NONSTATIONARITIES IN HYDROLOGIC AND ENVIRONMENTAL TIME SERIES

Water Science and Technology Library

VOLUME 45

The titles published in this series are listed at the end of this volume.

NONSTATIONARITIES IN HYDROLOGIC AND ENVIRONMENTAL TIME SERIES

by

A. RAMACHANDRA RAO

School of Civil Engineering,
Purdue University, West Lafayette, Indiana, U.S.A.

KHALED H. HAMED

Irrigation and Hydraulics Department,
Faculty of Engineering, Cairo University, Giza, Egypt

and

HUEY-LONG CHEN

Department of Environmental Engineering,
Lan-Yang Institute of Technology, Touchen, Ilan, Taiwan

KLUWER ACADEMIC PUBLISHERS
DORDRECHT / BOSTON / LONDON

A C.I.P. Catalogue record for this book is available from the Library of Congress.

ISBN 1-4020-1297-7

Published by Kluwer Academic Publishers,
P.O. Box 17, 3300 AA Dordrecht, The Netherlands.

Sold and distributed in North, Central and South America
by Kluwer Academic Publishers,
101 Philip Drive, Norwell, MA 02061, U.S.A.

In all other countries, sold and distributed
by Kluwer Academic Publishers,
P.O. Box 322, 3300 AH Dordrecht, The Netherlands.

The illustration on the cover is taken from figure 6.25.

Printed on acid-free paper

TABLE OF CONTENTS

5. Time-Frequency Analysis

6. Time-Scale Analysis

7. Segmentation of Non-Stationary Time Series

8. Estimation of Turbulent Kinetic Energy Dissipation

9. Segmentation of Observed Data

10. Linearity and Gaussianity Analysis

11. Bayesian Detection of Shifts in Hydrologic Time Series

List of Tables

List of Figures

Acknowledgements

The work discussed in this book grew out of the authors' work at the School of Civil Engineering, Purdue University, over the last decade. We would like to thank the support of our colleagues at Purdue University, Professors M. Hondzo, D.A. Lyn, G.S. Rao and G.D. Jeong. Dr. Khaled H. Hamed would like to thank his colleagues at the Faculty of Engineering, Cairo University, for their support and encouragement. Dr. Miki Hondzo contributed the lake temperature data discussed in this book. We would like to express our gratitude to him. It would not have been possible to develop this work without the support of our families. We are deeply indebted to their support. Mrs. Dinah Hackerd produced several versions of the manuscript with her usual cheer and good humor. We would like to thank her for all that she has done.

1. Introduction

Hydrologic time series analysis plays an important role in managing water resources. Practical applications include the design and operation of hydraulic structures, flood control, water supply management, hydropower generation, and irrigation systems design among others. The task of modeling streamflow can be achieved by following either of two approaches. The first approach is deterministic in nature, which involves the physical simulation of the physical system through numerical models using rainfall as input. The second is stochastic in nature and involves the study of the important statistical properties of streamflow sequences. Besides the difficulties in modeling the relationship between rainfall and runoff, determining future flows requires forecasts of rainfall. One is thus forced either to resort to more complicated meteorological modeling, or to apply stochastic modeling to rainfall. On the other hand, river flows, as stochastic processes, suffer from a smaller degree of randomness than rainfall since the storage in river basins smooths out some of the randomness. Also, large fluctuations in rainfall are dampened by flow passing through river basins. It is therefore desirable to study streamflow series using the stochastic approach. Several attempts have also been made to establish some physical foundation for stochastic streamflow models of different types (see Salas et al., 1988, Chapter 1).

The concept of random variables has been used in the field of hydrology since the beginning of last century (Hazen, 1914; Sudler, 1927; Hurst, 1951). Since that time, many models have been proposed in the literature. These include, among others , autoregressive models (AR) (Thomas and Fiering, 1962; Yevjevich, 1963; Matalas, 1967); autoregressive moving average models (ARMA) (Carlson et al., 1970, O'Connell, 1971); fractional Gaussian noise (FGN) (Mandelbrot and Wallis, 1968; Matalas and Wallis, 1971); Markov mixture models (Jackson, 1975); and ARMA-Markov models (Lettenmaier and Burges, 1977).

However, despite the increasing sophistication of time series modeling techniques, practical hydrologic time series modeling is still performed by using simple methods. As technology advances, and with the evolution of more powerful devices for recording, storing, and retrieving hydrologic data, a wealth of hydrologic as well as climatic data are available for analysis. With the advent of powerful computers the use of more advanced analysis methods on a larger number of data sets is becoming more and more accomplishable, and the use of simplified methods can be justified only to a limited extent. The analysis of a large number of data sets also gives more credibility to the obtained results and allows for better generalization of these results.

A wide range of time series models developed during the past fifty years depend on the two key assumptions of linearity and stationarity. However, both of these assumptions are mathematical idealizations which, in some cases, may be valid approximations to the real situation (Priestley, 1988). In practice, it is usually assumed that the departure of the observed time series from such assumptions is not large enough to seriously affect the results of analysis within the time interval of interest. Stationarity implies that the statistical properties of a time series do not change with time. Stationarity in its strict sense implies that for a set of given times $t_1, t_2, ..., t_n$ and any integer k, the joint probability distribution of the variables $\{X(t_1), X(t_2), ..., X(t_n)\}$ is identical with the joint probability distribution of the variables $\{X(t_1+k), X(t_2+k), ..., X(t_n+k)\}$. A number of methods of time series analysis require a weaker type of stationarity, known as 'weak stationarity' or 'second order stationarity'. In this case it is assumed that the first and second moments of $X(t)$ are independent of time. In addition, the expected value of [X

$(t)\, X(s)]$ will be a function only of the lag time $(t\text{-}s)$ and hence the covariance of $X(t)$ and $X(s)$ will be a function of the lag time $(t\text{-}s)$ but not of either t or s separately.

Recently, more attention has been paid towards nonstationary time series modeling. In these models, the basic idea is to extend the classical approaches to accommodate certain types of nonstationarity. On the other hand, if the type of nonstationarity can be identified and modeled, one can proceed to remove it to arrive at a stationary time series. However, according to the above definition of stationarity, even in the case of "weak stationarity", nonstationarity can take various forms. In this book, several methods for identifying nonstationarity in hydrologic and climatic time series are considered. The first method for identifying nonstationarity is trend analysis, which deals with nonstationarity in the mean of the time series. The second method deals with periodic components in the observed hydrologic and climatic time series, which can be studied through spectral analysis. The third method is based on a newly introduced analysis technique known as evolutionary spectral analysis. Evolutionary spectral analysis, also known as time-frequency analysis, is concerned with the evolution of spectral properties in time, i.e. changes in the shape and composition of the spectra of time series with time. The fourth method considered here is based on time-scale analysis. Time-scale analysis, also known as multi-resolution analysis, is also a newly introduced technique for analyzing time series as well as other signals and is similar in some aspects to time-frequency analysis. In time-scale analysis, time series are decomposed into components of different scales, from which inferences can be made about the composition of the time series.

New analysis methods have been recently introduced, which help accomplish the task of identifying nonstationarity in time series. In spectral analysis, two recently introduced methods which are aimed at increasing resolution and decreasing power leakage are the Multi-Taper Method (MTM) (Thomson, 1982) and the Maximum Entropy Method (MEM) (Childers, 1978). In the field of time-frequency analysis, evolutionary spectral analysis (Priestley, 1965) offers a method for detecting nonstationarity in time series based on the variations in the spectra of the time series at different time instants. This method, combined with the improved spectral estimates provided by the MTM and MEM methods is a powerful tool for identifying nonstationarity in time series data. In the field of time-scale analysis the newly introduced Wavelet analysis (Daubechies, 1992) is a very useful tool for identifying different types of nonstationarity in time series based on multi-resolution analysis. These methods of identifying nonstationarity in hydrologic and climatic time series are discussed in this book .

Previous methods described deal mainly with nonstationarity in the mean, or the spectral and scale properties of the entire series. The fifth method of analysis of nonstationary time series considered in this book is to apply a model of piecewise-stationary time series with successive stationary segments. Statistical properties of each of the stationary segments are described by using stationary stochastic models. In order to establish the piecewise-stationary time series, segmentation algorithms have been developed. Tests often used in segmentation algorithms without requiring *a priori* spectral information are mainly based on AR models.

The first example considered in this book, of applying the segmentation algorithm, is to determine stationary segments from temperature gradient profiles which are measured from vertically rising instruments in oceans or lakes. In the area of environmental hydraulics, the kinetic energy dissipation rate is one of the key intrinsic fluid flow parameters. It is obtainable through direct and indirect methods. In an indirect method, the kinetic energy dissipation rate is estimated from the Batchelor spectrum (Dillon and Caldwell, 1980; Oakey, 1982). The Batchelor spectrum is a theoretical spectrum derived by Batchelor (1959). In order to compare the estimated temperature gradient spectrum to the theoretical Batchelor spectrum, nonstationary temperature gradient profiles are partitioned into stationary segments. Segmentation algorithms using tests based on AR models (Lovell and Boashash, 1987; Imberger and Ivey, 1991) have been proposed to partition nonstationary temperature gradient profiles to stationary segments. Due to different assumptions and hypotheses made in different tests, segmentation results of different tests would be different. Therefore, a general comparison of segmentation results with respect to different tests is necessary. Such a comparison of tests in presented herein. Besides, two important assumptions usually made in linear stochastic models, Gaussianity and linearity, are tested to investigate the application of these tests which are based on AR models. Results from these tests are presented.

The second example of applying the segmentation algorithm is to examine stationarity in standardized hydrologic time series. In areas of stochastic hydrology and water resource systems, a conventional practice is to transform seasonal or monthly time series into standardized time series with zero mean and unit variance. Often the standardized time series are assumed to be weakly stationary. Therefore, stationary stochastic models such as AR, MA or ARMA models are frequently used to model the standardized time series. Commonly used hydrologic time series are streamflow, precipitation, temperature and Palmer's drought index, which may be obtained from national databases such as the United States Geological Survey (USGS), the Historical Climatology Network (HCN), and the National Climatic Data Center (NCDC) databases. However, this standardization procedure does not assure stationarity of the transformed time series. Besides, the other two important assumptions, Gaussianity and linearity, may not be valid either. These aspects are discussed in some detail.

The use of Bayesian statistics to investigate the changes in the characteristics of time series is another area of growing interest. Examples of using the Bayesian statistics are discussed in the last chapter of this book.

Many of these techniques have been discussed in journals and conference proceedings. One of the purposes of this book is to bring these techniques together in a single source. It is hoped that this endeavor would facilitate further research in this important area.

2. Data Used in the Book

2.1. HYDROLOGIC AND CLIMATIC DATA

Hydrologic data discussed in this book include flow, rainfall, temperature, tree-ring data and Palmer's Drought Severity Index (PDSI) Series. Flow data are obtained from the United States Geological Survey (USGS, 1992) and Earth Info (1993) data bases. Both rainfall and temperature data are obtained as two different sets. One set is obtained from the Historical Climatology Network (HCN) database (Nelson and Boden, 1993) for individual stations. Another set is obtained from the National Climatic Data Center (NCDC) database (NCDC, 1991) for state divisions. Precipitation and temperature data are also taken from Earth Info (1993) database. Tree-ring data are obtained from the National Oceanic and Atmospheric Administration database (NOAA, 1997). The PDSI data are obtained from National Climatic Data Center, NOAA.

Flow time series from 30 stations in the study region are obtained from the USGS and Earth Info (1993) databases. These consist of 6 stations in Iowa, 7 stations in Illinois, 7 stations in Indiana, and 7 stations in Ohio and one each in Michigan, Missouri and Wisconsin. The station information and length of the time series from each station as well as starting and ending years are given in Table II.1. The flow time series from the first 26 stations are shown in Fig. 2.1.

The first 26 series in Table II.1 are used in the analyses discussed in the first six chapters of the book. The last four series in Table II.1 are used in the analyses discussed in the following chapters. Stations included in the HCN database are believed to be not influenced by artificial changes in local environment. All stations in the database were quality controlled by NCDC using outlier and areal edits and corrected for time-of-observation differences, instrument changes, instrument moves, station relocation, and urbanization effects. Also, data missing from some records were estimated by using data from neighboring stations. Rainfall and temperature time series from 36 stations were obtained from the HCN database. These consist of 9 stations in Iowa, 10 stations in Illinois, 8 stations in Indiana, and 9 stations in Ohio. In addition, rainfall data from five stations and temperature data from four stations from Earth Info (1993) were used in this book. The lengths of data from the Earth Info (1993) data base and the HCN database are different. The station information as well as the lengths of the time series for each station are given in Table II.2. HCN Rainfall and Temperature data are available up to the year 1987. Rainfall and temperature data from the first 36 stations in Table II.2 are used in the analysis discussed in chapters 3-6 in this book. Data from Stations 37-42 in Table II.2 are used in the analyses discussed in the chapters following chapter 6. The rainfall and temperature time series, respectively, from the HCN stations are shown in Figs. 2.2 and 2.3.

NCDC rainfall and temperature data are available as divisional averages. Each state is divided into a number of divisions: 9 divisions in Iowa, 9 divisions in Illinois, 9 divisions in Indiana, and 10 divisions in Ohio. The divisional average data are obtained by averaging the records from a number of stations within that division. NCDC Divisional average data are available for all divisions from 1895 to 1992 for rainfall data, and from 1895 to 1995 for temperature data. The rainfall and temperature time series, respectively, from the 36 NCDC divisions are shown in Figs. 2.4 and 2.5. The Earth Info (1993) data, numbers 37 - 42 in Table II.2, are from individual stations.

For tree ring data, 22 time series were chosen from the database. These consist of 9 series in Iowa, 9 series in Illinois, one series in Indiana, and 3 series in Ohio. The station information and the length of each tree ring series, the starting and ending years, and the name of the major researcher who collected the information are given in Table II.3. The 22 chosen tree-ring time series are shown in Fig. 2.6.

The monthly PDSI series analyzed in this book include six series from stations in Indiana, Illinois, and Ohio. The PDSI data are obtained from the United States National Climatic Data Center. The value of PDSI ranges from -4.0 to 4.0. PDSI is a drought indicator which is evaluated based on evapotranspiration, antecedent moisture and runoff. The details of monthly PDSI series are summarized in Table II.4. The regions in Indiana, Illinois and Ohio states, for which the PDSI series are analyzed in this study, are shown in Fig. 2.7.

A separate set of hydrologic and air pollution data was used in the analysis discussed in chapter 11. The details of these data are found in chapter 11.

2.2. SYNTHETIC AND OBSERVED ENVIRONMENTAL DATA

The synthetic environmental data used in this book include: (1) two sets of synthetic series generated for examining the variations of three test statistics based on autoregressive (AR) models with respect to the AR order p, of which the first is generated from an AR model, and the second set is generated by sampling from the Batchelor spectrum; (2) four sets of synthetic series consisting of multiple segments generated by sampling from the Batchelor spectrum in order to compare the performances of different test statistics in detecting change points in time series; (3) two other sets of multiple-segment synthetic series generated by sampling from the Batchelor spectrum for comparing the segmentation algorithms with and without boundary optimization; and (4) three sets of synthetic series generated by sampling from the Batchelor spectrum with white noise added to examine Batchelor curve fitting. Each synthetic data set contains ten series. The observed environmental data consist of ten profiles of vertical temperature gradient measured in lakes.

The details of sampling from the Batchelor spectrum are given in Section 2.2.1. Specifications and examples of synthetic series generated by sampling from the Batchelor spectrum and the AR model are given in Sections 2.2.2 and 2.2.3, respectively.

2.2.1. *Synthetic Data Sampling from Batchelor Spectrum.* Synthetic data are generated by sampling from the Batchelor spectrum, and from AR models. A one dimensional Batchelor spectrum of temperature gradient fluctuations is given by Eq. 2.1 (Dillon and Caldwell, 1980; Oakey, 1982)

$$S(k) = \left(\frac{q}{2}\right)^{1/2} \frac{x^{T}}{k_B D_T} S_N(\alpha) \qquad [(\text{degC/m})^2/\text{cycle/m})] \tag{2.1}$$

where $q = 3.46$ (Dillon and Caldwell, 1980) is a universal constant, D_T is the molecular diffusivity of heat $[m^2/s]$, k is the wave number [cycle/m], k_B is the Batchelor wave number.

$$k_B = \left(\frac{\varepsilon}{D_T^2 \nu} \right)^{\frac{1}{4}} \qquad \text{[cycle/m]} \tag{2.2}$$

In Eq. 2.2, ε is the kinetic energy dissipation rate $[\mathrm{m}^2/\mathrm{sec}^3]$, v is the kinematic viscosity $[\mathrm{m}^2/\mathrm{s}]$, and x_T is the dissipation of temperature variance given by Eq. 2.3,

$$x_T = 6D_T \int_0^\infty S(k)dK = 6D_T \overline{\left(\frac{\partial T'}{\partial z} \right)^2} \qquad [(\mathrm{degC})^2/\mathrm{sec}] \ , \tag{2.3}$$

where T' is the temperature fluctuation, and z is water depth. The nondimensional shape of the spectrum is given by the normalized spectrum in Eq. 2.4.

$$S_N\,(\alpha) = 2\pi\alpha \left[\exp\left(-\frac{\alpha^2}{2} \right) - \alpha \int_\alpha^\infty \exp\left(-\frac{x^2}{2} \right) dx \right] , \tag{2.4}$$

In eq. 2.4 α is a nondimensional wave number given by

$$\alpha = 2\pi (2q)^{\frac{1}{2}} \frac{k}{k_B} . \tag{2.5}$$

Variations of the Batchelor spectrum due to the kinetic energy dissipation rate ε and the dissipation of temperature variance x_T are shown in Figs. 2.8 and 2.9 respectively. The effect of changing ε is to shift the Batchelor Spectrum along the slope of -1 in the log-log space (Fig. 2.8) and the spectrum is amplified proportionally by increasing x_T (Fig. 2.9).

A discrete weakly stationary series $\{\eta_t\}$ which has the same spectral properties as the theoretical Batchelor spectrum may be generated by using Eq. 2.6 (Luketina and Imberger, 1998),

$$\eta_t = \sum_{j=1}^{n_f} A_j \sin\left(2\pi t \frac{j}{2n_f} + \theta_j \right) , \tag{2.6}$$

where θ_j is a uniformly distributed random variable ranging from 0 to 2π, and n_f is the number of Fourier components used to determine the amplitude A_j from the Batchelor spectrum. In order to prevent aliasing in the spectrum of the synthetic series, the components above the Nyquist frequency 0.5 are not included in Eq. 2.1. The magnitude of A_j is

$$A_j = (4S(j\delta)\delta)^{\frac{1}{2}} , \tag{2.7}$$

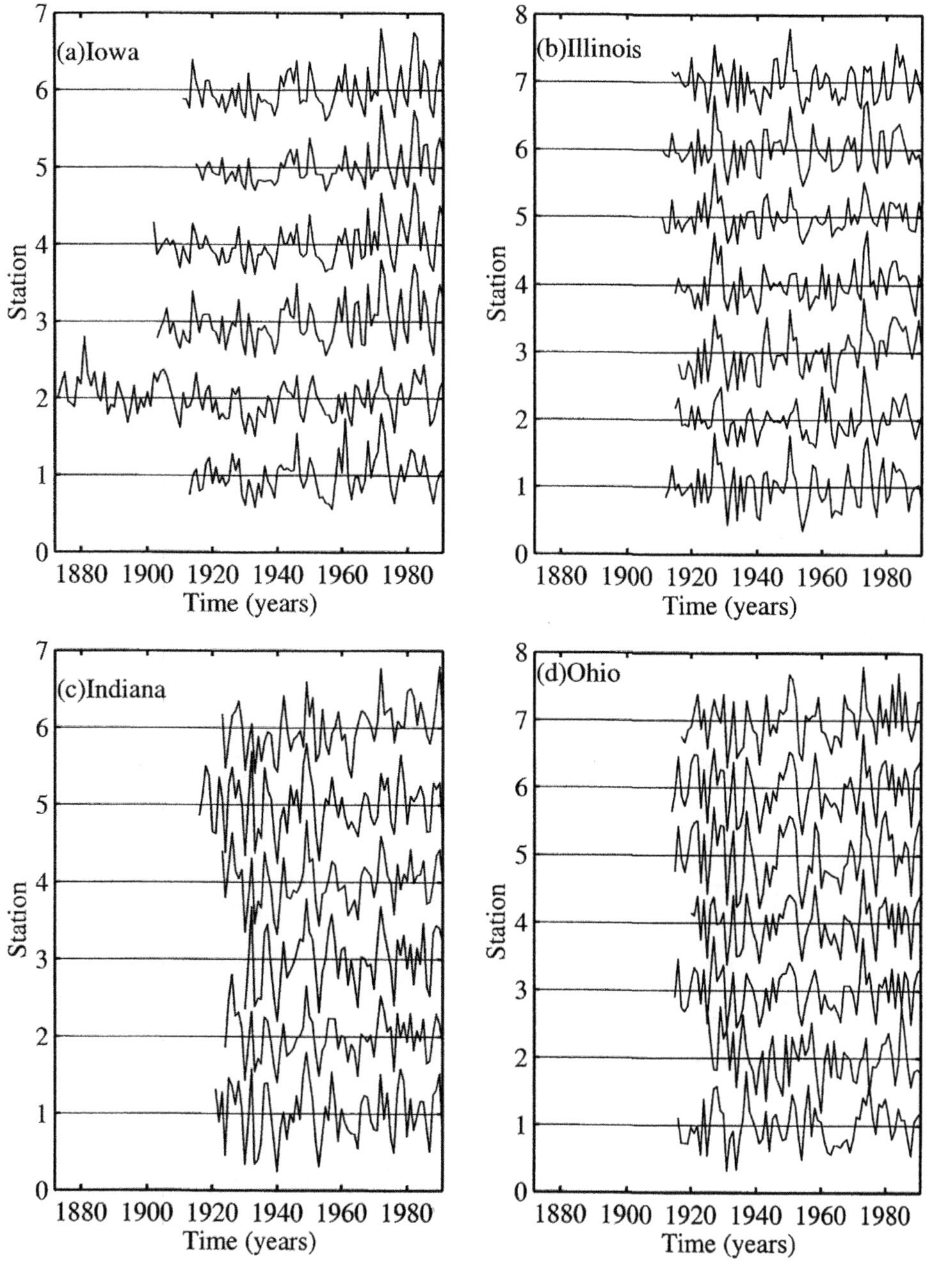

Figure 2.1. Flow time series from 26 USGS stations.

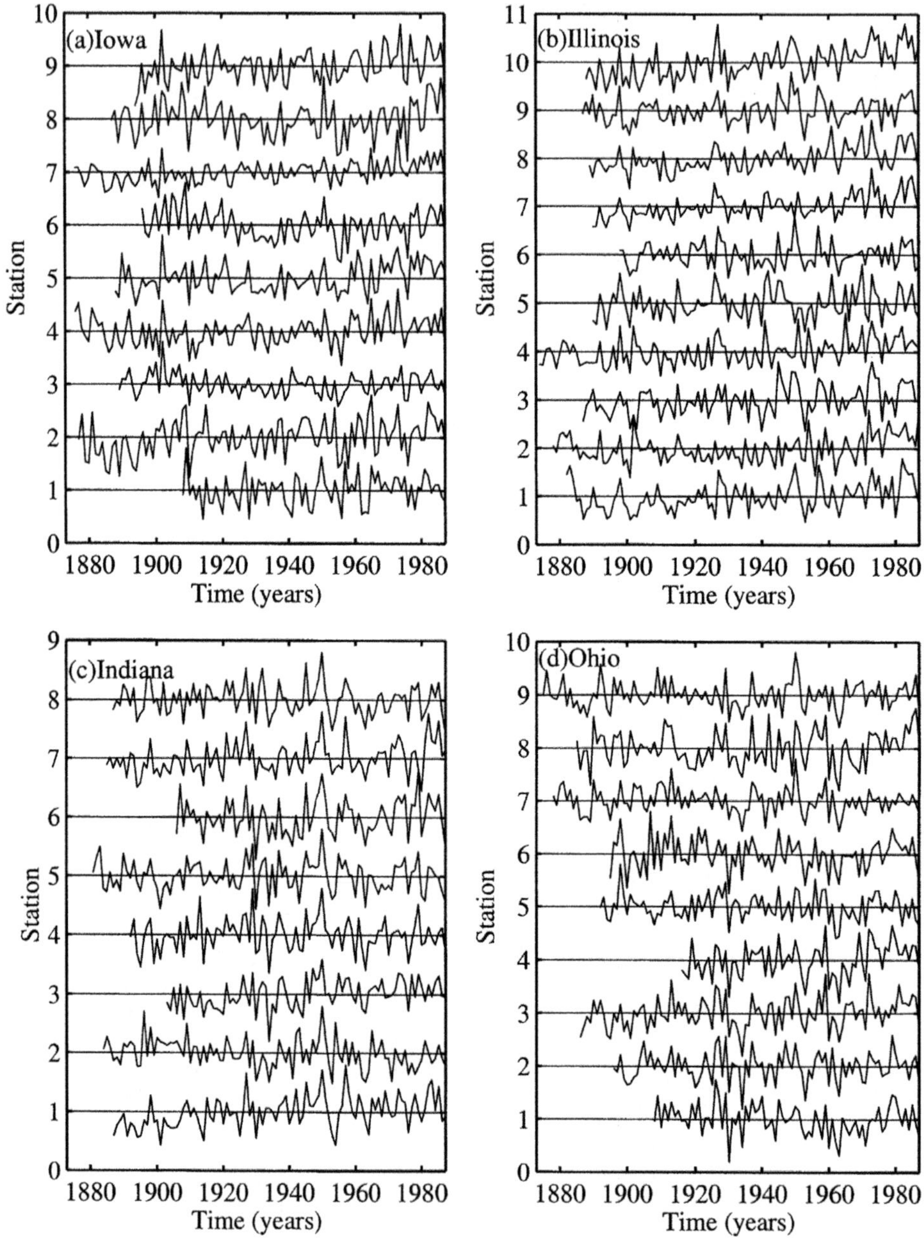

Figure 2.2. Rainfall time series from 36 HCN stations.

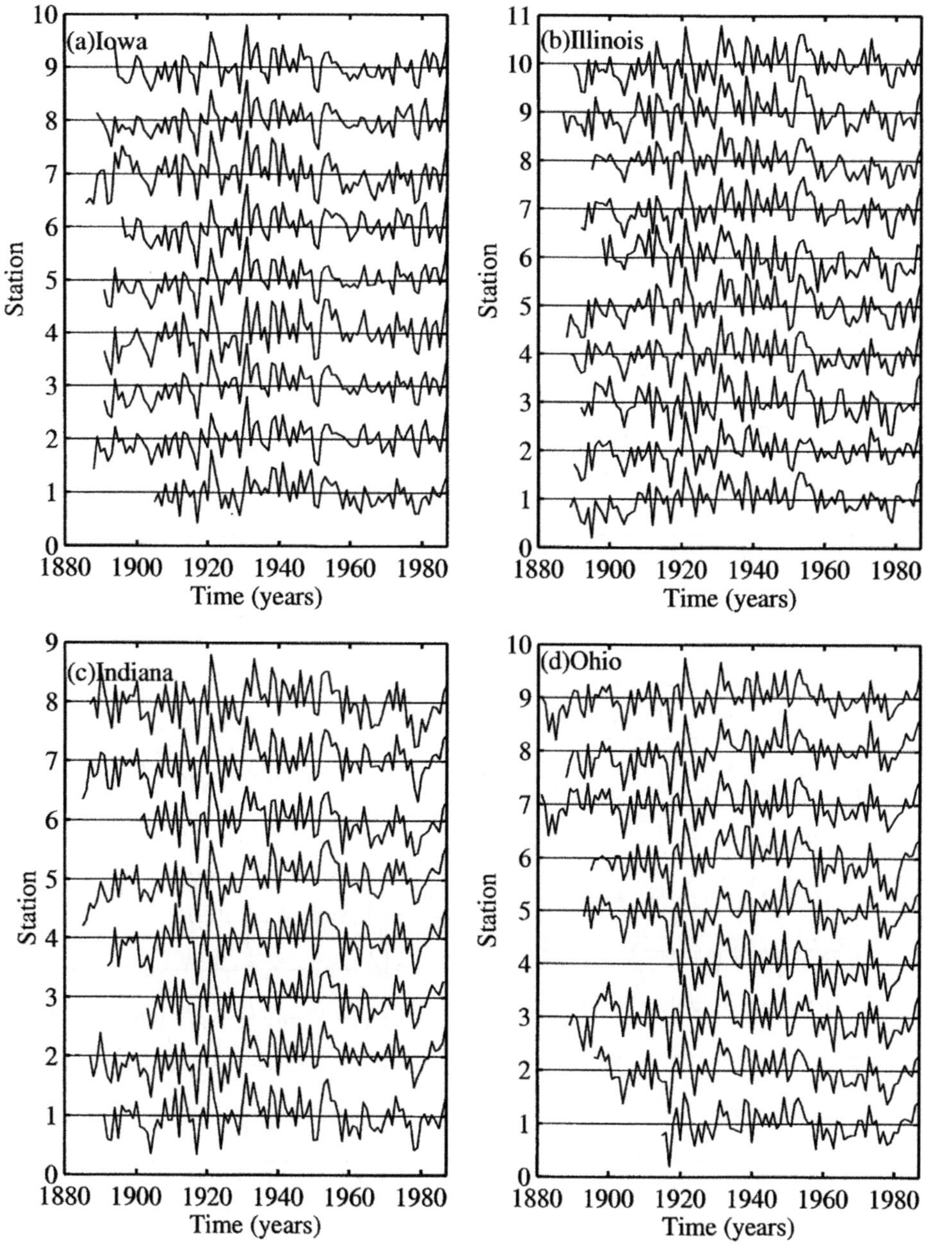

Figure 2.3. Temperature time series from 36 HCN stations.

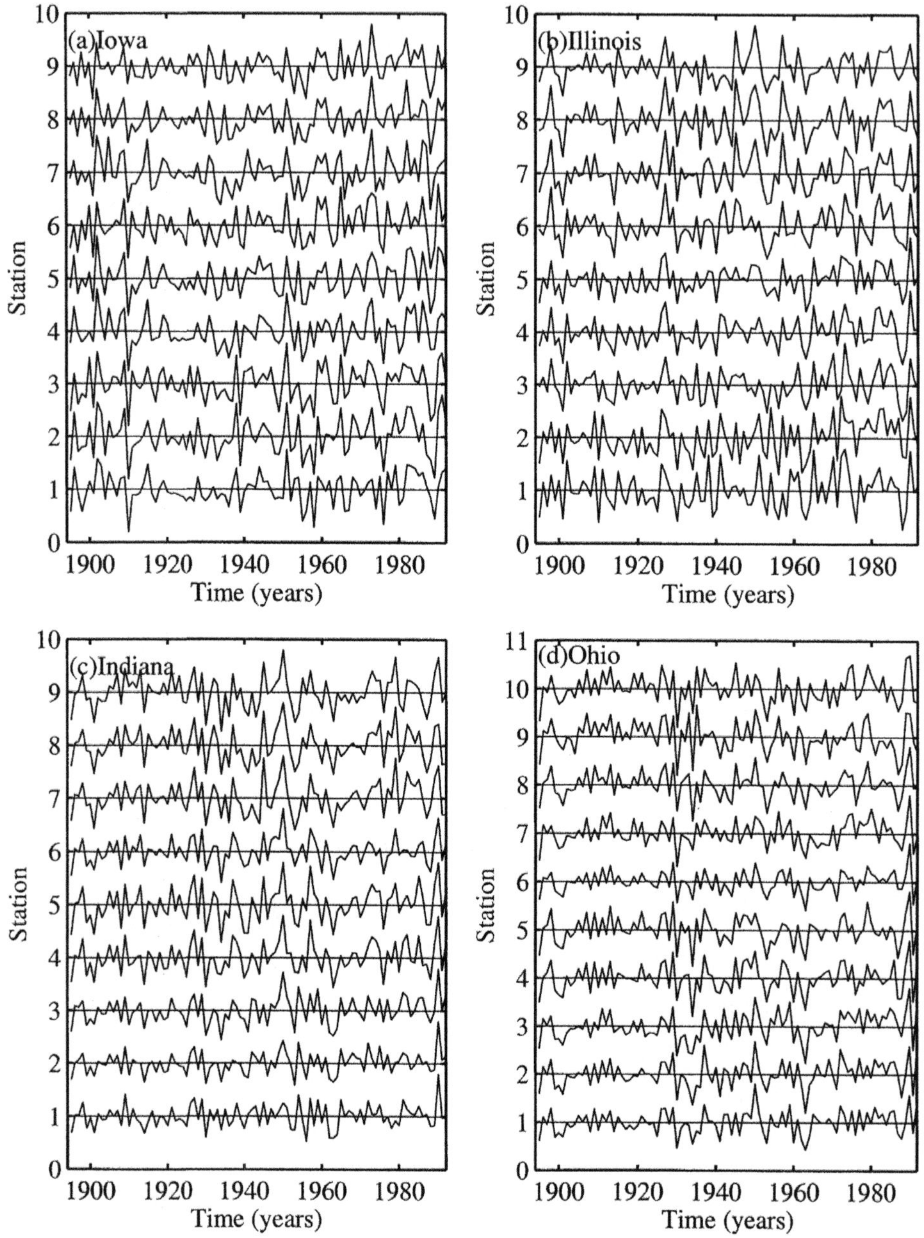

Figure 2.4. Average divisional rainfall time series from 37 NCDC divisions.

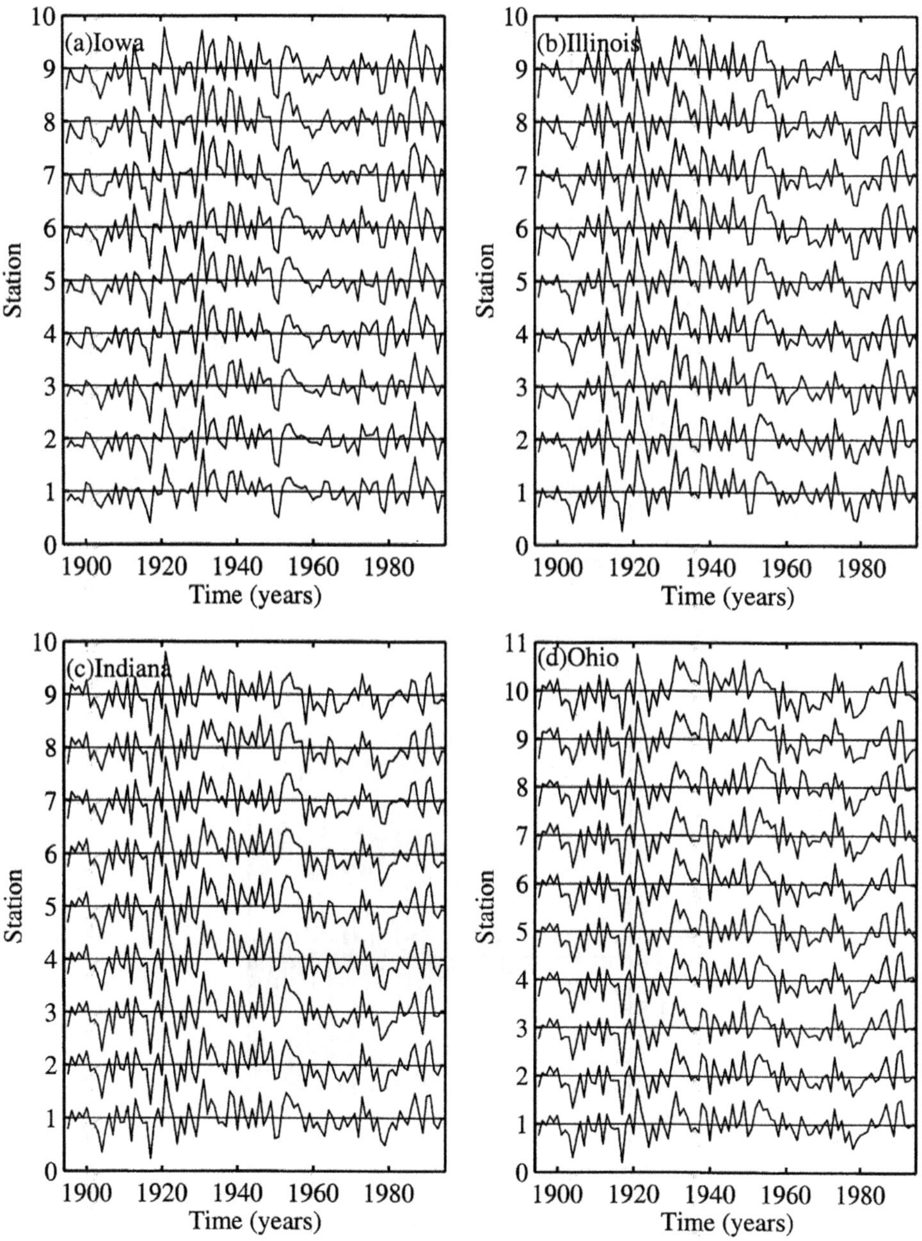

Figure 2.5. Average divisional temperature time series from 37 NCDC divisions.

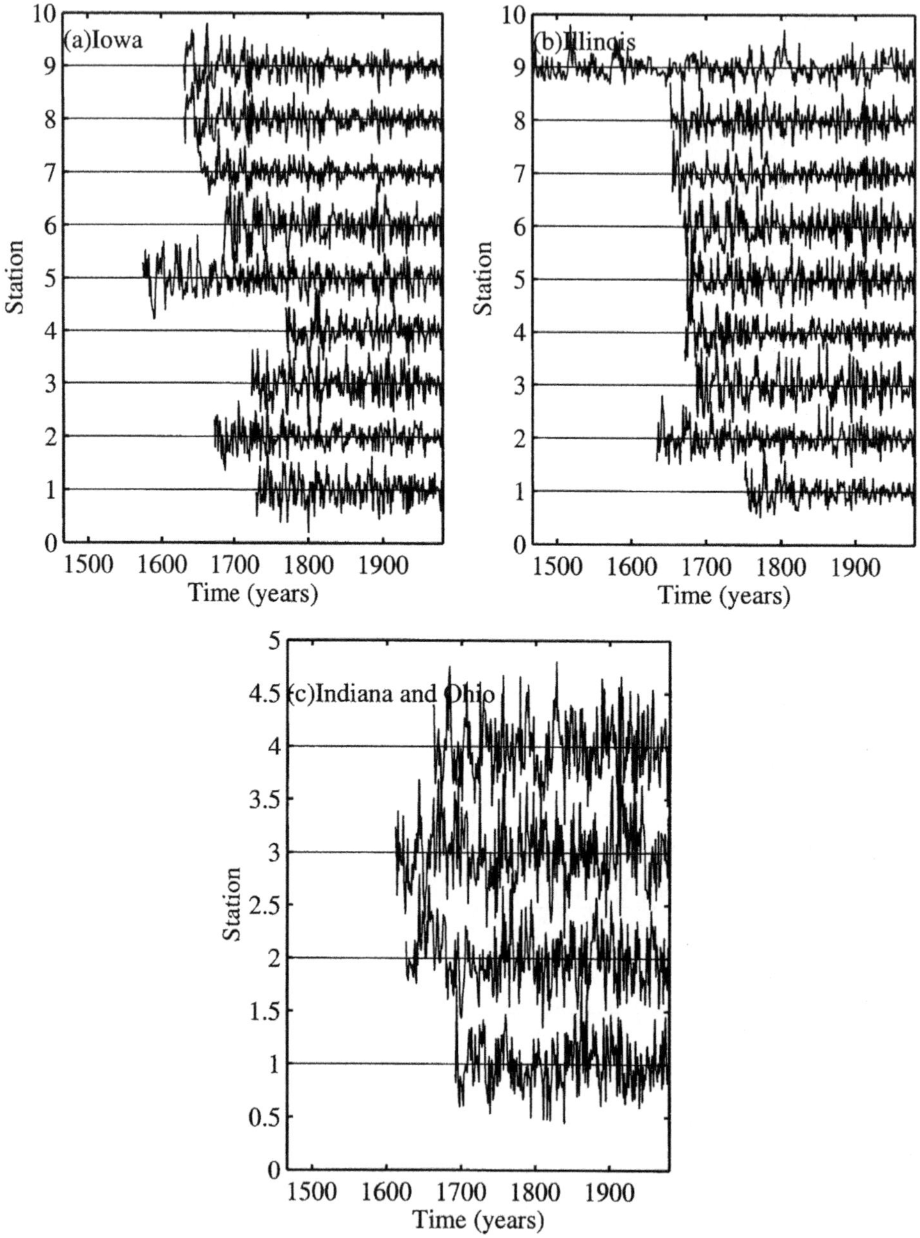

Figure 2.6. Tree-ring time series from 22 locations in the NOAA tree-ring database.

Table II.1. USGS flow data.

No	State	USGS ID Number	Station Name	Start Year	End Year	N
1	IA	05418500	Maquoketa River near Maquoketa, IA	1913	1992	79
2		05420500	Mississippi River at Clinton, IA	1873	1992	11
3		05454500	Iowa River at Iowa City, IA	1903	1992	89
4		05464500	Cedar River at Cedar Rapids, IA	1902	1992	90
5		05484500	Racoon River at Van Meter, IA	1915	1992	77
6		05490500	Des Moines River at Keosauqua, IA	1911	1992	81
7	IL	03345500	Embarasse River at Ste. Marie, IL	1912	1992	80
8		05435500	Pecatonica River at Freeport, IL	1915	1992	77
9		05527500	Kankakee River near Wilmington, IL	1916	1992	76
10		05570000	Spoon River at Seville, IL	1915	1992	77
11		05572000	Sangamon River at Monticello, IL	1911	1992	81
12		05592500	Kaskaskia River at Vandalia. IL	1911	1992	81
13		05597000	Big Muddy River at Plumfield, IL	1914	1992	78
14	IN	03276500	Whitewater River at Brookville, IN	1921	1992	71
15		03324500	Salmonie River at Dora, IN	1924	1992	68
16		03326500	Mississinewa River at Marion, IN	1930	1992	62
17		03335500	Wabash River at Lafayette, IN	1923	1992	69
18		03373500	East Fork White River at Shoals, IN	1915	1992	77
19		05518000	Kankakee River at Shelby, IN	1923	1992	69
20	OH	03109500	L. Beaver Cr. Near E. Liverpool, OH	1916	1992	76
21		03202000	Racoon Creek at Adamsville, OH	1925	1992	67
22		03261500	G. Maimi River at Sidney, OH	1915	1992	77
23		03262000	Loramie Creek at Lockington, OH	1920	1992	72
24		03269500	Mad River near Springfield, OH	1915	1992	77
25		03270500	G. Miami River at Dayton, OH	1914	1992	78
26		04191500	Auglaize River near Defiance, OH	1917	1982	75
27	IN	03275000	Whitewater River near Alpine, IN	1928	1992	54
28	MI	04108500	Kalamazoo River at Fennville, MI	1929	1993	64
29	MO	06934500	Missouri River at Humann, MO	1929	1993	64
30	WI	05395000	Wisconsin River at Merrill, WI	1903	1991	88

Table II.2. Details of rainfall and temperature time series.

No.	State	Station # & Division	Station Name	N Rainfall	N Temperature
1		0112-08	Albia	80	83
2		0600-06	Belle Plaine	111	100
3		1402-02	Charles City	99	97
4		2864-03	Fayette	112	97
5	IA	2977-02	Forest City 2NNE	100	97
6		4894-04	Logan	92	92
7		5952-03	New Hampton	112	102
8		8296-05	Toledo	101	99
9		8688-09	Washington	94	94
10		0338-02	Aurora	106	99
11		1280-06	Carlinville	110	98
12		3335-01	Galva	101	96
13		3717-06	Griggsville	114	99
14	IL	3879-09	Harrisburg	98	100
15		4108-06	Hillsboro	90	90
16		4823-03	La Harpe	98	96
17		8147-08	Sparta	99	93
18		8740-05	Urbana	101	101
19		8916-01	Walnut	100	98
20		0177-05	Anderson Sewage	101	97
21		0676-03	Berne	104	101
22		1229-06	Cambridge City	85	85
23	IN	1747-05	Columbus	96	96
24		5337-05	Marion 2N	107	103
25		6705-08	Paoli	82	87
26		7646-05	Rushville	103	103
27		8036-07	Shoals HIghway	101	101
28		2119-05	Delaware	80	73
29		2791-01	Findlay Sewage	92	92
30		3375-04	Greenville Water	102	99
31		5315-07	Millport 2NW	72	69
32	OH	5669-01	Napoleon	96	95
33		6781-09	Portsmouth	93	93
34		8534-02	Upper Sandusky	110	107
35		8822-01	Wauseon Water	103	100
36		8830-09	Waverly	113	107
37			Urbana	99	90
38	IL		Aledo	88	88
39	IN		Ft. Wayne	44	44
40	IN		Evansville		42
41	IN		Indianapolis		44
42	MN		Minneapolis	101	101

Table II.3. NOAA tree-ring time series.

No.	State	Location Lon.-Lat.	Location Name	Start Year	End Year	N	Researcher
1	IA	42N-91W	Backbone St. Park	1730	1980	251	Duvick, D.N.
2		40N-93W	Nine Eagles St. Park	1672	1980	309	Duvick, D.N.
3		40N-91W	Geode St. Park	1724	1980	257	Duvick, D.N.
4		41N-91W	Palisades-Kepler St. Park	1770	1980	211	Duvick, D.N.
5		41N-93W	Lake Ahquabi St. Park	1574	1980	407	Duvick, D.N.
6		42N-94W	Dolliver Memorial St. Park	1658	1980	296	Duvick, D.N.
7		43N-91W	Yellow river St. Forest	1651	1980	330	Duvick, D.N.
8		42N-91W	White Pine Hollow St. Pres.	1631	1980	350	Duvick, D.N.
9		42N-91W	Merritt Forest St. Preserve	1631	1980	350	Duvick, D.N.
10	IL	41N-88W	Sandwich White Oak	1752	1980	229	Duvick, D.N.
11		41N-89W	Starved Rock State Park	1633	1980	348	Duvick, D.N.
12		41N-88W	Kankakee River State Park	1686	1980	295	Duvick, D.N.
13		39N-89W	Lincoln's New Salem St.Park	1671	1980	310	Duvick, D.N.
14		39N-88W	Fox Ridge St. Park	1674	1980	307	Duvick, D.N.
15		40N-87W	Kickapoo St. Park	1670	1980	311	Duvick, D.N.
16		37N-88W	Ferne Clyffe St. Park	1655	1980	326	Duvick, D.N.
17		37N-89W	Giant City St. Park	1652	1980	329	Duvick, D.N.
18		37N-89W	Cache River	1468	1980	513	Stahle, D.W.
19	IN	41N-86W	Pulaski Woods	1692	1980	289	Cook, E.R.
20	OH	40N-81W	Andrew Johnson Woods	1626	1980	355	Cook, E.R.
21		41N-81W	Stebbin's Gulch Holden Arb.	1612	1980	369	Cook, E.R.
22		39N-84W	Davis-Purdue-Glen Helen	1662	1980	319	Cook, E.R.

Table II.4. Summary of monthly PDSI series.

Station	State	Period of record	N
Region 2	IL	1895 - 1993	98
Region 8	IL	1895 - 1993	98
Region 1	IN	1895 - 1993	98
Region 7	IN	1895 - 1993	98
Region 2	OH	1895 - 1993	98
Region 9	OH	1895 - 1993	98

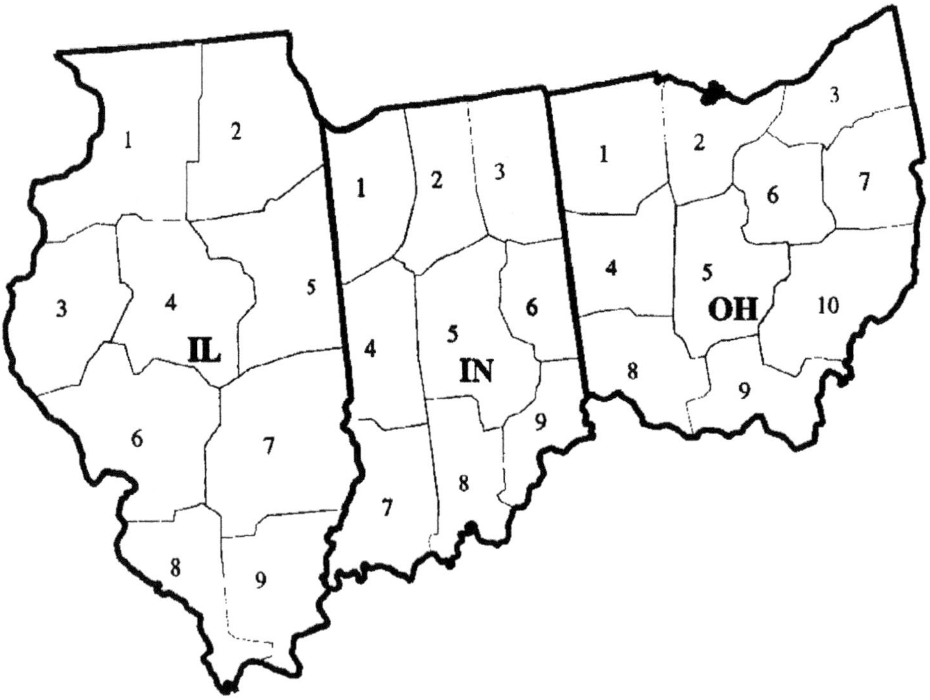

Figure 2.7. The regions in Indiana, Illinois and Ohio states, for which the PDSI series are analyzed in this study.

where δ is the interval between wave number ordinates transformed from the Fourier ordinates; i.e.

$$\delta = \frac{f_s}{2n_f u_s}, \tag{2.8}$$

where f_s and u_s are the scanning rate and the moving velocity of the probe of SCAMP (Self-Contained Autonomous Microstructure Profiler) (Carter and Imberger, 1986). $S(j\delta)$, the Batchelor spectrum given by Eq. 2.1, is sampled by specifying values of (ε, x_T), with fixed values of D_T and v which are $1.4*10^{-7}$ m^2/s and 10^{-6} m$^2/s$, respectively. The scanning rate f_s and velocity u_s of the probe are fixed at 100 Hz and 0.1 m/s, respectively. The number of Fourier components $n_f = 64$ is selected so that each time a synthetic series of 128 observations is generated. The procedures for generating a synthetic series by sampling the Batchelor spectrum are summarized as follows:

1. Specify the parameters in Table II.5, and assign values to variables (ε, x_T) in Table II.6.

2. Determine the amplitude A_j from the Batchelor spectrum at $n_f = 64$ ordinates between wave numbers corresponding to 0 and 0.5 Hz.

3. Use Monte Carlo simulation to sample θ_j, $j = 1, ..., n_f$ from uniform distribution $[0,2\pi]$.

4. Generate η_t, $t = 1,...,2n_f$ according to Eq. 2.6. That is, a synthetic series of $2n_f = 128$ observations.

Table II.5. Parameters used in sampling from the Batchelor spectrum.

Parameter	Value	Definition
n_f	64	Number of Fourier components.
f_s	100 Hz	Scanning rate of the probe.
u_s	0.1 m/s	Moving velocity of the probe.
q	3.46	Universal constant.
D_t	$1.4*10^{-7}$ m^2/s	Diffusivity of heat.
v	10^{-6} m^2/s	Kinematic viscosity.

Table II.6. Variables used in sampling from the Batchelor spectrum.

Parameter	Units	Definition
ε	[m^2/sec^3]	Kinetic energy dissipation rate.
x_T	[(degC)^2/sec]	Dissipation of temperature variance.

2.2.2. *Details of Data Generated by Sampling from the Batchelor Spectrum.*

1. A set of synthetic series (Model 2.1.1 in Table II.7) are generated by sampling from the theoretical Batchelor spectrum ($\varepsilon = 10^{-6}$ m^2/sec^3, $x_T = 10^{-6}$ (degC)^2/sec) to examine variations with respect to the AR order p of test statistics for segmentation. Each series contains 512 observations. A synthetic series generated from Model 2.1.1 is shown in Fig. 2.10.

Table II.7. Model of synthetic series generated by sampling from the Batchelor spectrum.

Model	Batchelor spectrum: B(ε [m^2/sec^3], x_T [(degC)^2/sec])	Number of observations
M. 2.1.1	B(10^{-6}, 10^{-6}).	512

2. Four sets of synthetic series (Models 2.1.2-2.1.5 in Table II.8) consisting of multiple segments are generated by sampling from the Batchelor spectrum in order to compare the performances of different tests for detecting change points in time series. Each of the first set (Model 2.1.2) of synthetic series consists of ten segments which are generated by sampling from the theoretical Batchelor spectrum with x_T ranging from 10^{-7} to 10^{-6}(degC)^2/sec and ε fixed at 10^{-6} m^2/sec^3. Each segment contains 256 observations. Model 2.1.2 is represented by B(10^{-6}, $1*10^{-7}$ to $10*10^{-7}$) in Table II.8. A synthetic series generated from Model 2.1.2 is shown in Fig. 2.11. The second set (Model 2.1.3) of synthetic series, each of which consists of three segments, are generated by sampling from the theoretical Batchelor spectrum with x_T ranging from $(10/3)*10^{-7}$ to 10^{-6} (degC)^2/sec and ε fixed at 10^{-6} m^2/sec^3. Each segment contains 768 observations. Model 2.1.3 is represented by B(10^{-6}, $(10/3)*10^{-7}$ to $10*10^{-7}$) in Table II.8. The next set (Model 2.1.4) of synthetic series, each of which consists of ten segments, are generated by sampling from the theoretical Batchelor spectrum with ε ranging from 10^{-7} to 10^{-6} m^2/sec^3 and x_T fixed at 10^{-6} (degC)^2/sec. Each segment contains 256 observations. The last set (Model 2.1.5) of synthetic series are generated by sampling from the theoretical Batchelor spectrum with ε ranging from $(10/3)*10^{-7}$ to 10^{-6} m^2/sec^3 and x_T fixed at 10^{-6}(degC)^2/sec. Each series of this set is composed of three segments. Each segment contains 768 observations.

Table II.8. Models of synthetic series consisting of multiple segments generated by sampling from the Batchelor spectrum.

Model	Batchelor spectrum: B(ε [m^2/sec^3], x_T [(degC)^2/sec])	Number of observations
M. 2.1.2	10 segments: B(10^{-6}, $1*10^{-7}$ to $10*10^{-7}$).	256/segment
M. 2.1.3	3 segments: B(10^{-6}, $10/3*10^{-7}$ to $10*10^{-7}$).	768/segment
M. 2.1.4	10 segments: B($1*10^{-7}$ to$10*10^{-7}$, 10^{-6}).	256/segment
M. 2.1.5	3 segments: B($10/3*10^{-7}$ to$10*10^{-7}$, 10^{-6}).	768/segment

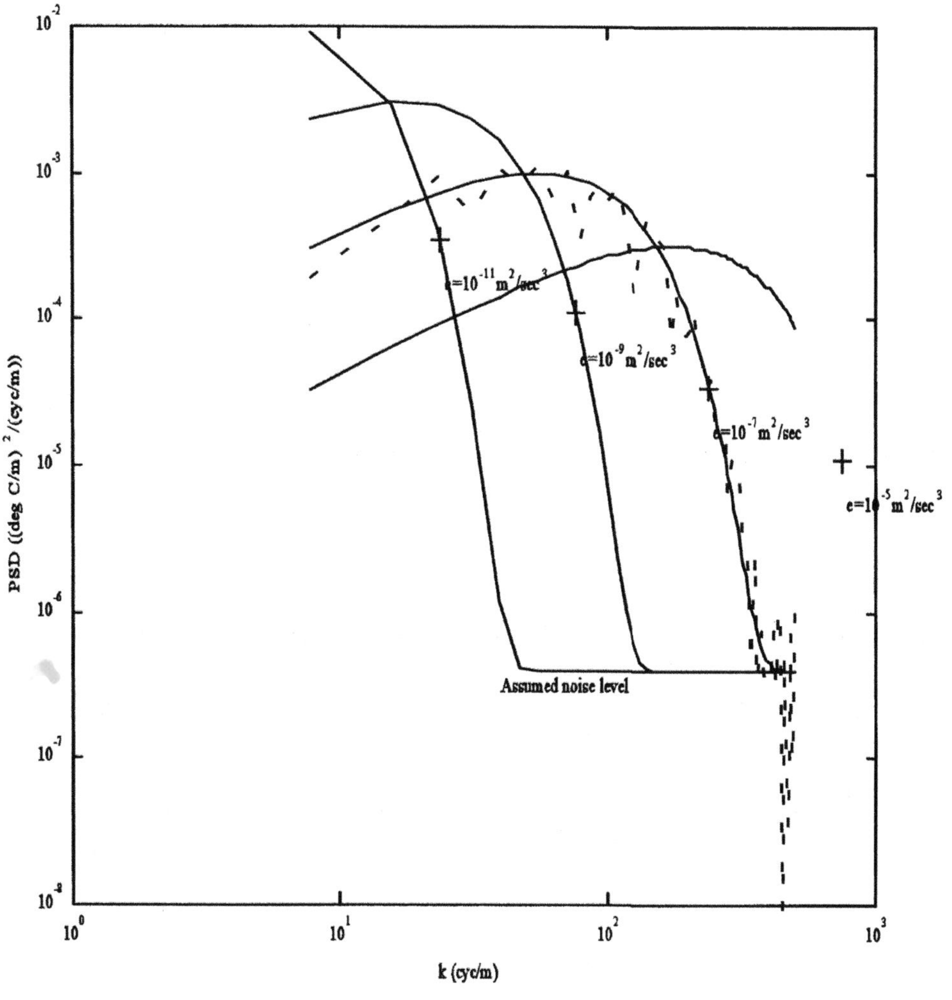

Figure 2.8. Batchelor spectra corresponding to different values of ε.

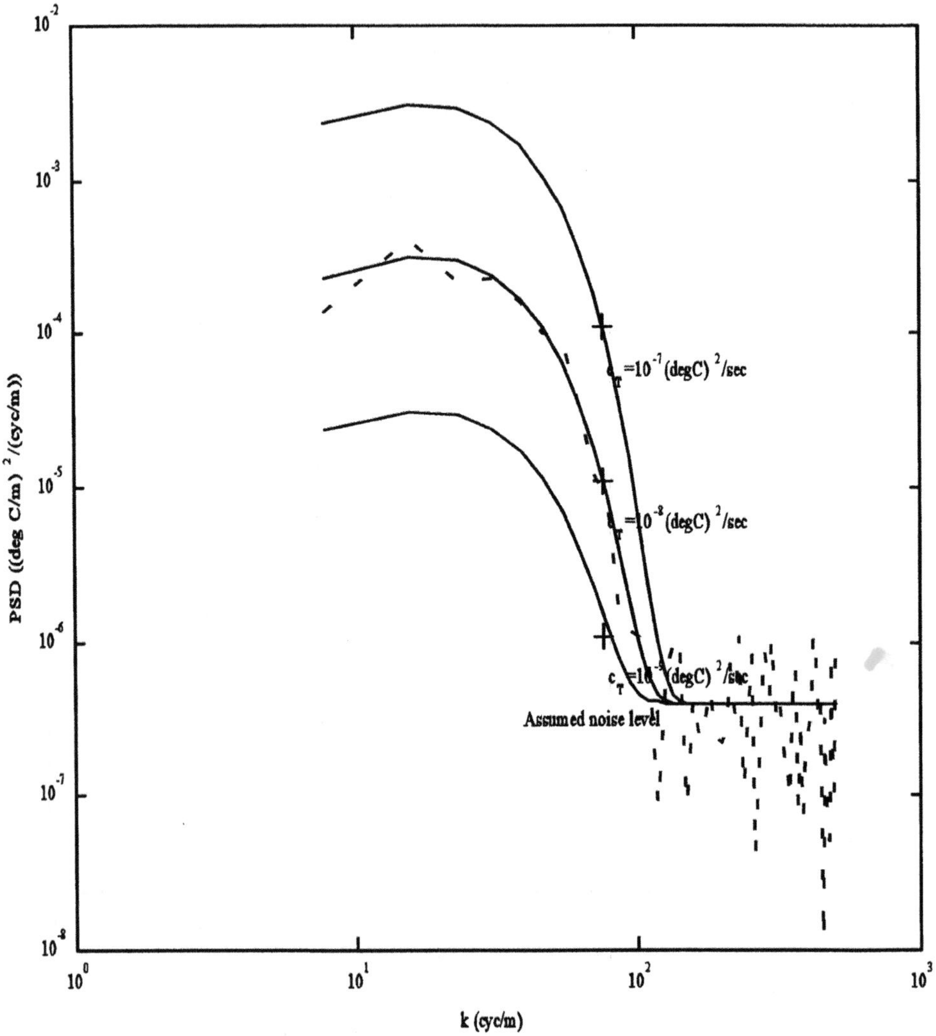

Figure 2.9. Batchelor spectra corresponding to different values of x_T.

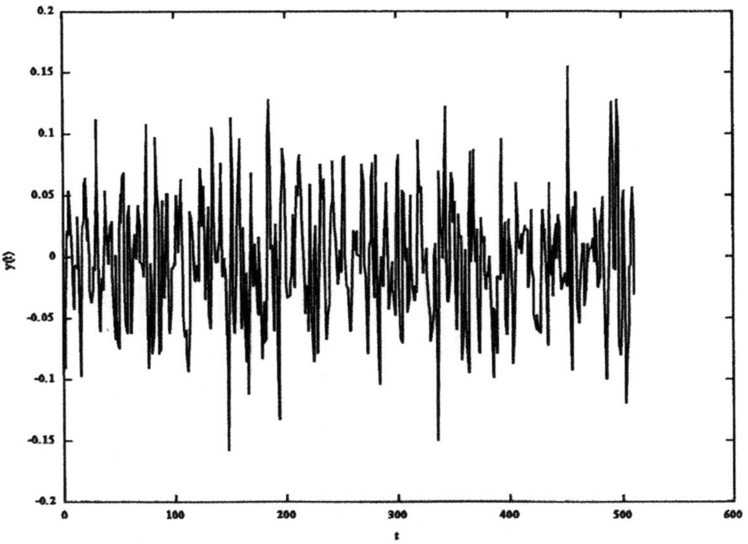

Figure 2.10. A synthetic series generated from Model 2.1.1.

3. Two sets of synthetic series (Models 2.1.6 and 2.1.7 in Table II.9) consisting of multiple segments are generated for comparing segmentation algorithms with and without boundary optimization. Each of the first set (Model 2.1.6) of synthetic series is composed of two stationary segments generated by sampling from the theoretical Batchelor spectrum ($\varepsilon = 10^{-6}$ m^2/sec^3, $x_T = 10^{-6}$ (degC)^2/sec), and a "non-stationary" segment in between. The nonstationary segment is composed of 3 sub-segments of white noise with the noise level $\sigma_{wn}{}^2 = 2.4*10^{-8}$, $2.4*10^{-7}$, and $2.4*10^{-8}$ (degC/m)^2/(cycle/m) respectively. Each segment generated by sampling from the Batchelor spectrum contains 512 observations, and each segment of white noise contains 64 observations. A synthetic series generated from Model 2.1.6 is shown in Fig. 2.12. For the second set (Model 2.1.7) of synthetic series, each series is composed of 4 segments generated by sampling from the theoretical Batchelor spectrum with (ε, x_T) ranging from 10^{-5} to 10^{-8} m^2/sec^3 and 10^{-5} to 10^{-8} (degC)^2/sec respectively. Each segment contains 512 observations.

4. Three sets of ten synthetic series (Models 2.1.8-2.1.10 in Table II.10) were generated to test Batchelor curve fitting, by sampling from a theoretical Batchelor spectrum ($\varepsilon = 10^{-7}$ m^2/sec^3, $x_T = 10^{-7}$ (degC)^2/sec) and three levels of white noise ($\sigma_{wn}{}^2 = 0$, $4*10^{-7}$ and $4*10^{-5}$ (degC/m)^2/(cyc/m)). The noise level $\sigma_{wn}{}^2 = 4*10^{-7}$ (degC/m)^2/(cyc/m) is equivalent to 10 dB signal-to-noise ratio at the wave number equivalent to 30 Hz, and the noise level $\sigma_{wn}{}^2 = 4*10^{-5}$ (degC/m)^2/(cyc/m) is equivalent to -10 dB signal-to-noise ratio. Each series contains 256 observations. A synthetic series generated from Model 2.1.10 is shown in Fig. 2.13.

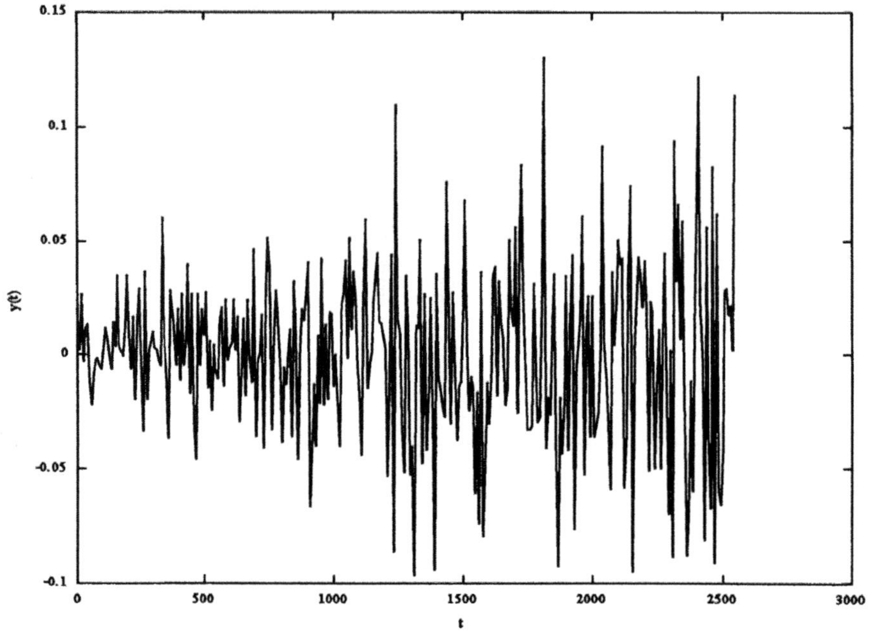

Figure 2.11. A synthetic series generated from Model 2.1.2.

Table II.9. Models of synthetic series consisting of multiple segments generated by sampling from the Batchelor spectrum.

Model	Batchelor spectrum: B(ε [m^2/sec^3], x_T [(degC)^2/sec]) White noise: W(σ_{wn}^2 [(degC/m)^2/(cyc/m)])	Number of observations
M. 2.1.6	5 segments: B(10^{-6}, 10^{-6}), W($2.4*10^{-8}$), W($2.4*10^{-7}$), W($2.4*10^{-8}$), B(10^{-6}, 10^{-6}).	B: 512/segment W: 64/segment
M. 2.1.7	4 segments: B(10^{-5}, 10^{-5}), B(10^{-6}, 10^{-6}), B(10^{-7}, 10^{-7}), B(10^{-8}, 10^{-8}).	512/segment

Each series consists of 512 observations. A synthetic series generated from Model 2.1.11 is shown in Fig. 2.14.

2.2.3. *Synthetic Data from AR Model.* A set of synthetic series (Model 2.1.11) are generated from a stationary AR model, in order to examine the variation in the AR order p corresponding to three test statistics based on AR models:

$$y_t = 1.1\, y_{t-1} - 0.3\, y_{t-2} + a_t, \qquad a_t \sim N(0,1) \text{ (Model 2.1.11).}$$

Table II.10. Models of synthetic series generated by sampling from the Batchelor spectrum with white noise added.

Model	Batchelor spectrum: B(ε [m^2/sec^3], x_T [(degC)^2/sec]) White noise: W(σ_{wn}^2 [(degC/m)^2/(cyc/m)])	Number of observations
M. 2.1.8	B(10^{-7}, 10^{-7}) + W(0).	256
M. 2.1.9	B(10^{-7}, 10^{-7}) + W(4*10^{-7}).	256
M. 2.1.10	B(10^{-7}, 10^{-7}) + W(4*10^{-5}).	256

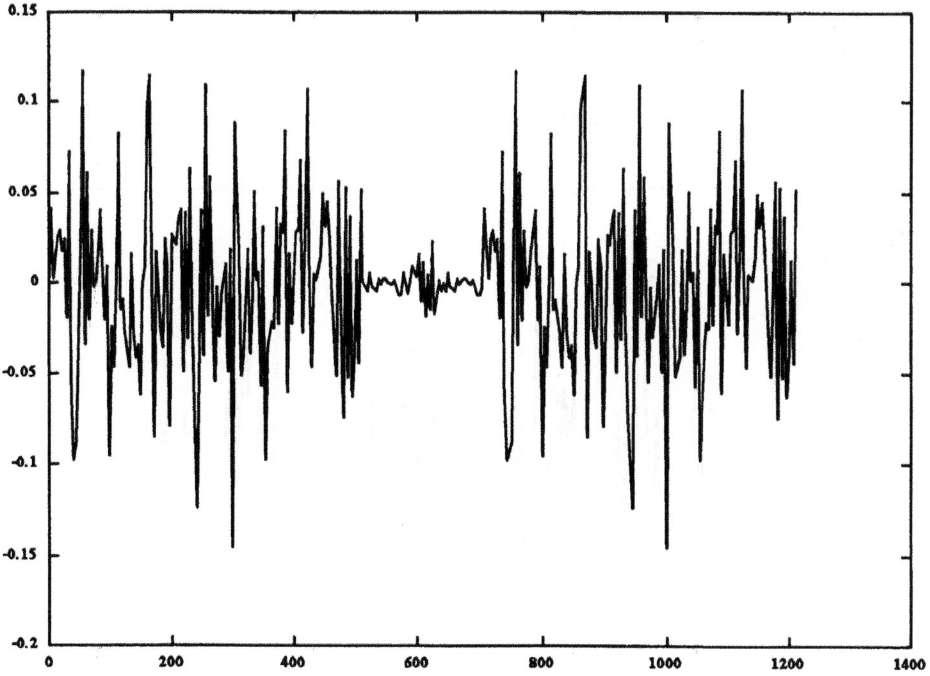

Figure 2.12. A synthetic series generated from Model 2.1.6.

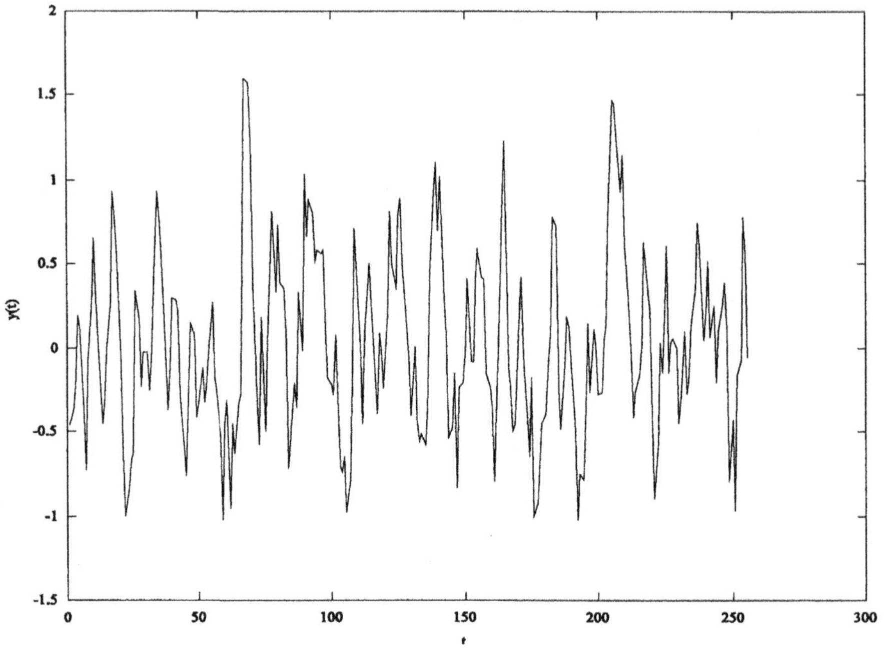

Figure 2.13. A synthetic series generated from Model 2.1.10.

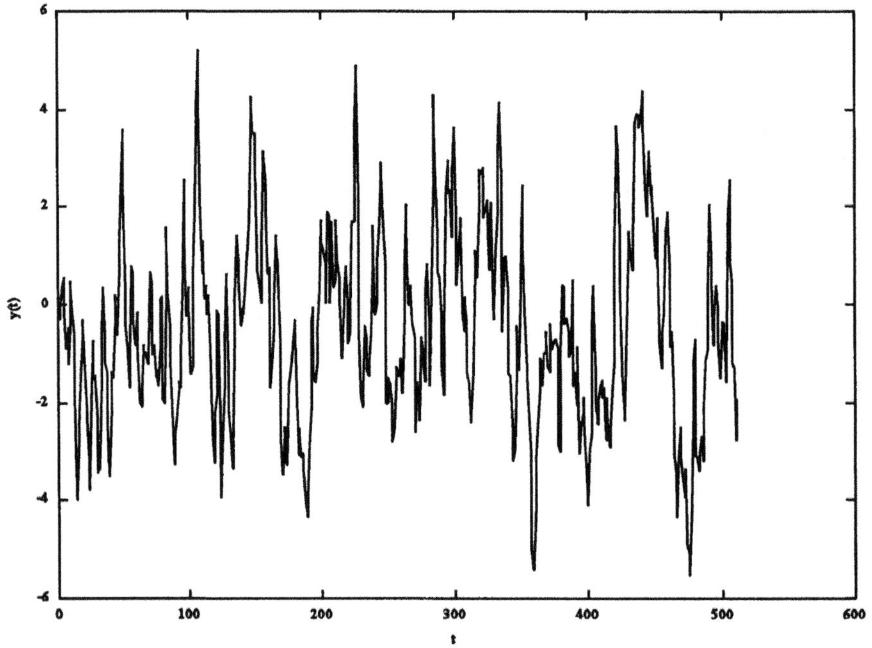

Figure 2.14. A synthetic series generated from Model 2.1.11.

2.3. OBSERVED DATA

Ten profiles of temperature gradient time series are analyzed and discussed in this book.

2.3.1. Measured Temperature Gradient Profiles. Ten profiles of temperature gradient are collected in three inland lakes located in the state of Minnesota. These lakes are Lake Calhoun (24m), Lake Waconia (9m) and Lake Christmas (19m). Each profile is collected with the probe of SCAMP vertically rising at $u_s = 0.1$ m/s and $f_s = 100$ Hz. D_t and ν are specified as $1.4*10^{-7}$ m^2/s and 10^{-6} m^2/s. The temperature gradient is detrended by subtracting the mean of a window of 128 elements of data around each point. This SCAMP profiling instrument is manufactured by Precision Measurement Engineering of Encinitas, CA. Information regarding recorded dates and profile length are given in Table II.11. As an example of the temperature series and temperature gradient series, Profile 1210 collected at 12:10 p.m., Aug. 15, 1997, in Lake Waconia is shown in Fig. 2.15.

Table II.11. Temperature gradient profiles analyzed in this study.

Lake	Profile (time)	Number of observations	Recorded date
Calhoun	1111	24043	Aug 10 1997
	1127	21665	Aug 10 1997
	1228	22428	Aug 10 1997
	1455	13083	Aug 10 1997
Waconia	1210	12062	Aug 15 1997
	1239	9005	Aug 15 1997
	1256	9005	Aug 15 1997
	1326	9005	Aug 15 1997
	1349	10024	Aug 15 1997
Christmas	1502	20050	Jul 24 1998

Twelve arbitrary segments of Profile 1210 are selected at different depths for illustrating spectral estimation. Each segment contains 128 data points (Table II.12). The numbers at different depths in Fig. 2.15 indicate the corresponding segments.

Table II.12. Segments of profile 1210.

Segment	1	2	3	4	5	6	7	8	9	10	11	12
Beginning	1	129	257	385	4001	4129	4257	4385	7001	7129	7257	7385
End	128	256	384	512	4128	4256	4384	4512	7128	7256	7384	7512

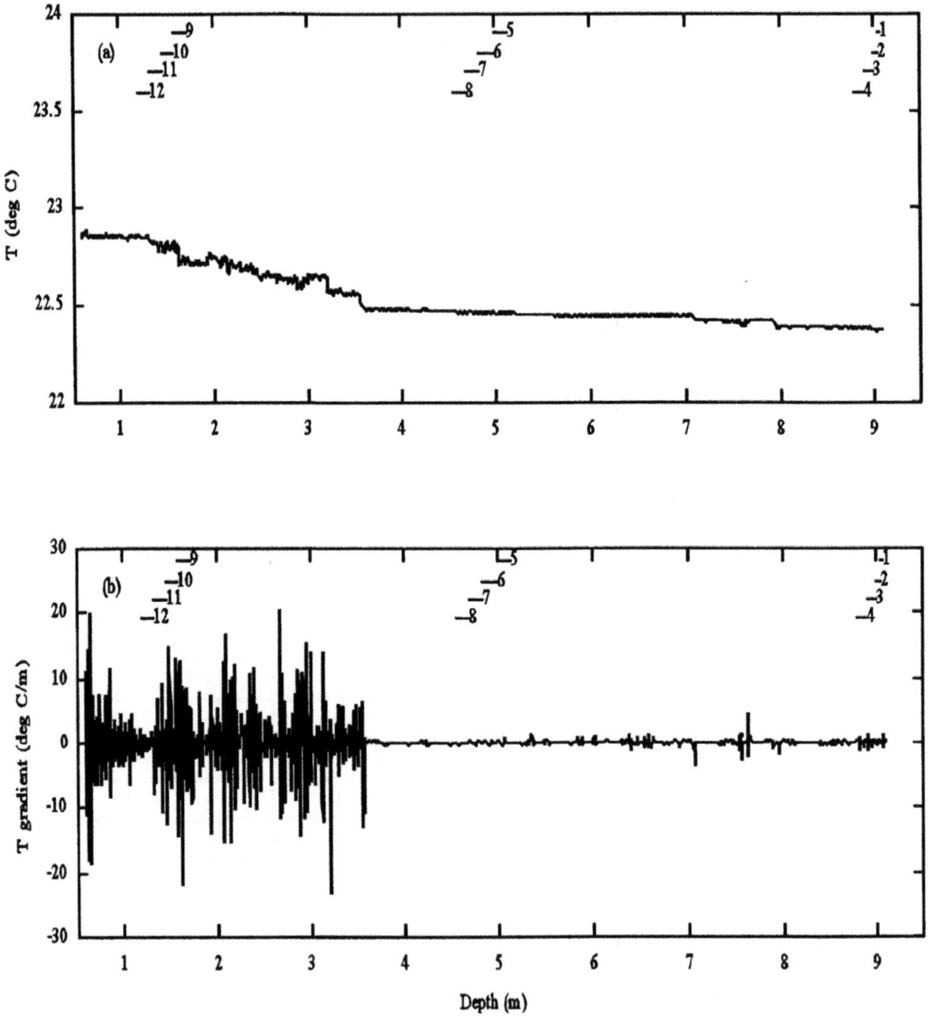

Figure 2.15. (a) Temperature series and (b) temperature gradient series of Profile 1210. The numbers at different depths indicate the corresponding segments.

3. Time Domain Analysis

3.1. INTRODUCTION

Nonstationary behavior in time series can take different forms. One of the most common types of nonstationarity is of the form in Eq. 3.1,

$$X_t = \mu_t + Z_t \tag{3.1}$$

where X_t is the observed time series, μ_t is a deterministic time dependent function, and Z_t is a zero- mean stationary stochastic process. Typical examples are time series with trend components when μ_t is an increasing or decreasing function of time in the form of a polynomial, and seasonal time series when μ_t is represented as the sum of a number of trigonometric functions.

Recently, more attention has been paid towards nonstationary time series modeling (cf. Priestley, 1988). In these models, the basic idea is to extend the classical approaches to accommodate certain types of nonstationarity. On the other hand, if the type of nonstationarity can be identified, one can proceed to remove the non-stationarity and then fit conventional models to the residuals. The most common method used in stochastic hydrology is to model the deterministic component μ_t and eventually remove it to arrive at the stationary time series Z_t. Another approach is to difference the series, a method which is commonly used for seasonal time series. However, in some cases, seasonal adjustment may obscure some of the properties of a time series (Newbold, 1980), and it would be more appropriate to incorporate seasonality within the model fitted to the data. Also, the deterministic component μ_t must be removed carefully to avoid damaging effects, especially when studying spectral properties of the time series (Currie and O'Brien, 1988).

Different methods have been proposed for detecting nonstationarity in time series. Time domain methods include, in addition to visual inspection, a number of statistical significance tests for trends either directly or through changes in the mean or median of the time series. On the other hand, frequency domain methods are more convenient for studying cyclic time series.

Analysis of trends in water resources data has received considerable attention in the past few years. A number of studies have been conducted to investigate trends in water quality time series from streams, lakes, and groundwater data (Yu et al., 1993, Bouchard and Haemmerli 1992, Zetterqvist 1991, Taylor and Loftis 1989, Hipel et al. 1988, Lettenmaier 1988 and 1976, Van Belle and Hughes 1984, Hirsch et al. 1982). Several parametric and non-parametric methods are used in these studies for trend detection. These methods can also be used for analyzing trends in hydrologic as well as climatic time series.

Trend analysis of hydrologic time series has not been extensively studied, especially for river flow data. In a study by Mitosek (1992), 150 long-term hydrologic time series from rivers in Asia, Africa, Europe, Australia, and North America were investigated. Results from these studies, based on a number of parametric and non-parametric tests, indicate that the hypothesis of stationarity of the overall record cannot be rejected at 5% significance. It was also found that various tests gave different results, and no regional pattern of trends was detected. Chiew and McMahon (1993) studied trends in flow data from a number of Australian rivers and streams. No significant trends were detected in most of the Australian river flow data at the 5% level. However, this was attributed by Chiew and McMahon to the high variability in river flow time series from the southern hemisphere, which is almost twice as high as the variability in the northern hemisphere.

Large scale trends in temperature and rainfall data have been studied on the global and regional scales (Boden et al. 1994). General results of trend analysis in temperature data indicate a warming trend for the first part of the century, followed by a cooling trend in the middle part, and finally a warming trend in the past 20 to 30 years. The latter warming trend is commonly referred to as the global warming due to the "greenhouse effect" produced by the increased concentration of carbon monoxide and trace gases in the atmosphere (WMO, 1995). However, it remains unclear whether global warming is the result of the "greenhouse effect" or some other phenomena (Boden et al. 1994). Global warming is believed to have a considerable effect on characteristics of hydrologic regimes. However, the effect may not be similar in different climatic regions (Beran and Arnell 1996, Burn 1994, Georgiadi 1991). For global rainfall data, records indicate a decreasing trend during the first half of the century, and an increasing trend in the last half of the century (Boden et al. 1994).

In this chapter, data from different stations in the study region are analyzed for the purpose of detecting trends using time domain methods. Riverflow, rainfall, and temperature data, as well as tree-ring data as a proxy for hydrologic data are used. The results from different time series are compared and the relationship between them are investigated.

3.2. VISUAL INSPECTION OF TIME SERIES

Time domain analysis ranges from visual inspection of a plot of the time series, or of some of its properties or transformations, to formal statistical tests of the hypothesis of stationarity. Objectives of such analyses are to identify different types of non-stationarity and test their significance. Visual inspection is the first step in the analysis, which can reveal or suggest possible types of nonstationarity in the time series. This, in turn, makes it easier to choose the appropriate tests of statistical significance (Klemeš, 1978). In many cases, the existence of the mean and the large variations of the observations around the mean make it difficult to identify nonstationarity in raw data. By rescaling the data and summing the departures around the mean, the situation can be improved (Garbrecht and Fernandez, 1994). The resulting quantities of such rescaling and summation of deviations from the mean are called the Rescaled Adjusted Partial Sums (RAPS). A plot of RAPS can reveal information about trends and fluctuations in the data. However, in some cases spurious periodicities may be observed (Feller, 1966; Klemeš and Klemeš, 1988; Yule, 1926). This is due to the fact that higher order RAPS converge rapidly to a sine wave. Thus if the process itself is a cumulative process, such as the water level in a reservoir, the application of RAPS may reveal a sine wave with a period of the sample size n (Klemeš and Klemeš, 1988). Consequently, RAPS plots should be interpreted with caution and used only as a guide but not as a replacement for formal statistical tests of significance (Garbrecht and Fernandez, 1994). The RAPS are defined as in Eq. 3.2,

$$S_k = \sum_{t=1}^{k} \frac{X_t - \bar{X}}{D_X} \quad ; \qquad k=1,2,\ldots,n \qquad (3.2)$$

where S_k is the kth term of the RAPS, X_t is the observed time series at time t, $\bar{X}$ is the mean of X_t, D_X is the standard deviation of X_t, and n is the number of observations. It is assumed that X_t follows a normal distribution or can be transformed into one (Legates, 1991). Garbrecht and Fernandez (1994) give the expected value and variance of S_k for the cases of independent and identically distributed X_t in addition to the cases of a sudden jump and uniform monotonic linear trend, as well as a general trend. Sudden jumps in X_t appear as a broken linear trend in S_k, while a linear trend in the data shows as a parabolic trend in S_k. For a monotonic positive trend, the values at the beginning of the record will, in general, be below the overall average and their adjusted partial sums will accumulate to a large negative value towards the middle of the record. The values toward the end of the record will be above-average and will balance S_k back to zero at the end of the record. This will result in the parabolic shape of a plot of S_k with negative ordinates when a positive linear trend exists in the data. For negative trends, the situation is reversed and the parabolic shape results with positive ordinates.

3.3. STATISTICAL TESTS OF SIGNIFICANCE

There are two types of trend that are commonly found in hydrologic and meteorologic data. The first type is monotonic trend, in which the location parameter (mean or median) changes monotonically over time, but not necessarily continuously or linearly. The other type is the step trend, where a shift in the location parameter occurs at a certain point in time. Generally, monotonic trend tests are used unless there is a strong belief that a sudden change is present in the data due to, for example, the relocation of a station or the existence of a large gap within the record (Hirsch et al., 1991). Trend testing is performed by using parametric or non-parametric methods. Parametric methods are usually more powerful than non-parametric methods but require stronger assumptions about the statistical distribution of the data. Parametric methods usually require that the data be independent and normally distributed. Non-parametric methods, on the other hand, often use the ranks of the observations rather than their actual values, which relaxes the requirements concerning the distribution of the data. Also, non-parametric methods are less sensitive to outliers in the data, since the rank, rather than the actual value, is used in these tests. The following tests are used to detect trends and changes in the mean or median of time series data.

3.3.1. Parametric Tests

a) The Linear Model Test

In this test the data are assumed to follow a linear trend in the form of Eq. 3.3,

$$X_t = \tau t + \gamma + \varepsilon_t \qquad (3.3)$$

where ε_t is normally distributed $N(0,\sigma_\varepsilon^2)$, τ is the trend magnitude, and γ is a constant. Estimates of ε_t, τ and γ are obtained by using standard least square regression methods. The variance of $\hat{\tau}$ is given by (Lettenmaier, 1976) as in Eq. 3.4.

$$var\ (\hat{\tau}) = \frac{12\ \sigma_\varepsilon^2}{n(n+1)(n-1)} \tag{3.4}$$

Based on the normality of $\hat{\tau}$, and for testing the null hypothesis that $\hat{\tau}=0$, the test statistic is

$$Z = \left[\frac{n(n+1)(n-1)}{12}\right]^{0.5} \frac{\hat{\tau}}{\sigma_\varepsilon} \tag{3.5}$$

The test statistic Z is compared to the t-distribution with n-2 degrees of freedom at the desired significance level. This is a parametric test and requires the data to be random and normally distributed.

b) Cumulative Deviation Test (Buishand, 1982)

This test is used to detect a change in the mean by studying the cumulative deviation from the mean. Given a time series X_t, the values of the Rescaled Adjusted Partial Sums (RAPS) are calculated as in Eq. 3.2. The test statistic Q is given by Eq. 3.6.

$$Q = \max |S_k| \tag{3.6}$$

Critical values of Q are given by Buishand (1982). This is a parametric test and requires the data to be random and normally distributed.

c) Worsley Likelihood Ratio Test (Worsley, 1979)

This test is designed to detect a change in the mean of a time series. The values of S_k from Eq. 3.2 are weighted according to their positions in the time series to form the sequence Z_k^* given in Eq. 3.7.

$$Z_k^* = [k(n-k)]^{-0.5}\ S_k \tag{3.7}$$

The test statistic W is then calculated as in Eq. 3.8,

$$W = (n-2)^{0.5}\ V/(1-V^2)^{0.5} \tag{3.8}$$

where

$$V = \max |Z_k^*| \tag{3.9}$$

Worsley (1979) gives critical values of W. This is a parametric test and requires the data to be random and normally distributed.

3.3.2 *Non-parametric Tests*

a) Mann-Kendall Test (Kendall, 1970)

This is a test of randomness against monotonic trend. Given a time series, $X = x_1, x_2, \ldots, x_n$, if $R_1, R_2, \ldots, R_n$ are the corresponding ranks of the elements of the series, the statistic S is calculated as in Eq. 3.10.

$$S = \sum_{i<j} sign\,(R_j - R_i) \tag{3.10}$$

Under the null hypothesis of randomly ordered observations, S is asymptotically normally distributed with mean zero and variance given by Eq. 3.11,

$$var(S) = \frac{1}{18}\left[n\,(n-1)\,(2n+5) - \sum_t t(t-1)(2t+5)\right] \tag{3.11}$$

where t is the number of X values involved in a given tie, and the summation is over all ties. The standardized normal test statistic is calculated as in Eq. 3.12.

$$Z(n) = S/[var(S)]^{1/2} \tag{3.12}$$

Kendall (1970) proposes a correction for continuity which results in Z given by Eq. 3.13.

$$Z = \begin{cases} \dfrac{S-1}{[var(S)]^{0.5}} & S>0 \\ 0 & S=0 \\ \dfrac{S+1}{[var(S)]^{0.5}} & S<0 \end{cases} \tag{3.13}$$

A positive value of Z indicates that the trend is positive and vice versa. This is a non-parametric test since the ranks of the observations, rather than their actual values, are used. The data need not be normally distributed, but they are required to be random.

b) Kruskal-Wallis Test (Sneyers, 1975)

The Kruskal-Wallis statistic tests for equality of subperiod means. If a time series is divided into m subsamples with lengths n_j, $j=1, 2, \ldots, m$, and R_{ij} is the rank of the ith observation of the jth subsample in the ordered complete sample, the test statistic is given by Eq. 3.14,

$$XS = 12 \sum_{j=1}^{m} \left[(R_j^2 / n_j) / \{ n(n+1) \} \right] - 3(n+1) \tag{3.14}$$

where R_j is the sum of ranks in the jth sample and n is the total sample size. Under the null hypothesis of equal subperiod means, the statistic XS is Chi-square distributed with $(m-1)$ degrees of freedom. This is also a non-parametric test and no assumptions are made about the distribution of the observations.

c) Distribution-free CUMSUM Test (McGilchrist and Woodyer, 1975)

This test is designed to detect a change in the median of a process. In this test the statistic V_k is defined as in Eq. 3.15,

$$V_k = \sum_{i=1}^{k} sign(X_t - X_m) \tag{3.15}$$

where X_m is the sample median. The test statistic KS is calculated as in Eq. 3.16.

$$KS = (2/n) \max |V_k| \tag{3.16}$$

The statistic KS has the same null distribution as the Kolmogorov-Smirnov two-sample statistic with sample size of $n/2$. This is also a non-parametric test and no assumption is made about the distribution of observations.

The World Meteorological Organization (WMO, 1988) selected the Mann-Kendall test, the Kruskal-Wallis test, the cumulative deviation test, and the Worsley likelihood test as standard procedures for detecting trends or changes in long time series of hydrologic data. The former two tests are non-parametric, while the latter two are parametric.

In general, parametric tests are more powerful than non-parametric ones, as long as the condition that the data are normally distributed is satisfied. However, even for small departures from normality, non-parametric procedures perform equally well and are sometimes better than parametric tests (Hirsch et al. 1991). In the case of water quality data, it is preferable to use non-parametric techniques (Hirsch et al. 1991) . This is due to the fact that water quality data are highly non-normally distributed. Hydrologic data share the same property of non-normality with water quality data but to a lesser extent. Accordingly, results obtained by using parametric and non-parametric tests are not expected to be identical in the case of hydrologic time series.

3.4. TESTING AUTOCORRELATED DATA

Some features of hydrologic data such as the existence of autocorrelation between the observations and seasonality affect both the accuracy and power of parametric and non-parametric tests as well. Accuracy is related to the empirical significance of the trend test, which is the percentage of samples rejected under the null hypothesis of no trend. If the data are uncorrelated, the empirical significance level should be equal to the significance level α at which the test is conducted. Positive serial correlation between the data, however, will increase the chance of detecting significant trends even in the absence of trends, and vice versa (Cox and Stewart 1955), which will result in inaccurate significance levels. For example, for samples with length $n=60$ and AR(1) dependence with $\varphi = 0.6$ using the Mann-Kendall trend test with significance levels $\alpha = 0.01, 0.02, 0.05$, and 0.10, the function of rejected samples are 0.11, 0.14, 0.20, and 0.26, respectively (Hamed and Rao 1998). Power of the trend test, on the other hand, is the percentage of samples rejected when a trend exists in the data. The power of a good test should be as large as possible for small deviations from the null hypothesis of no trend, while maintaining an accurate empirical significance level under the null hypothesis. For the time series used in this study, flow data have lag-one autocorrelation $\rho(1)$ of 0.11 to 0.39, temperature data have $\rho(1)$ of 0 to 0.28, and rainfall data have $\rho(1)$ of -0.25 to 0.24. These large values of lag-one autocorrelations affect the results of both parametric and non-parametric tests.

Autocorrelation in the data may be dealt with by using one of two different approaches. The first approach is to incorporate the effect of autocorrelation in the test itself. The second approach is to remove the autocorrelation from the data by using a suitable model, and then perform the tests on the uncorrelated residuals. The former approach requires the modification of each test separately, thus performing the test on the undisturbed data, while in the latter approach reformulation of tests is not needed, but the structure of the data will be altered.

One of the most common problems in testing the significance of trends is the difficulty in distinguishing between trends and long-term persistence associated with large autocorrelations. However, there is a strong belief that most annual flow time series have characteristics of AR(1) or ARMA(1,1) processes with $0.0<\rho(1)<0.6$ (Hirsch and Slack 1984). Also, rainfall data usually have smaller autocorrelations. Therefore, with the relatively small autocorrelation values in the data analyzed in this book ($\rho(1) <= 0.39$) the effect of long-term persistence is of minor concern.

To accommodate the effect of autocorrelation in the data, a modification to the Mann-Kendall trend test that accounts for the existence of autocorrelation in the data (Hamed and Rao, 1998) is used in addition to the previously mentioned trend tests. In this modified Mann-Kendall test, the autocorrelation between the ranks of the data is first calculated after removing the apparent trend. The variance of the statistic S in Eq. 3.11 is then adjusted to account for the autocorrelation as in Eq. 3.17,

$$var(S) = \frac{1}{18}\left[n\,(n-1)\,(2n+5)\right] * \frac{n}{n_S^*} \tag{3.17}$$

with the value of n/n_S^* given by Eq. 3.18,

$$\frac{n}{n_S^*} = 1 + \frac{2}{n(n-1)(n-2)} \sum_{i=1}^{p} (n-i)(n-i-1)(n-i-2)\,\rho_s(i) \tag{3.18}$$

where n is number of observations in the sample, n_S^* is the "effective" number of observations, $\rho_s(i)$ is the autocorrelation between the ranks of the observations for lag i. Examination of the autocorrelation plots for the various data used in this book indicates that the autocorrelation becomes insignificant after a lag of three. Therefore, the value of p in Eq. 3.18 is taken as 3 to account only for the significant autocorrelation in the data. The power of this test is comparable to that of the original Mann-Kendall test while providing much more accurate significance levels (Hamed and Rao 1998).

The second approach to deal with autocorrelation is to remove it from the data. This can be accomplished in two steps. First, data are transformed to normality, and then the autocorrelation is removed by fitting AR(p) models to the data. As stated before, $p = 3$ was found to be sufficient to account for the autocorrelation in the data. In transforming data to normality, considering the large amount of data to be processed, it was necessary to follow a systematic approach. The approach used here is the inverse of the monotonic transformation technique, which is originally used to obtain correlated non-Gaussian series from Gaussian ones (Grigoriu 1995). The transformed Gaussian time series in this case would be as in Eq. 3.19,

$$y_i = \Phi^{-1}\left[F(x_i)\right] \tag{3.19}$$

where y_i is the transformed Gaussian series, Φ^{-1} (.) is the inverse normal cumulative distribution function, and $F(x_i)$ is the cumulative distribution function of x_i, the original non-Gaussian time series. Since the actual distribution function $F(x_i)$ is not known, it is estimated from the data. A plotting position estimator for $F(x_i)$ given in Eq. 3.20 is used,

$$F(x_i) = \frac{R_i - 0.5}{n} \tag{3.20}$$

where R_i is the rank of the observation x_i and n is the total number of observations. This formula is due to Hazen (1914) and may be used for skewed data. Investigation of other plotting position formulae indicated that the choice of a plotting position formula did not have a significant effect on trend test results. AR(3) models are then fitted to the detrended data and the tests are performed on the AR-filtered data (the uncorrelated residuals) in addition to the trend component. The normal transformation allows for the investigation of the performance of parametric tests and the comparison of their results with the results from non-parametric tests.

3.5. APPLICATION OF TREND TESTS TO HYDROLOGIC DATA

3.5.1. Visual Inspection of Data. Visual inspection of the time series plots in Figures 2.1 through 2.6 does not reveal adequate information about possible trends in the data. Periods of below-average and above-average values can be noticed in most of the data series. However, these variations can be attributed to short-term dependence (persistence) that is commonly observed in hydrologic and other data. The Rescaled Adjusted Partial Sums (RAPS) are calculated and plotted for different time series in the study region in Figures 3.1 through 3.6.

The RAPS plots of the USGS flow data are shown in Fig. 3.1. A number of stations exhibit a parabolic shape indicating the possibility of linear trends. Data from stations 3 to 6 in Iowa (Fig. 3.1a), station 3 in Illinois (Fig. 3.1b), station 6 in Indiana (Fig. 3.1c), and station 7 in Ohio (Fig. 3.1d) indicate a positive trend. Data from station 2 in Iowa (Fig. 3.1a), on the other hand, indicate a negative trend for the first part of the record up to around 1940. Data from the remaining stations do not indicate trends in the data.

The RAPS plots for the HCN rainfall data are shown in Fig. 3.2. A larger number of stations in Figure 3.2 exhibit the parabolic shape than in Figure 3.1 for flow data. Data from stations 2, 4, 5 , 7, and 9 in Iowa (Fig. 3.2a); stations 1 to 4, 7, 8, and 10 in Illinois (Fig. 3.2b); stations 1, 3, and 7 in Indiana (Fig. 3.2c); and stations 3 and 4 in Ohio (Fig. 3.2d) all indicate the possibility of a positive trend. On the other hand, data from station 3 in Iowa (Fig. 3.2a), station 2 in Indiana (Fig. 3.2c), and station 1 in Ohio (Fig.3.2d) indicate the possibility of negative trends.

The RAPS plots for the NCDC divisional rainfall data are shown in Fig. 3.3. It can be noticed that the plots in Fig. 3.3 are much more uniform than those in Fig. 3.2 for the individual rainfall stations. This is mainly due to the fact that divisional data are the spatial averages of a number of individual stations, and hence random fluctuations due to spatial variability are smoothed by the averaging process. In general, the RAPS plots in Fig. 3.3 indicate possible positive trends for most of the divisions, especially after 1930. However, the plot for division 9 in Ohio (Fig. 3.3d) indicates a negative trend.

The RAPS plots for the HCN temperature data are shown in Fig. 3.4. Unlike flow and rainfall data in Figs. 3.1 to 3.3, temperature data in Fig. 3.4 indicate two types of trend. For most of the stations there is an indication of a positive trend for the first part of the record from around 1890 till 1950. Data in the period from around 1950 till the end of the record in 1987, on the other hand, indicate a negative trend.

RAPS plots for the NCDC divisional temperature data are presented in Fig. 3.5. As was the case with divisional rainfall data, the plots in Figure 3.5 are more uniform than the plots for the individual stations in Figure 3.4. The plots in Figure 3.5 indicate, in general, a possible positive trend for the first part of the record between 1900 and 1950, and a possible negative trend from around 1950 to the end of the record in 1992.

The RAPS plots for the NOAA tree-ring data are given in Fig. 3.6. It can be seen that most of the data in Iowa (Fig. 3.6a) and Illinois (Fig. 3.6b) indicate no trends. However, station 9 in Illinois (Fig. 3.6b) as well as the three stations in Indiana and Ohio (Fig. 3.6c) indicate periods of positive and negative trends ranging from 30 to 100 years.

The above interpretation of visual inspection results is highly subjective. Nevertheless, some indications of the type and extent of trends in the data are obtained by comparing the results from different sites in the region. Next, statistical tests are used to evaluate the significance of trends in the data. Both the overall length and subperiods of the records are investigated.

3.5.2. Statistical Trend Tests. The trend tests are performed at significance level α of 5%, which is commonly used. A trend test for an individual station merely indicates whether the trend at that station is significant or not at the 5% level. However, data from a number of stations in a region can be considered as different realizations of the same process, thus providing more information about trends. For a region with m sites, the expected number of significant sites under the null hypothesis of no trend would be 0.05 m. A confidence interval for that expected number at the 5% significant level can be obtained by using the binomial distribution with m observations and a probability of failure p of 0.05. The null hypothesis of no-trend should therefore be rejected at the 5% level if the number of significant stations falls outside the specified confidence interval.

For flow data from 26 stations, the expected number of significant sites at 5% significance is one site and a 95% confidence interval is zero to 3. We will use the notation 0(1)3 for the confidence interval where the number in parenthesis is the mean value. For HCN temperature and rainfall data from 36 stations, the confidence interval is 0(2)4. For state division temperature and rainfall data from 37 stations, the confidence interval is also 0(2)4. For tree-ring data from 22 sites, the confidence interval is 0(1)3.

The total number of sites with significant trends (positive and negative) are given in Tables III.1 through III.6 for each of the six data sets. The number of sites with negative trend, where such information is given by a test (MK, MM, and LM tests) is shown as a superscript to the total number of stations with significant trends. Numbers in bold typeface in Tables III.1 through III.6 indicate that the number of stations with significant trends is within the confidence interval for that group of sites at 5% significance. The trend tests are performed on the original, normalized, and normalized-AR-filtered data. Normalized data are obtained by applying the transformation in Eq. 3.19 to the data. AR filtering is achieved by fitting an AR(3) model to the detrended data and considering the residuals from the model in addition to the trend component. The following abbreviations are used for the test names:

CD = Cumulative Deviation Test, WL = Worsely Likelihood Test, LM = Linear Model Test, MK = Mann-Kendall Test, MM = Modified Mann-Kendall Test, CU = Non-parametric CUMSUM Test, KW = Kruskall-Wallis Test which is performed using 5, 10, 20, and 30-year sub-periods. The LM, MK, and MM tests in Tables III.1 through III.6 are separated from the other tests by a thick line. These tests use direct measures of trend, i.e. the test statistic is directly related to the slope of the trend. Therefore negative trends can be distinguished from positive ones when these three tests are used.

The results in Tables III.1 through III.6 indicate that the effect of non-normality of the data on various trend test results is very small. This is reflected by the small differences in the number of significant stations indicated by parametric tests, which are sensitive to non-normality of the data, when using the original versus normalized data. This is to be expected, since the departure from normality is considered small for flow, rainfall and temperature series.

Table III.1: Trend tests on USGS flow data. Given is the number of significant stations at 5% significance. Total number of stations is 26. The 95% C.I. is 0(1)3.

	Parametric			Non-Parametric						
	CD	WL	LM	MK	MM	CU	KW5	KW10	KW20	KW30
Original	7	7	6	6	5	11	5	7	7	4
Normalized	5	5	6	6	5	11	5	7	7	4
Nor.-Filtered	**2**	**2**	6	4	4	11	**0**	4	5	**2**

* Bold is within 95% CI.

Table III.2: Trend tests on HCN rainfall data. Given is the number of significant stations* at 5% significance. Total number of stations is 36. The 95% C.I. is 0(2)4.

	Parametric			Non-Parametric						
	CD	WL	LM	MK	MM	CU	KW5	KW10	KW20	KW30
Original	17	19	18^{1}	19^{1}	19^{1}	21	5	15	15	14
Normalized	18	19	19^{1}	19^{1}	19^{1}	21	5	15	15	14
Nor.-Filtered	15	16	18^{1}	18^{1}	18^{1}	20	8	12	17	13

* Number is total significant, superscript is negative trend. Bold is within 95% CI.

Table III.3: Trend tests on NCDC divisional rainfall data. Given is the number of significant stations at 5% significance. Total number of stations is 37. The 95% C.I. is 0(2)4.

	Parametric			Non-Parametric						
	CD	WL	LM	MK	MM	CU	KW5	KW10	KW20	KW30
Original	**2**	**2**	7^{0}	6^{0}	13^{1}	16	**0**	**0**	**2**	**1**
Normalized	**2**	**2**	6^{0}	6^{0}	13^{1}	16	**0**	**0**	**2**	**1**
Nor.-Filtered	8	**3**	10^{1}	10^{1}	10^{1}	19	**0**	**0**	**3**	**2**

* Number is total significant, superscript is negative trend. Bold is within 95% CI.

Table III.4: Trend tests on HCN temperature data. Given is the number of significant stations at 5 % significance. Total number of stations is 36. The 95% C.I. is 0(2)4.

	Parametric			Non-Parametric						
	CD	WL	L M	MK	MM	CU	KW5	KW10	KW20	KW30
Original	22	24	16_3	16^3	11^3	27	28	22	14	36
Normalized	22	25	16_3	16^3	11^3	27	28	22	14	36
Nor.- Filtered	9	8	8^1	8^1	8^1	17	**2**	**1**	**1**	16

* Number is total significant, superscript is negative trend. Bold is within 95% CI.

Table III.5: Trend tests on NCDC divisional temperature data. Given is the number of significant stations* at 5% significance. Total number of stations is 37. The 95% C.I. is 0(2)4.

	Parametric			Non-Parametric						
	CD	WL	LM	MK	MM	CU	KW5	KW10	KW20	KW30
Original	6	5	**0**	**0**	**0**	23	20	12	14	8
Normalized	**4**	6	**0**	**0**	**0**	23	20	12	14	8
Nor.- Filtered	**0**	**0**	**0**	**0**	**0**	5	**0**	**0**	**0**	**0**

* Number is total significant, superscript is negative trend. Bold is within 95% CI.

Table III.6: Trend tests on NOAA tree-ring data. Given is the number of significant stations* at 5% significance. The number of series is 22. The 95% C.I. is 0(1)3.

	Parametric			Non-Parametric						
	CD	WL	LM	MK	MM	CU	KW5	KW10	KW20	KW30
Original	**2**	6	**0**	**0**	**0**	6	15	6	5	4
Normalized	**0**	5	**0**	**0**	**0**	6	15	6	5	4
Nor.- Filtered	**0**	**3**	**0**	**0**	**0**	**0**	**0**	**2**	**0**	**0**

* Bold numbers indicate that the value is within 95% CI.

The effect of serial correlation, on the other hand, is found to be substantial for data with large autocorrelations. The results in Table III.2 for the HCN rainfall data, which are the least autocorrelated of the data, indicate that the number of significant stations does not change very much between the case of original autocorrelated data (19 stations, MK test) and the case of uncorrelated AR-filtered data (18 stations, MK test). In fact, based on the results for divisional rainfall data in Table III.3 the number of stations with significant trends (10 stations, MK test) is underestimated when the original data are used (6 stations, MK test), due to negative autocorrelation in some of the time series. For the USGS flow data in Table III.1, the number of stations with significant trend obtained by using the AR-filtered data (4 stations, MK test) is slightly smaller than that indicated by testing the original data (6 stations, MK test). This is due to the existence of positive autocorrelation in the original data. For the HCN temperature data in Table III.4 the number of stations with significant trend drops from 16 stations when using the MK test to 8 stations after accounting for autocorrelation in the data either by AR-filtering or by using the MM test.

In general, the test results in Tables III.1 through III.6 indicate that the MM test, which is designed to account for autocorrelation in the data, performs better than the LM and MK tests when autocorrelated data are tested. For example, the number of stations with significant trends in Table III.1 from the MK test for original auto-correlated data is 6 stations, while the number given by the MM test is 5. After accounting for autocorrelation through AR-filtering, the number of stations drops to 4 when either the MK or MM test is used. Thus, the MM test gives a more reasonable number of stations by accounting for autocorrelation without the need for filtering the data. The same can be observed in Table III.4, where the MK test indicates 16 stations with significant trends. The MM test indicates 11 stations, while both tests indicate 8 stations using uncorrelated data.

As can be seen from the formulation of the trend tests in section 3.3, the LM, MK, and MM tests, which are enclosed between the two thick lines in Tables III.1 through III.6, use direct measures of trend, i.e. the test statistic is directly related to the slope of the trend. On the other hand, the CD, WL, CU, and KW tests evaluate trends by considering changes in the mean (or median) of the time series and therefore tend to be more sensitive to non-monotonic fluctuations in the data in addition to monotonic trends. For example, if a time series exhibits an increasing trend followed by a decreasing one, direct trend tests will not show an overall significance, while the other tests will reflect the fluctuations in the mean. The CU test appears to be the most sensitive of the tests for fluctuations in data as seen in Tables III.1 to III.6. For example, in Table III.4, 27 stations are identified as having significant changes in the mean, while other trend tests indicate only 11 stations have significant trends. The CD and WL tests give comparable results to the direct LM, MK and MM tests. For example, on the last line in Table III.2, 18 stations are identified by the LM, MK, and MM tests as having significant trends. The same is true on the last lines of Tables III.3 and III.4 with 10 and 8 stations, respectively. The KW test with 5, 10, 20 and 30-year sub-periods gives results that depend, as might be expected, on the length of the sub-period. Sub-periods of 20 and 30 years seem to give the largest number of significant sites, which gives an idea of the scale of significant fluctuations. Results from the direct trend tests LM, MK, and MM agree for normalized-filtered data, with the CD test being the closest to direct tests among indirect tests.

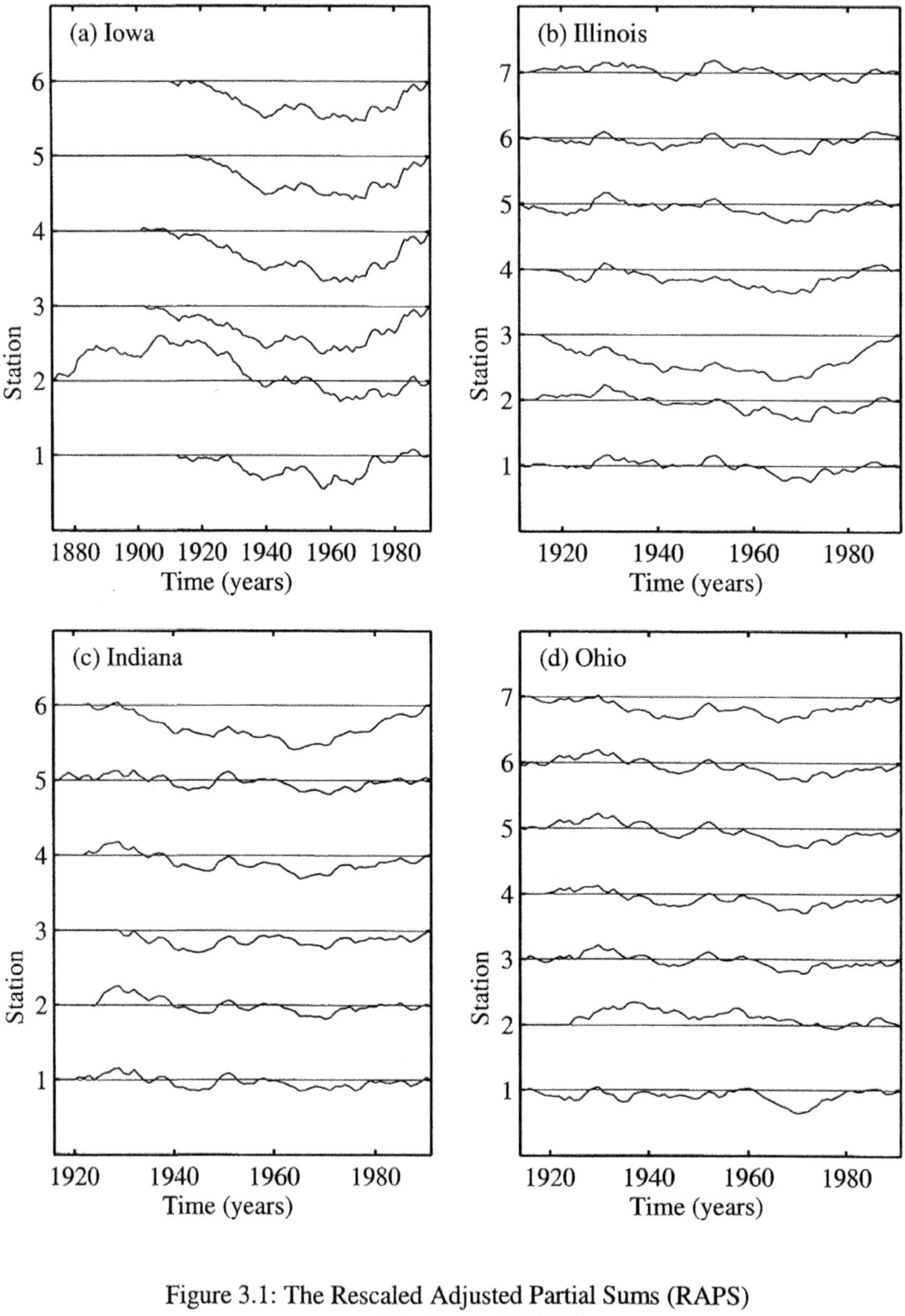

Figure 3.1: The Rescaled Adjusted Partial Sums (RAPS) plots for the USGS flow.

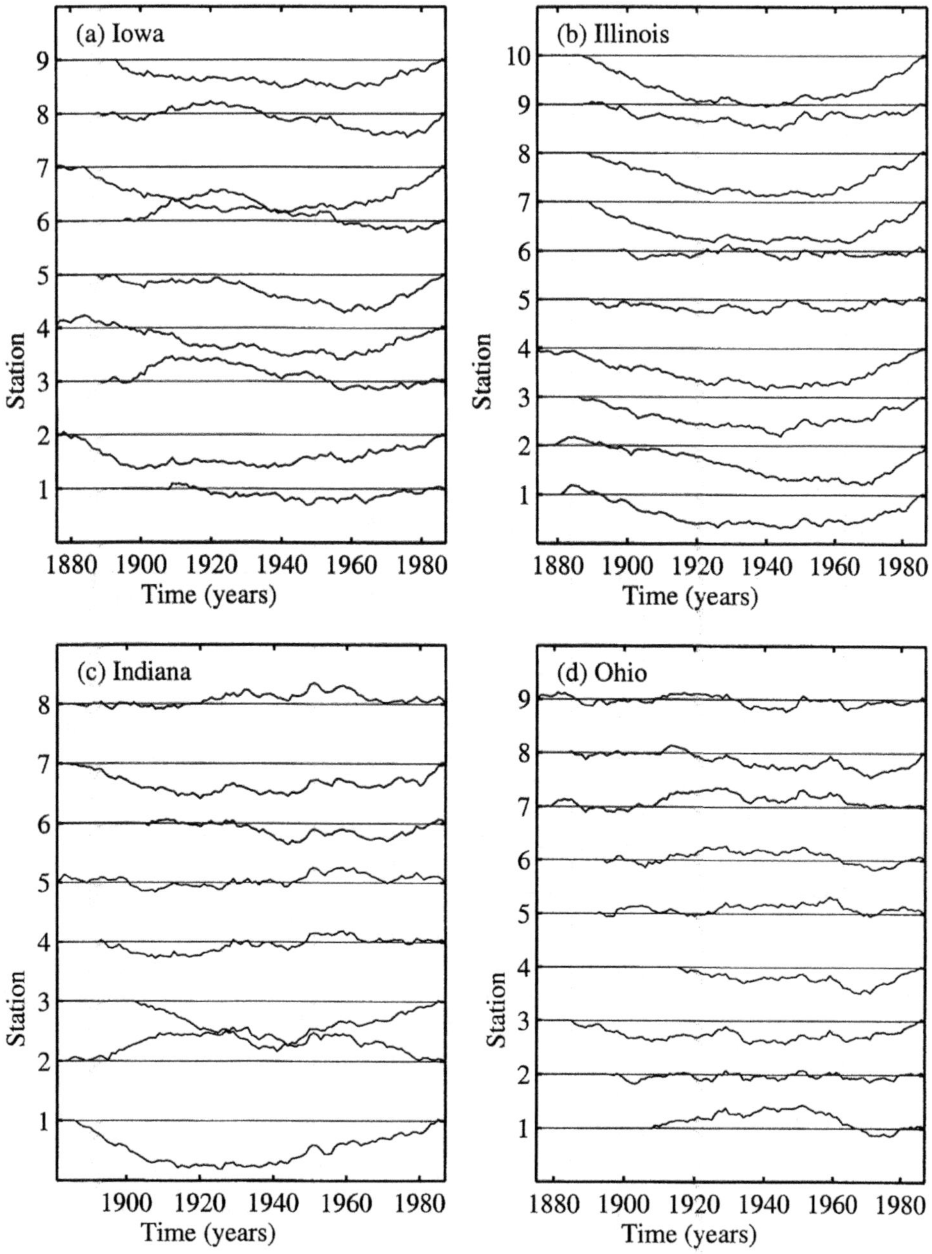

Figure 3.2: The Rescaled Adjusted Partial Sums (RAPS) plots for the HCN rainfall data.

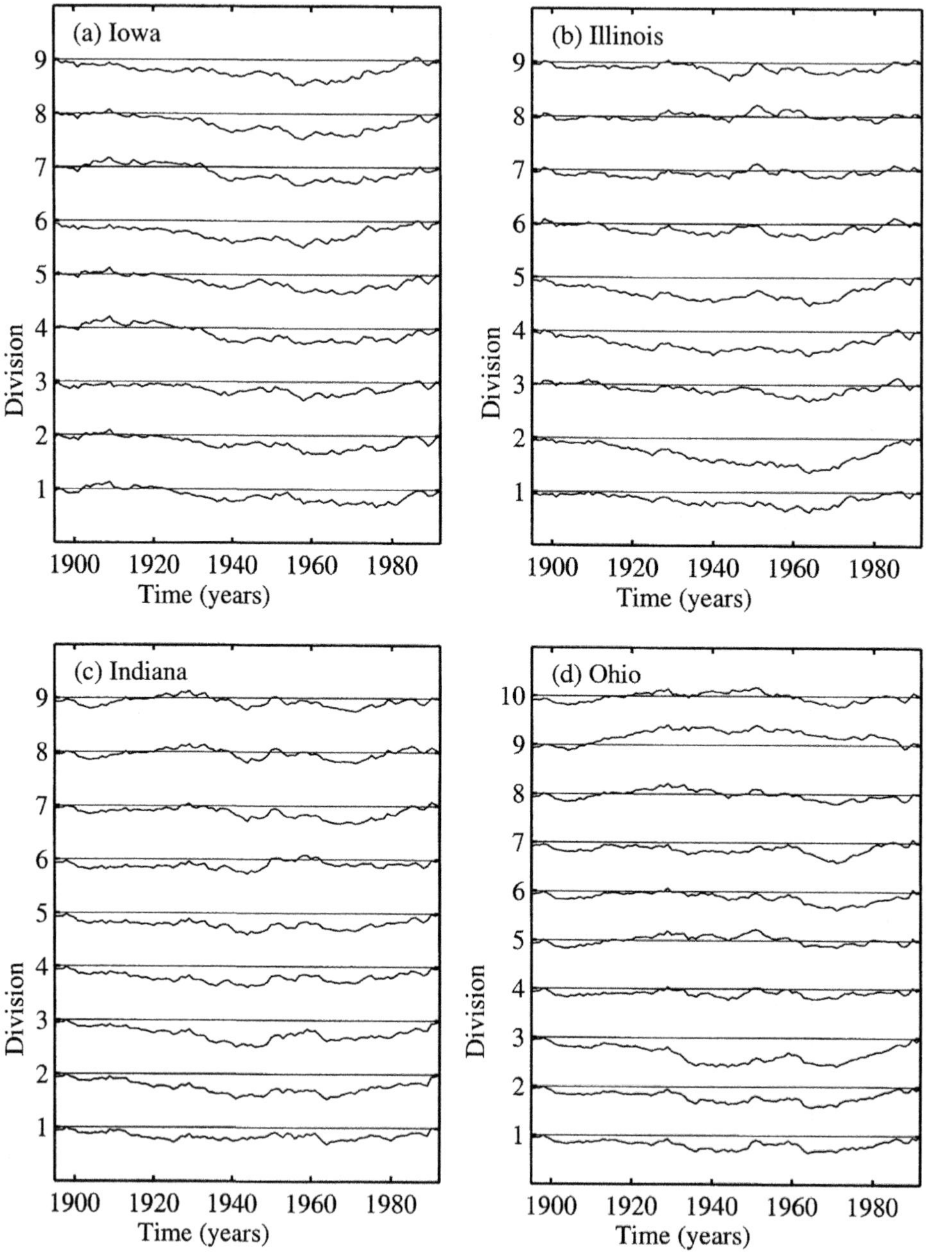

Figure 3.3: The Rescaled Adjusted Partial Sums (RAPS) plots for the NCDC divisional rainfall data.

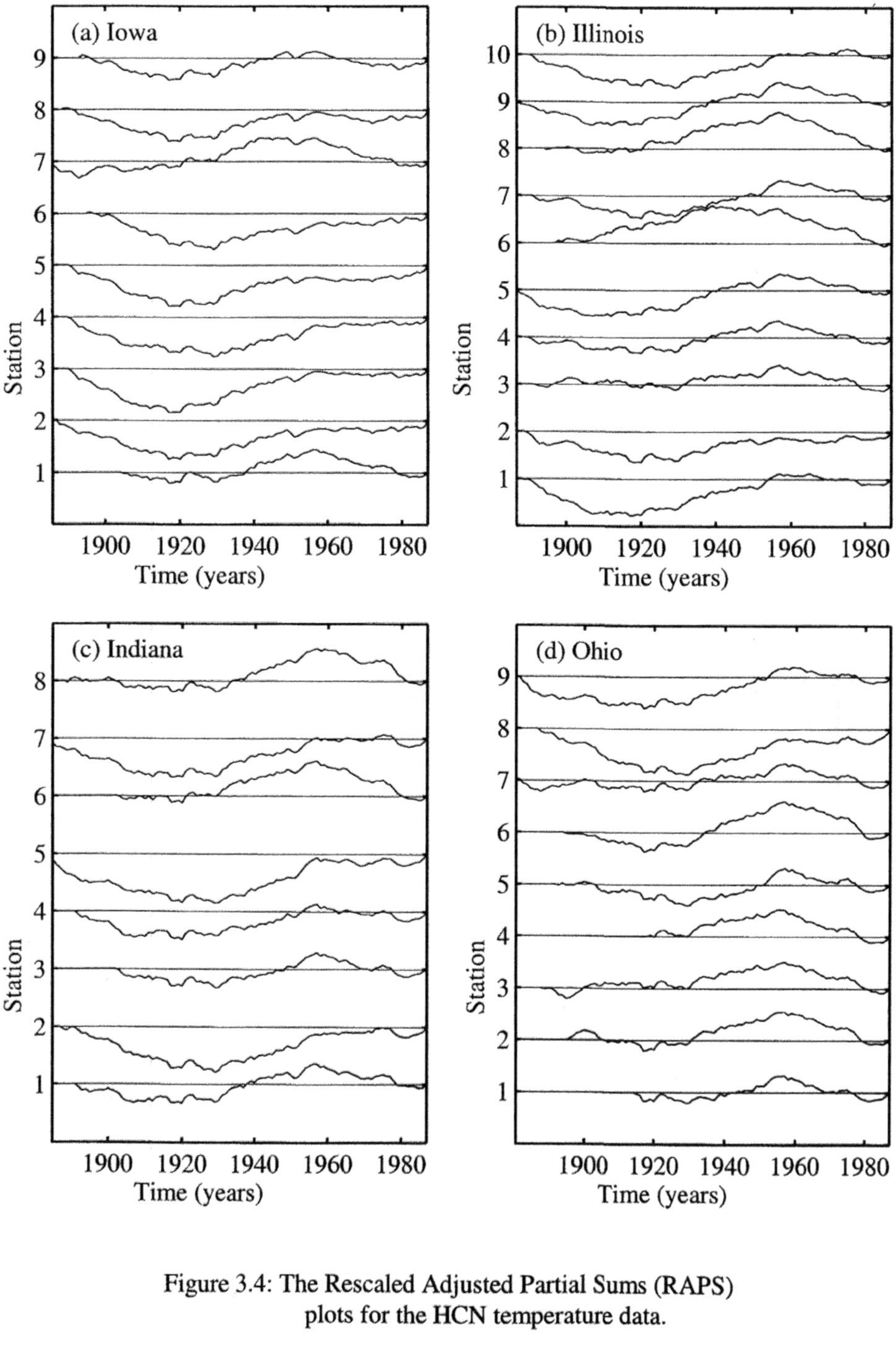

Figure 3.4: The Rescaled Adjusted Partial Sums (RAPS) plots for the HCN temperature data.

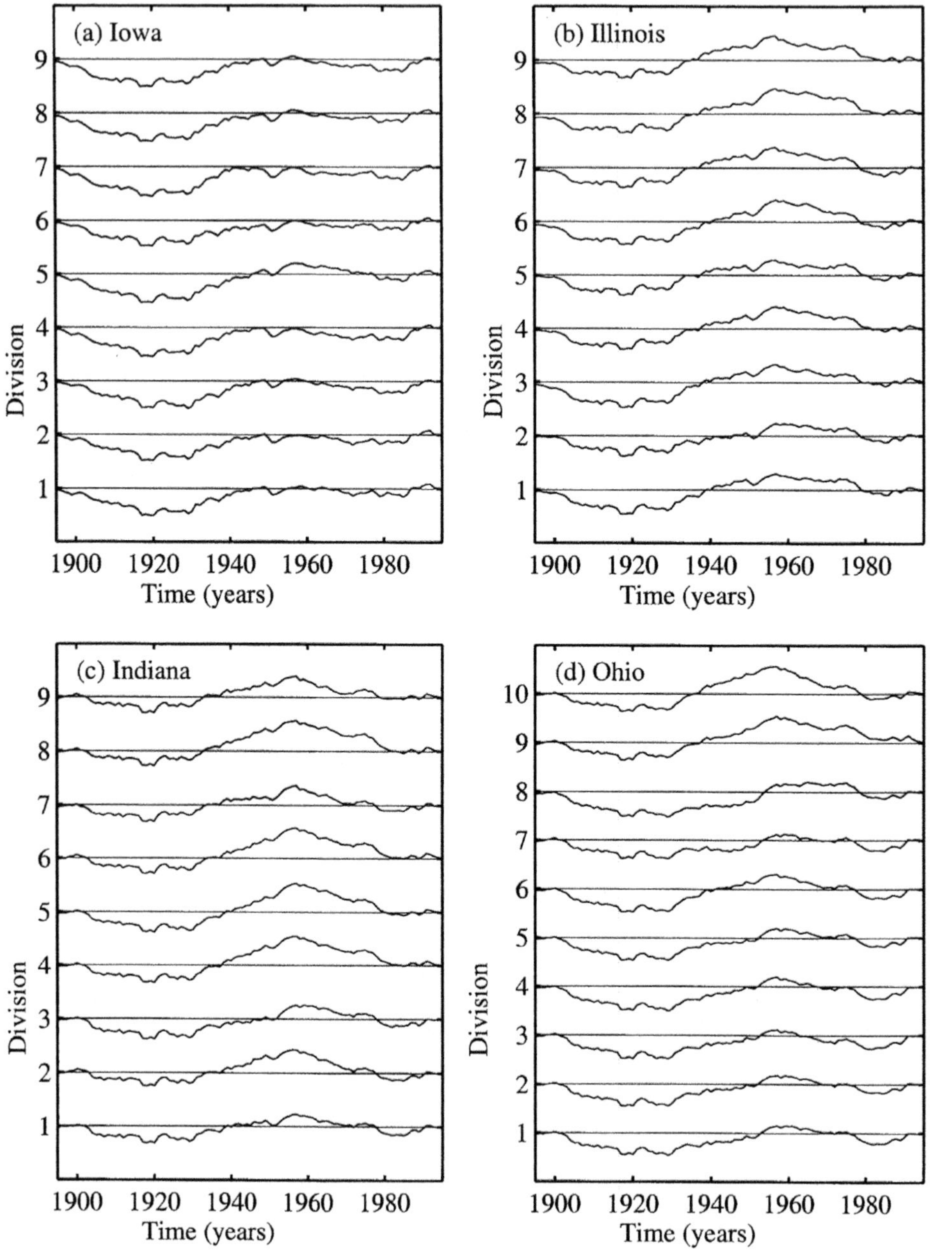

Figure 3.5: The Rescaled Adjusted Partial Sums (RAPS) plots for the NCDC divisional temperature data.

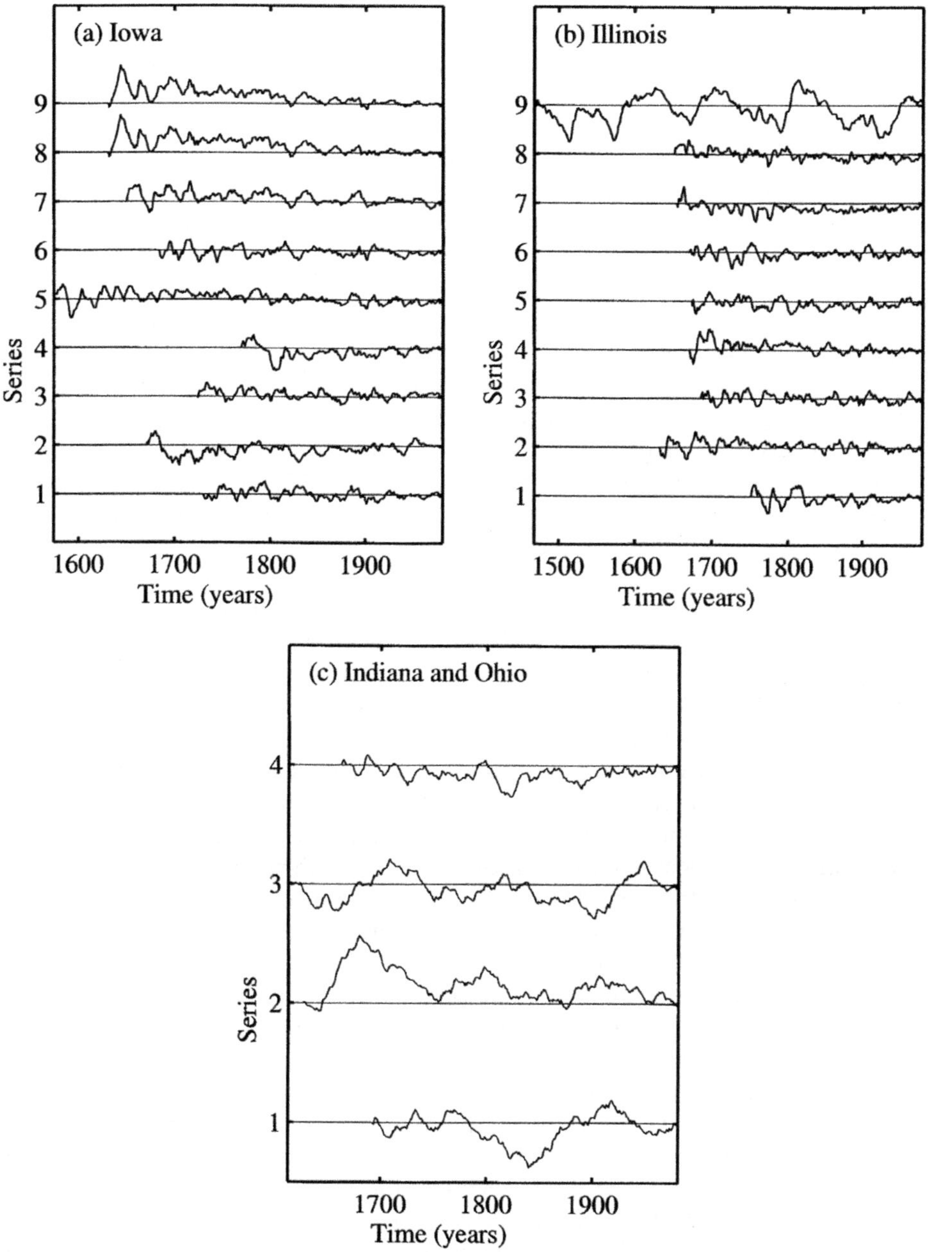

Figure 3.6: The Rescaled Adjusted Partial Sums (RAPS) plots for the NOAA tree-ring data.

The results from testing the USGS flow data in Table III.1 indicate that about four of the 26 stations have significant positive trends at the 0.05 level. This number is slightly outside the 95% confidence interval of 0(1)3, and the null hypothesis of no overall trend must be rejected, but the evidence is not overwhelming. In addition, the CU test indicates that 11 stations suffer from significant fluctuations in the mean at the 0.05 level. For the HCN rainfall data, the results in Table III.2 indicate 18 stations out of 36 have significant trends, 17 of which exhibit positive trends. With a confidence interval of 0(2)4 the test results give stronger support for rejection of the null hypothesis of no overall trend. The results from the NCDC divisional rainfall data in Table III.3 also support the rejection of the null hypothesis. The number of time series with significant trends in rainfall data in Table III.3 is 10 out of 37, of which nine series exhibit positive trends, which is well outside the 95% confidence interval of 0(2)4. In addition, the CU test indicates that 19 stations suffer from significant fluctuations in the mean at the 0.05 level. The results for the HCN temperature data in Table III.4 indicate eight significant stations, seven of which have positive trends. With a confidence interval of 0(2)4, these results further support the rejection of the null hypothesis. The CU test also indicates that 17 stations suffer from significant fluctuations in the mean at the 0.05 level. However, results from the NCDC divisional temperature data in Table III.5 show no evidence of significant trends. Results for the NOAA tree-ring data in Table III.6 also show no evidence of overall trend in the record. The tree-ring record is the longest of the data sets considered in this study with more than 250 years of data in each series.

3.5.3. Sub-period Trend Analysis. The previous analysis of the results in Tables III.1 through III.6 reveals inconsistencies between trends for different data types. These inconsistencies are most obvious when the results from HCN temperature data and those from state division data in Tables III.4 and III.5, respectively are compared. Also the tree-ring data, although closely related to temperature and rainfall data, do not show any significant overall trends. These inconsistencies in trend test results are partly due to the differences in the starting points and length of the record for different data groups. In order to further study the nature of trends in the data and resolve the inconsistencies, trends within sub-periods of the records are studied. The sub-period trend analysis is also motivated by the results obtained from the visual inspection of the RAPS plots in Figs. 3.1 through 3.6, which indicate the existence of trends in opposite directions at different times in some of the data series.

The sub-period trend analysis is performed by checking at each point of the time series for significant trends that span the largest number of observations. Only trends extending for more than 10 years are considered to be meaningful long term trends, as opposed to short term fluctuations common to autocorrelated series with moderate values of correlation. Starting at the first point of the time series, x_1, the sub-series $x_1 .. x_n$, $x_1 .. x_{n-1}$, ... x_1 ... x_{10} are formed and tested for significance of trends, and the longest significant trend is retained. Next, the analysis is advanced to the next point, x_2 and the longest significant trend over the remainder of the time series is retained. By doing this a significant trend spanning the largest number of points is obtained at each point. From these significant trends global trends which span the largest number of data can be easily identified.

In this analysis, data are first transformed to normality by applying the transformation in Eq. 3.19 in order to eliminate the effect of non-normality on the test results. To account for autocorrelation on trend significance, an AR(3) model is fitted to the detrended data and the test is performed on the residuals of the model in addition to the trend component. The MK test is used to test the significance of trends at this point in the analysis. The results of the sub-period trend analysis are presented in Figures 3.7 through 3.12 for the different data groups. Trends at various stations are depicted in these figures in image form, where black lines indicate significant positive trends at the 0.05 level, white lines indicate significant negative trends at the 0.05 level, and gray lines indicate the absence of significant trends. Each station is thus shown as a cluster of possibly overlapping black and white lines indicating periods of sustained positive or negative trends, respectively, on a gray background.

Figure 3.7 shows significant trends in the USGS flow data. The results in Fig. 3.7 indicate positive trends for most of the stations in the period from around 1930 to the end of record in 1992. Almost all stations in Figure 3.7 exhibit negative trends starting at around 1940 to 1960, except for stations 6 in Indiana, and 3 and 5 in Illinois. Although previous studies of flow data (Mitosek 1992, Chiew and McMahon 1993) do not indicate significant overall trends, it seems that their failure to detect trends is due to the existence of overlapping positive and negative trends at different times in the time series. Figure 3.7 clearly shows that an increasing trend in flow data exists starting from 1930 to 1992 in all stations except for stations 2 in Ohio, 1-3 in Indiana, and 7 in Illinois. On the other hand, the evidence for negative trends at the beginning of the record is not as strong. However, this may be mainly due to the lack of long flow data records. Indeed, data from the longest flow record (station 2 in Iowa) in Fig. 3.7 clearly indicate a negative trend extending from 1876 to 1940.

In Figure 3.8, for the HCN rainfall data some of the stations (stations 2, 4, 5, 7 and 9 in Iowa; 1-4 and 7-10 in Illinois; 1, 3, and 7 in Indiana; and 3 in Ohio) show a positive trend throughout the record starting roughly at 1900 until 1987. Another group of stations (stations 3 in Ohio, 6 and 7 in Indiana, and 6 in Illinois) exhibit positive trends starting from 1930 to 1987. Negative trends are observed in the period from 1910 to 1960 in stations 6 and 9 in Ohio; 6 and 8 in Indiana; 6 in Illinois; and 3, 6, and 8 in Iowa. Figure 3.9 for the state division rainfall data, on the other hand, shows more consistent trends than those in the HCN rainfall data in Figure 3.8. Significant positive trends are observed in many stations starting at around 1945 and extending towards the end of the record up to 1985 to 1992 especially in Illinois and Iowa. Significant negative trends are also observed mostly between 1900 and 1940 for a considerable number of stations (stations 6 in Ohio; 7, 8, and 9 in Indiana; and 9 in Illinois; and all stations in Iowa). These trends are similar to those observed in the USGS flow data series. The results also agree to a large extent with the results from studies by Groisman and Easterling (1994) and Karl et al. (1994b). The results from these studies indicate a decreasing trend from 1900 to 1935 and an increasing trend from 1935 to 1992 in United States national as well as some regional rainfall data.

The results from the HCN temperature data in Figure 3.10 indicate a significant positive trend at the beginning of the record around 1890 up to 1960 in almost all stations. A significant negative trend is also observed in the period from 1920 to 1980 in almost all stations. Another significant positive trend is observed starting at 1975 to the end of record at 1987. The results for the divisional temperature data in Figure 3.11

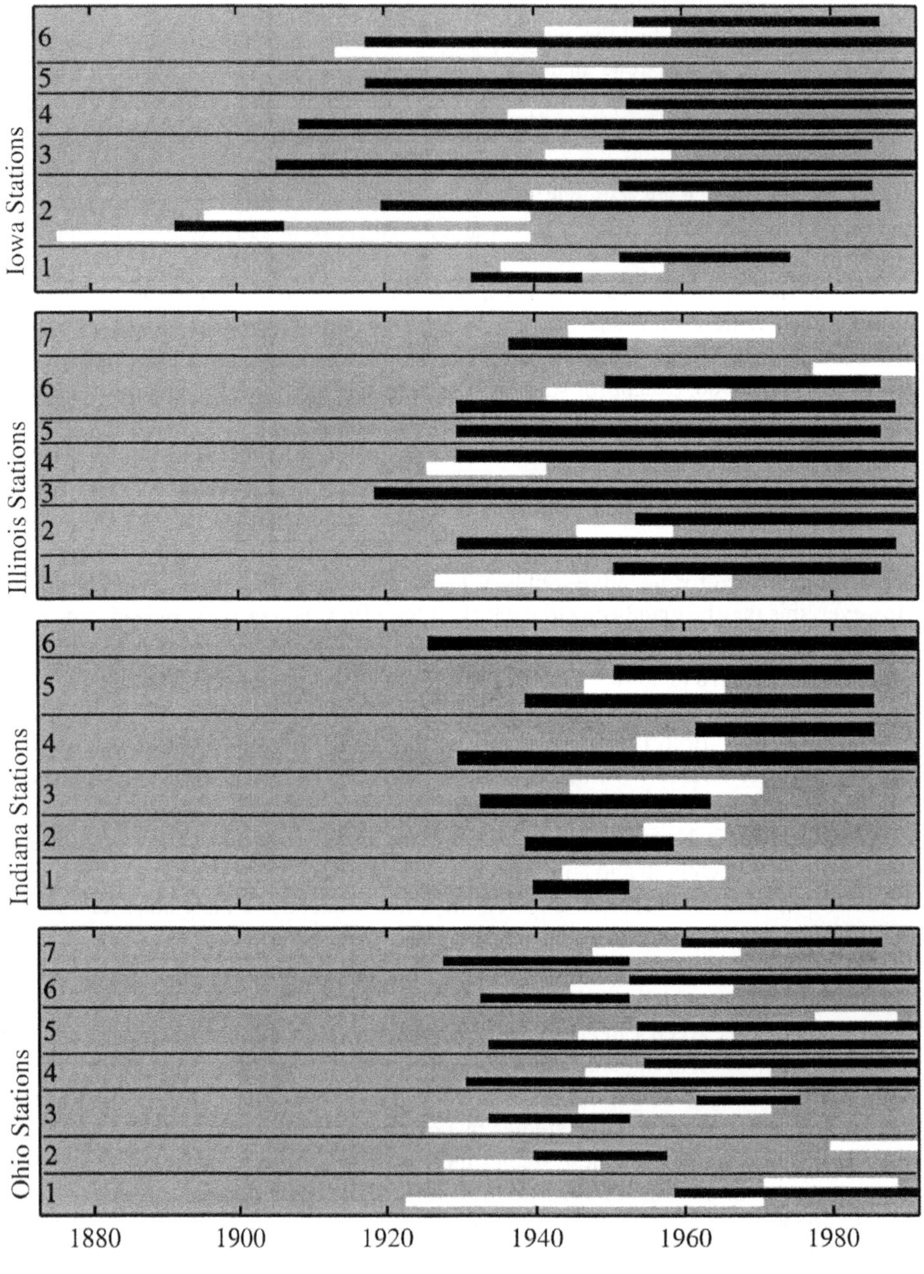

Figure 3.7: Significant trends in the USGS flow data. Black lines indicate positive trends, while white lines indicate negative trends.

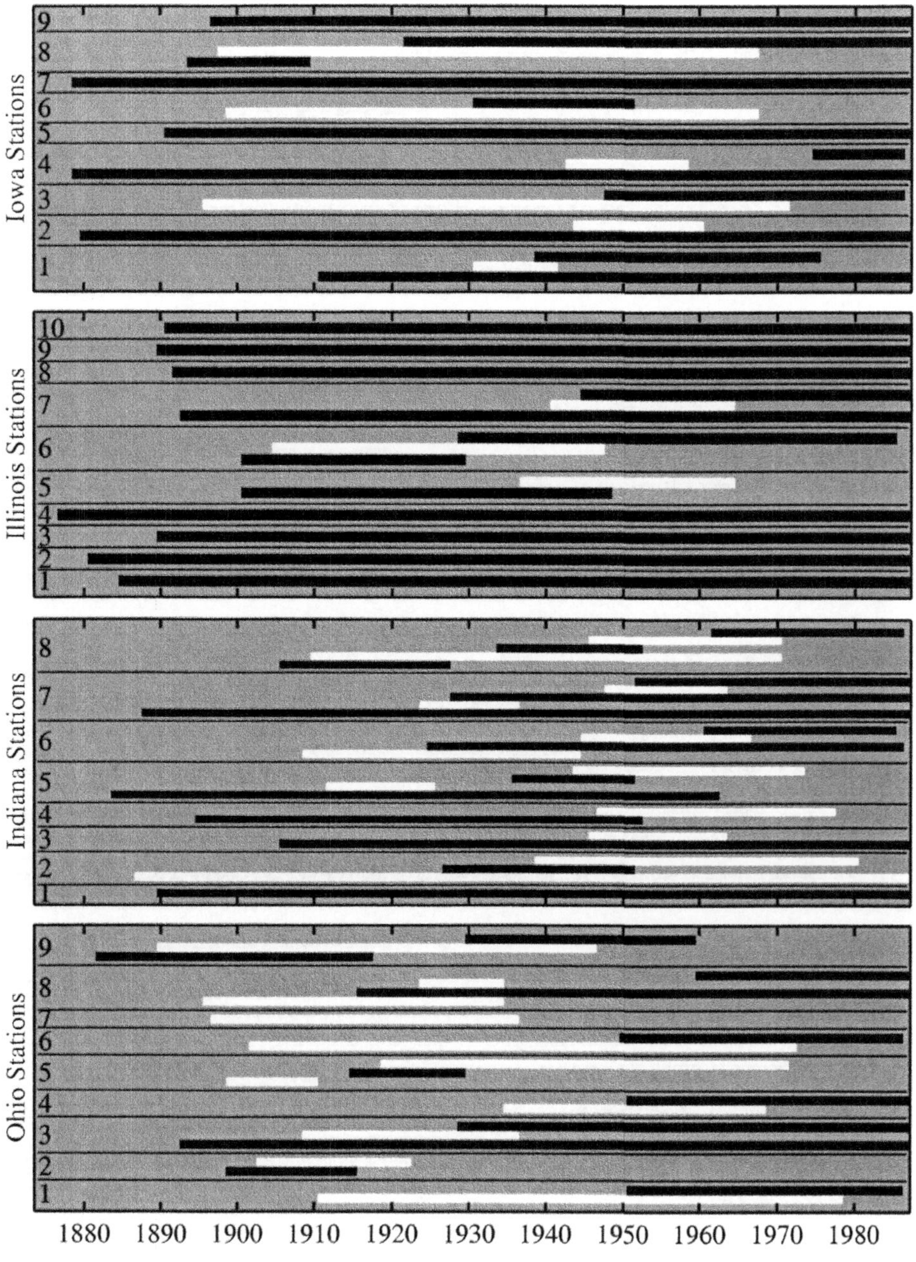

Figure 3.8: Significant trends in the HCN rainfall data. Black lines indicate positive trends, while white lines indicate negative trends.

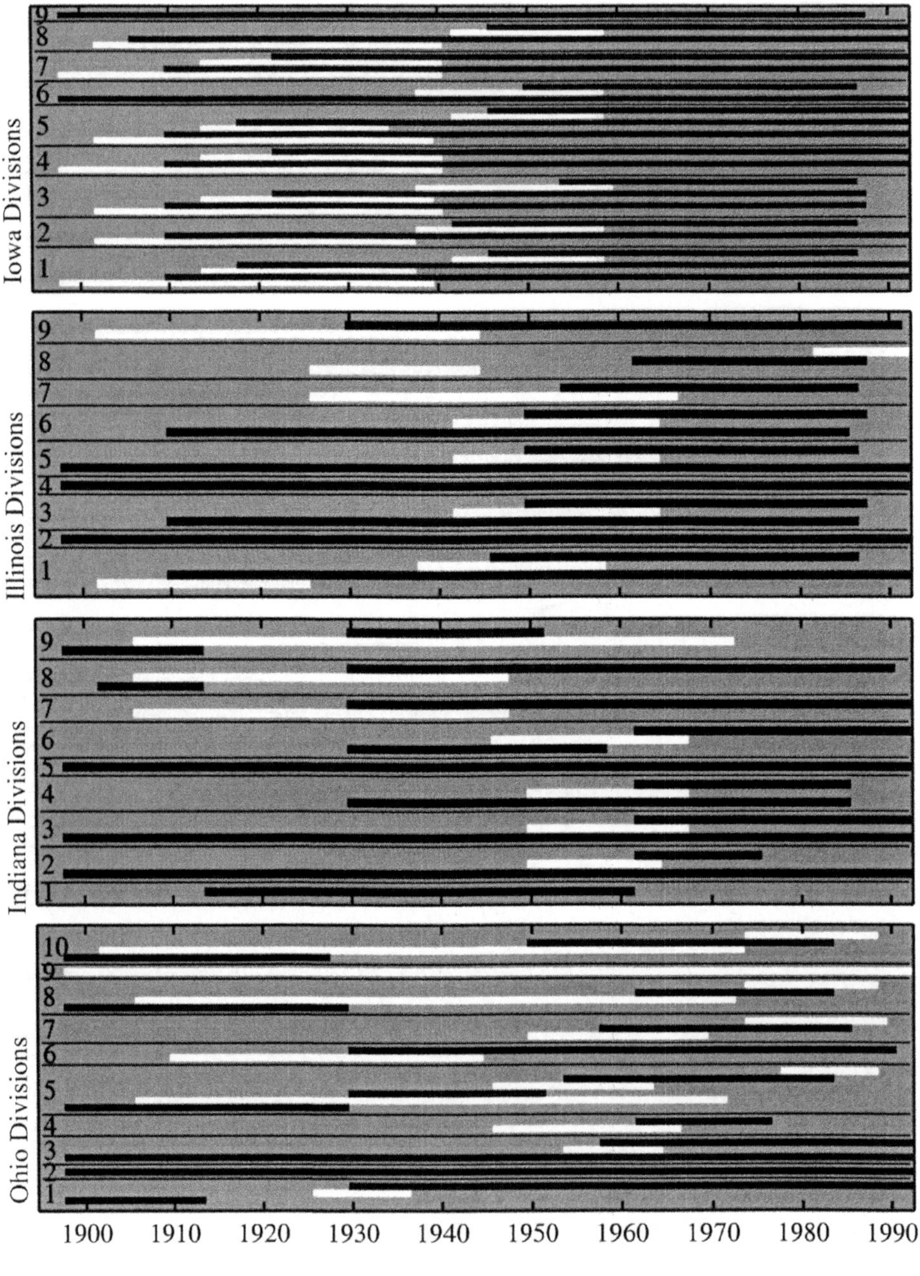

Figure 3.9: Significant trends in the NCDC divisional rainfall data. Black lines indicate positive trends while white lines indicate negative trends.

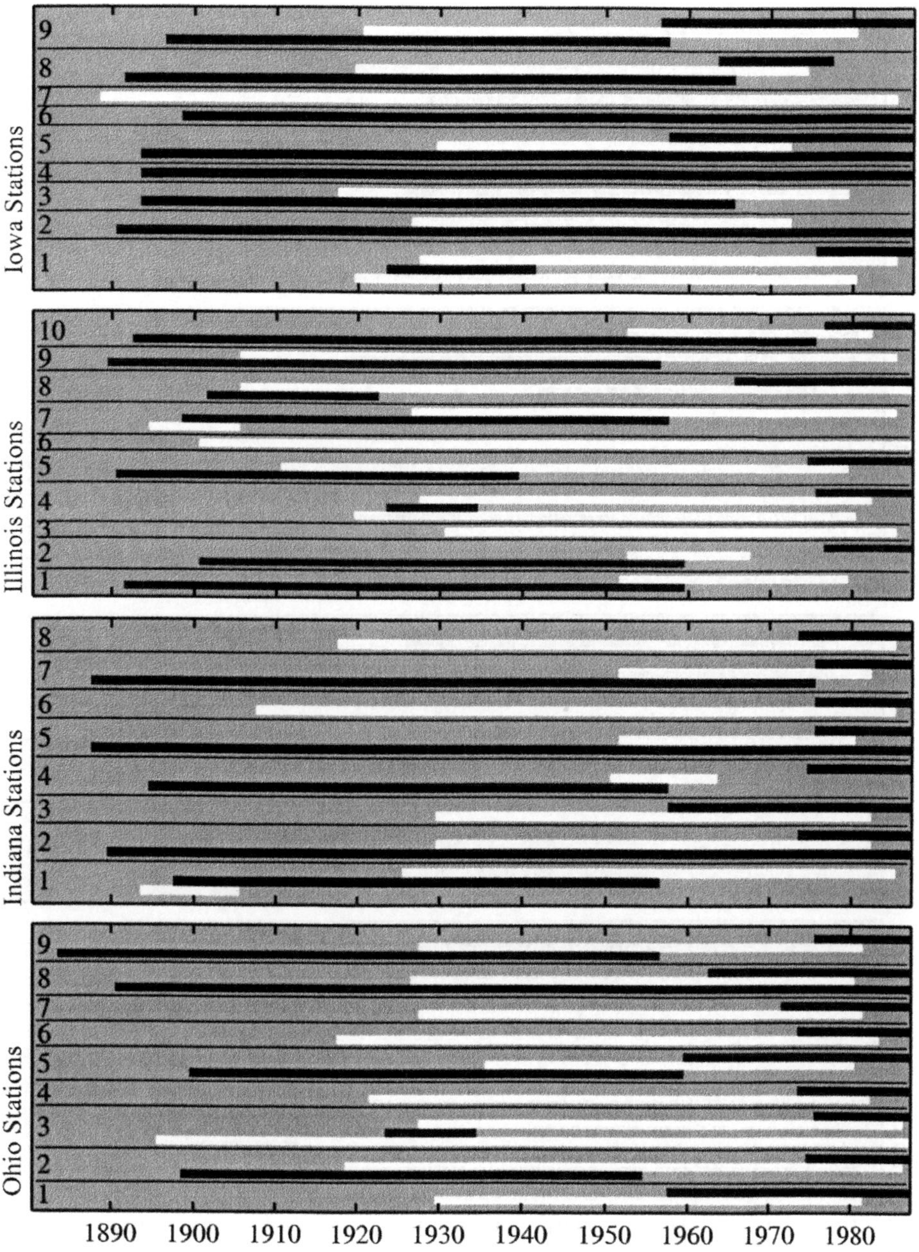

Figure 3.10: Significant trends in the HCN temperature data. Black lines indicate positive trends, while white lines indicate negative trends.

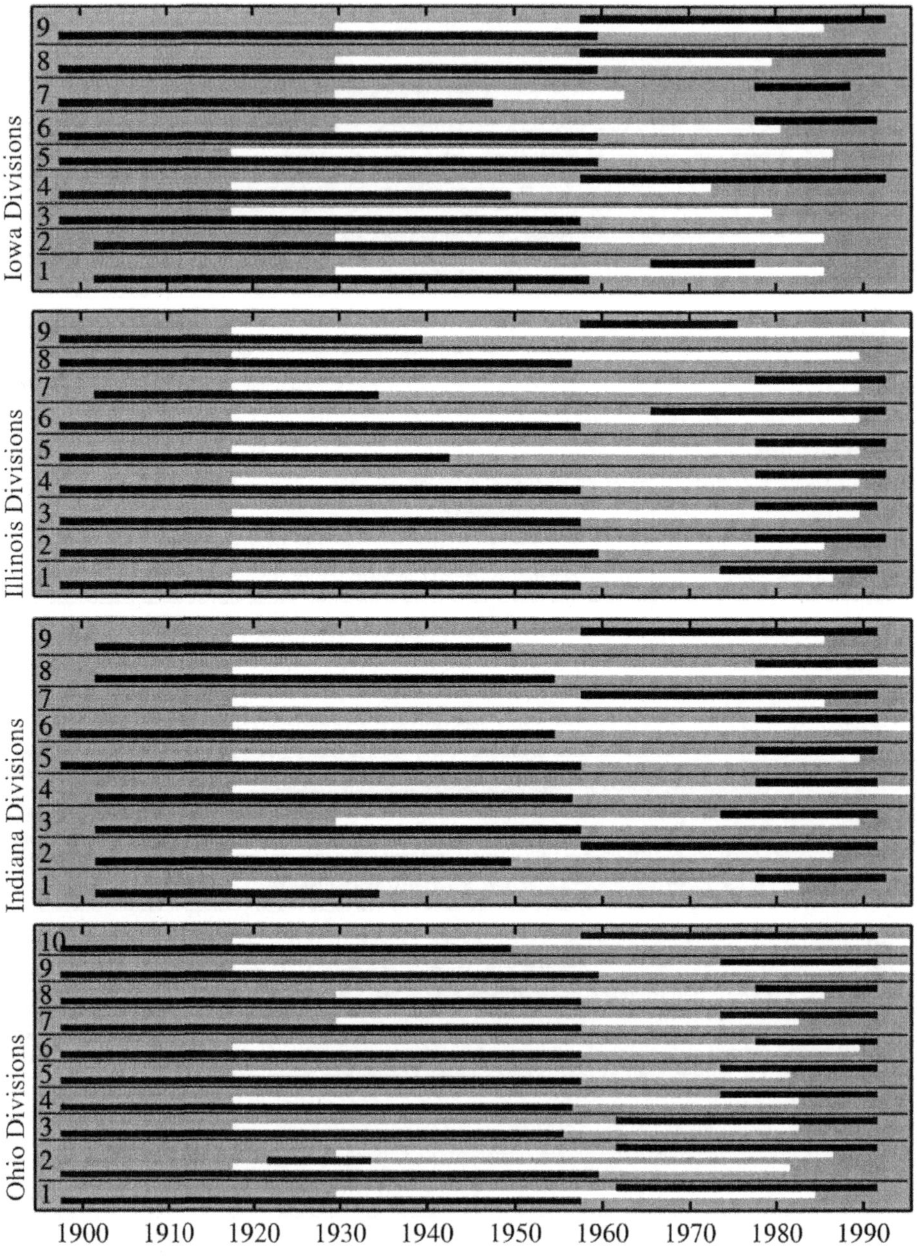

Figure 3.11: Significant trends in the NCDC divisional temperature data. Black lines indicate positive trends, while white lines indicate negative trends.

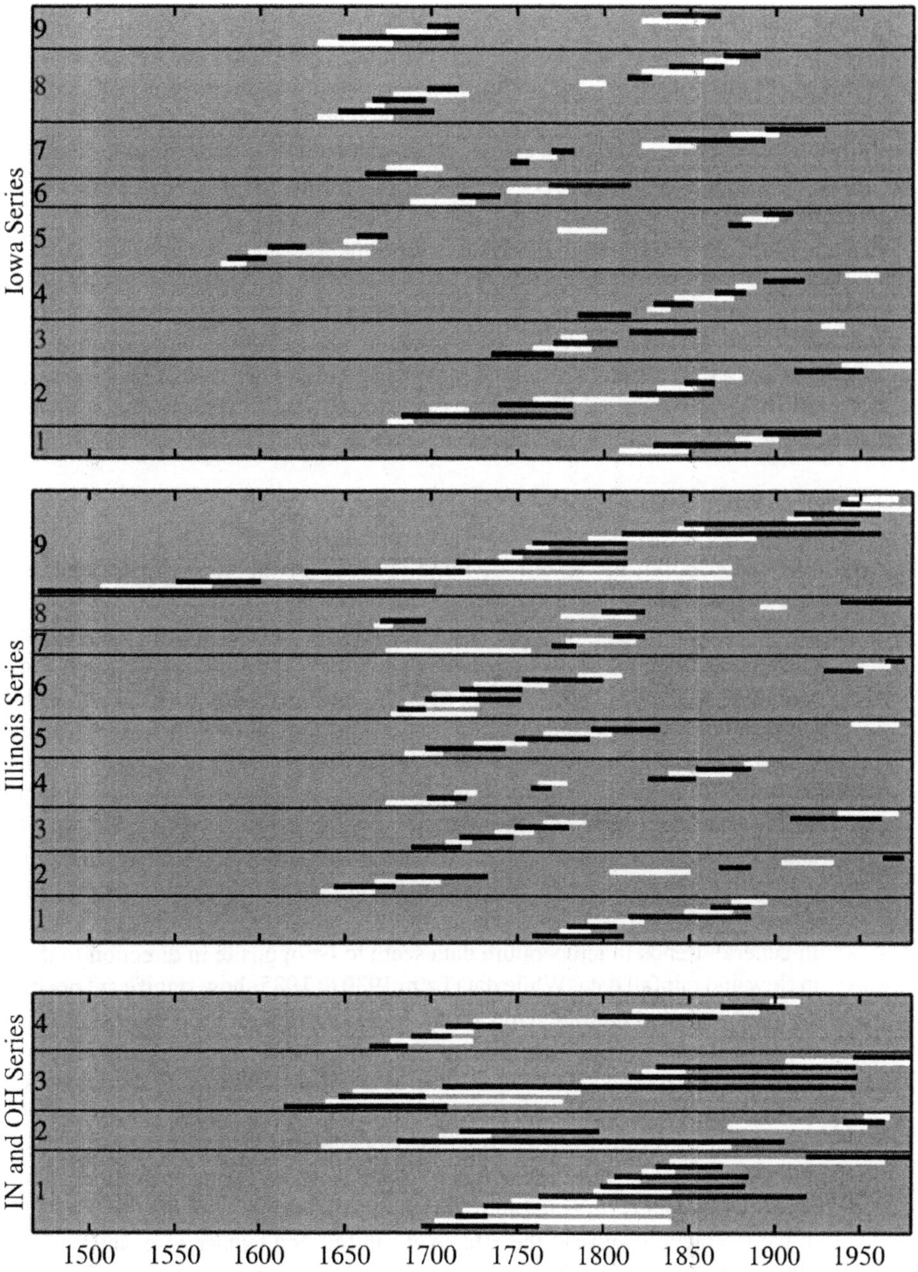

Figure 3.12: Significant trends in the NOAA tree-ring data. Black lines indicate positive trends, while white lines indicate negative trends.

indicate more consistent trends with a significant positive trend starting at around 1900 and extending to 1960. A significant negative trend is observed starting at 1928 and extends to around 1980. Another positive trend is observed starting at 1975 and extends to around 1992. These results are in good agreement with trends in global and United States national temperature data studied by Jones et al. (1994), and Wilson and Hansen (1994). The results from these two studies indicate a warming trend from 1900 to 1940 followed by a cooling trend from 1940 to 1965 and then followed by another warming trend from 1965 to 1993. Karl et al. (1994a) studied a number of United States national and regional data, from which they concluded that a cooling trend existed in the period from 1948 to 1987.

In Figure 3.12, the results from the NOAA tree-ring data show in general, cycles of alternating significant positive and negative trends. However, the fluctuation between positive and negative trends does not seem to be consistent among all time series. For the period from 1880 to 1980, which is concurrent to the available flow, rainfall, and temperature data, a few stations with significant positive trends and much fewer ones with negative trends can be observed. There is no evidence of trends in tree-ring data that are similar to those in flow, rainfall, and temperature data.

3.6. CONCLUSIONS

General conclusions from the results of trend analysis of the yearly hydrologic and meteorologic data discussed in this chapter are the following:

- There is evidence that trends are not monotonic within most of the records, and most records contain two or more opposite trends at different times, which are usually overlapping. This is a very important aspect as it implies some variability in the data rather than monotonic trends. This is the reason why tests on the full record may give different results for different data sets depending on the starting point and the extent of the data series. The disparity between the results from the HCN temperature data and the state division temperature data in Tables III.4 and III.5, respectively, is very clear. However, the sub-period analysis in Figures 3.10 and 3.11 shows more or less consistent periods of positive and negative trends.
- In general, trends in temperature data seem to be opposite in direction to those in flow and rainfall data. While data from 1930 to 1985 show significant positive trends in flow and rainfall, temperature data from 1920 to 1980 show significant negative trends. This relationship between trends in flow, rainfall and temperature data is not as clear near the beginning and the end of the record due to the edge effects, but is still partially observed.
- By comparing results from the tree-ring data in Figure 3.12 with the results from the other data it seems that not much of the observed trends in the rainfall and temperature data is reflected in tree-ring data. However, this may be attributed to the interaction between the effects of rainfall and temperature, and possibly other factors as well, in the formation of tree-rings, especially with the existence of opposite trends observed in rainfall and temperature data.
- The existence of cycles of positive and negative trends, which are clearly noticed in tree-ring data in Figure 3.12, with periods of 30 to 100 years suggests that the

available record length of hydrologic and meteorologic data may not be long enough to determine whether the observed trends are monotonic trends or merely a manifestation of long term variations with the overall process being stationary. With the available short records of around 100 years, at least one cycle consisting of a positive trend followed by a negative trend or vice versa is observed in most of the data series. Data series which are at least twice as long as the present ones are needed before definite conclusions about the nature of the observed trends in hydrologic time series can be drawn.

- In general, there is evidence of significant trends in hydrologic data. The results of trend analysis for flow, rainfall, and temperature data are more or less consistent. These trends are also consistent with national and global trends. However, the relatively short length of the available hydrologic time series dose not allow for firm conclusions about the nature of these trends and whether they are monotonic or cyclic.

4. Frequency Domain Analysis

4.1. INTRODUCTION

The results discussed in chapter 3 reveal the existence of trends in the data. The results from trend tests, however, suggest that this variability is more likely to be non-monotonic in nature; i.e. periodic or cyclic. Spectral analysis in the frequency domain is suitable for analyzing cyclic behavior in time series. The basic idea of spectral analysis is to represent the time series as a sum of sinusoidal components of different frequencies. The power spectrum of a time series, which is the squared amplitude of these sinusoids reflects the distribution of the variance of the stochastic process over these frequencies. By studying the power spectrum, those cyclic components which contribute most to the overall variability of the time series can be identified. The shape of the spectrum also reveals features of the process that are useful in selecting the types of models which are suitable for analyzing the observed data (Jenkins and Watts, 1968).

Conventional methods for spectral estimation include the correlogram and the spectrum, which is the Fourier transform of the autocovariance function of the data, and the periodogram, which is related to the Fourier transform of the time series itself. Because the raw spectrum estimates computed by using these non-parametric methods suffer from large variances, several smoothing techniques are used to reduce the variance of the estimated spectrum (Priestley, 1984). The smoothing of raw spectral estimates involves the use of different types of spectral averaging windows. However, smoothing often results in poor spectral resolution, and components which are close to each other can be hard to resolve. Another problem with conventional estimates is the spectral window leakage (Percival and Walden, 1993; Priestley, 1984). Spectral window leakage causes high power at one frequency to influence the power at other frequencies, thus making it difficult to isolate the effect of components with different frequencies. Power leakage can be reduced by using specialized conventional windows in some instances but at the expense of spectral resolution.

In the frequency domain, processes can be divided into two main categories according to the nature of their spectra (Priestley, 1984). The first category includes processes with discrete (line) spectra. These processes are deterministic periodic processes in which the energy of the process is divided among a discrete set of frequencies. The second category includes processes with continuous spectra. These are non-periodic stochastic processes such as random, autoregressive (AR), moving average (MA), and autoregressive moving average (ARMA) processes. In practice, however, many time series arise from processes with mixed spectra, i.e. with both discrete and continuous spectral components. Such time series will have a stochastic non-periodic component on which one or more periodic components are superimposed. Spectral analysis of time series with mixed spectra should take into account the effect of line components on the shape of the spectrum. Problems associated with conventional spectral analysis such as power leakage and window smoothing will obscure much of the information about the true composition of the spectrum.

Recently, new methods of spectral analysis have been introduced. These methods are aimed at increasing the resolution of spectral estimates while reducing power leakage at different frequencies. One of these new methods is the Multi-Taper Method (MTM)

(Thomson, 1982; 1990a and b). Similar to conventional spectral estimation methods, the MTM is a non-parametric method in which raw spectral estimates are smoothed by a number of special type windows designed to reduce leakage and increase spectral resolution. The advantage of using a non-parametric method for spectral estimation is that the analysis is not hindered by the limited capability of parametric models to take arbitrary forms. The MTM method can be used for both the estimation of the continuous part of the spectrum and the identification of periodic components and their frequencies. In addition, the MTM method provides a measure of the statistical significance of each periodic component. Thus significant components can be identified by using the MTM method and their contribution to the mixed spectrum can be isolated.

The Maximum Entropy Method (MEM) (Burg, 1975) is a parametric method of spectral analysis. The method is based on the maximum entropy principle and is closely related to autoregressive spectral estimation. In the MEM, autoregressive parameters are estimated by using Burg's algorithm (Burg, 1968) which results in more accurate spectral estimates than those obtained by using ordinary autoregressive spectral estimation methods. The MEM offers high resolution power spectra and provides great precision for the frequency of individual harmonics.

Conventionally, hydrologic processes have been regarded as continuous spectrum processes such as the autoregressive (AR), autoregressive-moving average (ARMA), and other models (Salas et al., 1988). No line components have been considered in the analysis of hydrologic data except for the well established, physically based daily and monthly cycles. However, recent studies suggest that other cycles with different frequencies may exist in hydrologic as well as temperature data (Currie, 1974, 1981, 1984, 1987, 1988, 1990; Rao and Yu, 1989; Labitzke and Van Loon, 1990; Mitra et al., 1991; Burrough, 1992; Mitra and Dutta, 1992; and Lall and Mann, 1995). These cycles represent the influence of a number of natural phenomena such as the sunspot cycle, the sun's magnetic (Hale) cycle, El-Niño/Southern Oscillations (ESNO), Quasi-Biennial Oscillations, and others. However, the analysis methods used in previous studies do not offer a measure of statistical significance for different cyclic components. The amplitude of line components is usually the only criterion for identifying such line components, giving the chance for spurious peaks to be mistaken for true cycles. In contrast, the MTM method offers a statistical test for assessing the significance of different components and provides a means for distinguishing between spurious and actual line components in the estimated spectrum, independent of their amplitude.

In this chapter, data series are analyzed by using the MTM as well as the MEM to identify periodic components in these time series and assess their statistical significance. Results from different time series are discussed and compared with previous studies. The relationships between results from different data sets are also investigated.

4.2. CONVENTIONAL SPECTRAL ANALYSIS

A stationary process X_t has a spectral representation of the form of Eq. 4.1 (Thomson, 1990a; Cramér, 1940),

$$x_t = \int_{-1/2}^{1/2} e^{i2\pi ft}\, dZ(f) \tag{4.1}$$

where f is frequency, $dZ(f)$ is a discrete orthogonal increment process (Brockwell and Davis, 1991). An important property of all stationary processes is that the expected value of $dZ(f)$ is zero. Also, the second moment of $dZ(f)$ for a stationary process is given by Eq. 4.2.

$$E\left\{ \left| dZ(f) \right|^2 \right\} = S(f)\, df \tag{4.2}$$

In Eq. 4.2, $S(f)$ is, by definition, the spectral density function of the stationary process X_t which is conventionally known as the power spectrum.

Two related non-parametric methods have been conventionally used for estimating the spectrum of a stationary process. The first method is the periodogram method. In its raw form, the periodogram is the squared magnitude of the Fourier transform of the time series, given by Eq. 4.3

$$I_n(f) = \frac{1}{n} \left| \sum_{t=1}^{n} x_t\, e^{-i2\pi tf} \right|^2 \tag{4.3}$$

However, this estimator of the spectrum is inconsistent in the sense that its variance does not decrease as the sample size n increases (Brockwell and Davis, 1991; Priestley, 1984). The periodogram also suffers from high variance and has an erratic and wildly fluctuating form (Priestley, 1984). In order to reduce the high variance of the periodogram, spectral windows are used for smoothing the periodogram. The smoothed periodogram in this case becomes as in Eq. 4.4,

$$\hat{S}(f_j) = \frac{1}{2\pi} \sum_{|k|<m} W_n(k)\; I_n(f_{j+k}) \tag{4.4}$$

where $W_n(k)$ is a sequence of weight functions known as a spectral window. A number of spectral windows with different properties have been developed such as the rectangular, Bartlett, Hanning, and Tukey windows. Using spectral windows to improve the properties of the periodogram, however, introduces bias in the spectral estimate and reduces the spectral resolution. As the width of the spectral window increases (more smoothing) the variance is reduced but the bias is increased, and vice versa. Also, the use of spectral windows introduces power leakage, when power at a given frequency affects the power at other frequencies (Priestley, 1984).

The second method for estimating the spectrum is by using the correlogram. The correlogram is the Fourier transform of the autocovariance function of the time series as in Eq. 4.5,

$$\tilde{S}(\omega) = \frac{1}{2\pi} \sum_{|k|<r} w(k/r)\,\hat{\gamma}(k)\, e^{-i2\pi kf} \tag{4.5}$$

where $\hat{\gamma}(k)$ is the autocovariance function of the time series at lag k, and $w(k/r)$ is a lag window. The lag window $w(k/r)$ in Eq. 4.5 is equivalent to the spectral window $W_n(k)$ in Eq. 4.4 and the relation between them is given in Eq. 4.6 (Brockwell and Davis, 1991).

$$W_n(f) = \frac{1}{n} \sum_{|h|\le r} w(h/r)\, e^{-i2\pi hf} \tag{4.6}$$

4.3. MULTI-TAPER METHOD (MTM) OF SPECTRAL ANALYSIS

The MTM method (Thomson, 1982) makes use of an extended version of the spectral representation in Eq. 4.1. The process x_t in this case may include a number of periodic components in addition to the underlying stationary process as in Eq. 4.7 (Percival and Walden, 1993),

$$\begin{aligned} x_t &= \sum_j C_j \cos(2\pi f_j t + \varphi_j) + \zeta_t \\ &= \sum_j \mu_j e^{i2\pi f_j t} + \mu_j^* e^{-i2\pi f_j t} + \zeta_t \end{aligned} \tag{4.7}$$

where ζ_t (the background continuum) is a zero mean stationary process with spectral density function $S(f)$ that is not necessarily constant, i.e. it can be colored noise, C_j and f_j are the amplitude and frequency of periodic or line component j, $\mu_j = (C_j/2)\, e^{i\varphi_j}$ is the complex amplitude corresponding to the real amplitude C_j , and (*) indicates complex conjugate. These types of processes are known as centered stationary or conditionally stationary processes and are often noted as having mixed spectra (Thomson, 1990a). For these processes, the expected value of the discrete orthogonal increment process $dZ(f)$ is no longer zero and is given by Eq. 4.8,

$$E\left\{dZ(f)\right\} = \sum_j \mu_j\, \delta(f-f_j)\, df \tag{4.8}$$

where δ is the Dirac delta function . The second central moment of $dZ(f)$ is given by Eq. 4.9.

$$E\left\{\left|dZ(f) - E\{dZ(f)\}\right|^2\right\} = S(f)\, df \tag{4.9}$$

That is, for processes with mixed spectra, the deterministic, or line, component of the spectrum is given by the first moment of $dZ(f)$, while the continuous non-deterministic part is given by the second central moment of $dZ(f)$. Conventional estimation, as was seen earlier, has been focused on the estimation of the second moment which is the continuous part of the spectrum. However, for processes with mixed spectra it is clear from Eq. 4.9 that the estimation of the continuous part of the spectrum requires the estimation of the deterministic part or line components as a first step. A major objection to conventional spectral analysis is that there is no separation between the first and second moments of $dZ(f)$, and spurious peaks in the spectrum can be identified as line components without an objective criterion for differentiating between real and spurious lines (Thomson, 1990b). In the MTM method, the multi-window approach provides a simple, yet very useful, likelihood ratio test for the significance of periodic components.

The basic idea in MTM spectral estimation is the use of multiple data windows known as "discrete prolate spheroidal sequences" or "Slepian sequences", which are defined as the solution of the symmetric Toeplitz matrix eigenvalue problem in Eq. 4.10 (Slepian, 1978),

$$\lambda_k v_n^{(k)}(N,W) = \sum_{m=0}^{N-1} \frac{\sin\{2\pi W(n-m)\}}{\pi(n-m)} v_m^{(k)}(N,W) \tag{4.10}$$

where N is the number of data points, W is the spectral band width, and λ_k are the eigenvalues associated with the Slepian sequences $v_n^{(k)}(N,W)$. Values of the Slepian sequences $v_n^{(k)}(N,W)$ can be calculated numerically by using methods given by Percival and Walden (1993) and Thomson (1990b). The Fourier transform of these Slepian sequences are called Slepian functions or discrete prolate spheroidal wave functions as given by Eq. 4.11.

$$V_k(f) = \sum_{n=0}^{N-1} v_n^{(k)}(N,W)\, e^{-i2\pi fn} \tag{4.11}$$

The Slepian functions in Eq. 4.11 have the maximum energy concentration in the interval $(f-W, f+W)$. The fraction of energy retained within the band $(f-W, f+W)$ by the kth function is given by λ_k in Eq. 4.10. Furthermore, the bias from all frequencies remote from the frequency of interest decreases exponentially as a function of the window width times the number of observations, and thus the use of these sequences is very effective in eliminating window leakage (Slepian, 1978). One of the important features of the Slepian sequences and functions is that their properties do not depend on the assumption of normality or stationarity of the time series (Thomson, 1990b).

As a first step in MTM spectral estimation the expansion or eigencoefficients of the time series X_t are calculated as in Eq. 4.12.

$$y_k(f) = \sum_{t=0}^{N-1} x_t \, v_t^{(k)}(N,W) \, e^{-i2\pi ft} \tag{4.12}$$

In Eq. 4.12, $v_n^{(k)}(N,W)$ are the Slepian sequences, $k = 0, 1, ..., K$-1. The band width W is usually chosen between $2/N$ and $20/N$, with $4/N$ a good initial choice. If W is too small, the resulting spectral estimate is unstable, but if W is too large the resolution is poor and fine features cannot be resolved (Thomson, 1990b). The number of data windows is usually taken as $K = 2NW$-1 which corresponds to the number of Slepian sequences with eigenvalues close to unity.

For a single periodic signal of amplitude μ and frequency f_o, considering Eq. 4.11 along with Eqs. 4.7 and 4.12, the expected value of $y_k(f)$ can be obtained as in Eq. 4.13 (Percival and Walden, 1993).

$$E\{y_k(f)\} = \mu \, V_k(f-f_o) + \mu^* \, V_k(f+f_o) \tag{4.13}$$

At $f = f_o$, Eq. 4.13 becomes Eq. 4.14.

$$\begin{aligned} E\{y_k(f_o)\} &= \mu \, V_k(0) + \mu^* \, V_k(2f_o) \\ &\approx \mu \, V_k(0) \end{aligned} \tag{4.14}$$

Assuming $2f_o > W$, the second term in Eq. 4.14 can be neglected since V_k is highly concentrated in the interval $(f-W, f+W)$. The complex amplitude μ in Eq. 4.14 can thus be estimated by minimizing the residual local squared error , i.e. when $f = f_o$, in Eq. 4.15 with respect to μ.

$$e^2(\mu, f) = \sum_{k=0}^{K-1} |y_k(f) - \mu(f) \, V_k(0)|^2 \tag{4.15}$$

The result of the minimization is given by Eq. 4.16.

$$\hat{\mu}(f) = \sum_{k=0}^{K-1} V^{*k}(0) \, y_k(f) \Big/ \sum_{k=0}^{K-1} |V_k(0)|^2 \tag{4.16}$$

This solution in Eq. 4.16 is simply a linear regression of $y_k(f)$ on $V_k(0)$. This fact allows the use of the usual regression F-test in the form of Eq. 4.17 to test for the significance of a line component with amplitude given by Eq. 4.16 at frequency f (Thomson, 1982).

$$F(f)=\frac{(K-1)\,|\hat{\mu}(f)|^2 \sum_{k=0}^{K-1} |V_k(0)|^2}{e^2(\hat{\mu},f)} \tag{4.17}$$

The location of the maximum value of $F(f)$ provides an estimate of the line frequency. This test works well for multiple lines if the lines are isolated, i.e. there is a single line in the range $(f-W, f+W)$. A more complex multiple line test may be used for resolving multiple line components that are spaced less than $2W$ (Thomson, 1990b). A good rule of thumb in implementing the F-test in Eq. 4.17 is not to consider significance levels less than $1-1/N$ (Thomson, 1990a).

By considering Eq. 4.7, the portion of the variance of X_t explained by a given line component can be calculated as in Eq. 4.18,

$$E_j = \frac{C_j^2}{2\sigma_x^2} = \frac{2\,|\mu_j|^2}{\sigma_x^2} \tag{4.18}$$

where σ_x^2 is the variance of the process X_t.

The MTM method also offers a better approach for the removal of the mean of a process. The usual practice is to subtract the mean and possibly the slope of the process and then treat the data as a sample from a zero mean process. This approach usually results in a hole in the spectrum around zero frequency and often leads to the appearance of a peak at a period commensurable with the length of the time series (Thomson, 1990a). A better approach is to include a line component at zero frequency and use the estimate of $\mu(0)$ from Eq. 4.16 as an alternative estimate of the mean. This will result in the preservation of the continuous part of spectrum at zero frequency. This is important when periods of interest are nearly as long as the data (Thomson, 1990a).

The continuous part of the spectrum can be estimated as in Eq. 4.19.

$$\hat{S}(f) = \frac{1}{K} \sum_{k=0}^{K-1} |y_k(f)|^2 \tag{4.19}$$

However, in the neighborhood of a significant line component with frequency f_o, the spectrum, according to Eq. 4.9 , must be reshaped by subtracting the contribution of the line component as in Eq. 4.20.

$$\hat{S}_r(f) = \frac{1}{K} \sum_{k=0}^{K-1} |y_k(f)-\hat{\mu}(f_o)\, V_k(f-f_o)|^2 \tag{4.20}$$

The correction in Eq. 4.20 is made for frequencies in the range (f_o-W, f_o+W) where the effect of the line component is concentrated. It should be noted that to prevent circular correlation, the Fourier transform in Eq. 4.12 should be obtained for a discrete mesh that is four to ten times the length of the time series by padding the observations with zeros. This is also essential since if a coarser mesh is used there is a high probability of the line frequency being far enough from the closest mesh frequency to be missed due to the narrowness of the F-test lines for strong periodic signals (Thomson, 1990b).

4.4. MAXIMUM ENTROPY SPECTRAL ANALYSIS

The use of window spectral estimators in the form of Eq. 4.5 implicitly forces the values of the autocovariance function to be zero for lags $k > r$. Burg (1968) suggested the use of another approach based on the maximum entropy principle. Maximum entropy spectral analysis as defined by Burg (1975) is based on choosing the spectrum which corresponds to the most unpredictable time series whose autocorrelation function agrees with the known values. The spectral density function (sdf) obtained by Burg is in the form of Eq. 4.21,

$$\tilde{S}(f) = \frac{\sigma_p^2}{\left|1 - \sum_{k=1}^{p} \varphi_{k,p} e^{-i2\pi fk}\right|^2} \tag{4.21}$$

where $\varphi_{k,p}$ and σ_p^2 are the solutions of the augmented Yule-Walker equations of order p. This is equivalent to the sdf of an AR(p) process. The difference between Maximum Entropy spectral analysis and ordinary AR spectral estimation lies in the method used for parameter estimation known as Burg's algorithm (Burg, 1968). Burg's algorithm can be regarded as a variation on the solution of the Yule-Walker equations via the Levinson-Durbin recursions. Burg's algorithm is based on the minimization of the backward and forward prediction errors given by Eqs. 4.22 and 4.23.

$$e_t^f(k-1) = X_t - \sum_{j=1}^{k-1} \varphi_{j,k-1} X_{t-j} \qquad k \le t \le n \tag{4.22}$$

$$e_{t-k}^b(k-1) = X_{t-k} - \sum_{j=1}^{k-1} \varphi_{j,k-1} X_{t-k+j} \qquad k+1 \le t \le n+1 \tag{4.23}$$

The minimization of these prediction errors results in the estimation of $\varphi_{k,k}$ as given by Eq. 4.24.

$$\hat{\varphi}_{k,k} = \frac{\sum_{t=k+1}^{n} e_t^{f2}(k-1) + e_t^{b2}(k-1)}{2\sum_{t=k+1}^{n} e_t^{f}(k-1)\, e_t^{b}(k-1)} \tag{4.24}$$

Initial values of the forward and backward errors are given by Eqs. 4.25 and 4.26.

$$e_t^{f}(0) = X_t \tag{4.25}$$

$$e_{t-1}^{b}(0) = X_{t-1} \tag{4.26}$$

Subsequent forward and backward errors are calculated by Eqs. 4.27 and 4.28.

$$e_t^{f}(k) = e_t^{f}(k-1) - \varphi_{k,k}\, e_t^{b}(k-1), \qquad k+1 \le t \le n \tag{4.27}$$

$$e_{t-k}^{b}(k) = e_{t-k}^{b}(k-1) - \varphi_{k,k}\, e_t^{f}(k-1), \qquad k+1 \le t \le n \tag{4.28}$$

The values of the AR coefficients $\varphi_{j,k}$, j=1, ..., k-1 can then be obtained by using Eq. 4.29.

$$\varphi_{j,k} = \begin{cases} \varphi_{j,k-1} + \varphi_{k,k}\varphi_{k-i,k-1} & j=1,\ldots,k-1 \\ \varphi_{k,k} & j=k \end{cases} \tag{4.29}$$

Burg's algorithm produces accurate AR spectral estimates for data which are truly AR (Kay, 1988). There are also many processes whose spectra can be well approximated by AR(p) spectra, and the maximum entropy method can be expected to work well for such processes (Percival and Walden, 1993).

Although the MEM is a method for estimating continuous spectra, processes with mixed spectra can also be analyzed and frequencies of line components can be estimated using the maximum entropy method. The estimation of sinusoidal components in this case is based on the fact that a randomly phased sinusoid in the form can be represented by a second order difference equation as in Eq. 4.30,

$$X_t = \varphi_{1,2} X_{t-1} + \varphi_{2,2} X_{t-2} = 2\cos(2\pi f)\, X_{t-1} - X_{t-2} \tag{4.30}$$

with the initial condition $X_o = C\cos(\varphi)$ and $X_1 = C\cos(2\pi f + \varphi)$. The process in Eq. 4.30 is called a pseudo-AR(2) process , because of the lack of an innovation term e_t (Percival and Walden, 1993). An important feature of the pseudo-AR(2) process in Eq. 4.30 is that the roots of its characteristic equation, Eq. 4.31, both lie on the unit circle.

$$z^2 - \varphi_{1,2} z - \varphi_{2,2} = 0 \tag{4.31}$$

Similarly, if X_t is composed of p sinusoids, it becomes a pseudo-AR($2p$) process with $2p$ roots on the unit circle in conjugate pairs. This leads to the fact that these sinusoids can be estimated through auto-regressive estimation procedures, including MEM, by looking at the roots of the characteristic equation of the estimated process. These roots will also correspond to sharp peaks in the estimated AR spectrum. However, noise in the observed data has the effect of moving the roots away from the unit circle , thus resulting in flatter peaks and equivalently poor spectral resolution. This can be remedied by increasing the AR order. The order of the fitted AR model should be chosen between $n/3$ and $n/2$, where n is number of observations (Ulrych and Ooe, 1983). However, it should be noted that spurious peaks may be introduced as the order of the AR process is increased (Kay, 1988). This can be handled by varying the order of the model and observing peaks which survive the change. Another way is to compare MEM and MTM peaks which, in general, should not share spurious peaks. The frequency of line components can be estimated based on either the locations of the largest peaks or the angles of the roots closest to unity. The latter approach gives more accurate results (Kay, 1988).

4.5. SPECTRAL ANALYSIS OF HYDROLOGIC AND CLIMATIC DATA

Both the MTM and MEM are used for analyzing the available data in the study region. For the sake of consistency, the number of frequency points used for all data is set at 2048, which is sufficient to prevent circular correlation and ensure accurate results of the F-test for the significance of peaks as discussed in section 4.3. Since the band width for various hydrologic and climatic time series is not known *a priori*, different values of the band width W were investigated for use in the MTM method. The significance of various harmonic components as well as their location and magnitude varied slightly as the band width changed. The results for $W = 2.5/N$ and $4/N$ are reported and discussed. These values of W correspond to using four and seven data tapers, respectively. For the MEM method, the order of the AR model was varied between $N/2$ and $N/3$ giving slightly variable results. The results for the cases of $N/2$ and $N/3$ are reported and discussed below.

4.5.1. *Results From MEM Analysis.* The MEM method offers no measure for testing the significance of different periodic components. However, the closeness of the roots of the characteristic equation to unity can be used as a guide for the significance of identified components. Bearing in mind that noise roots may be pushed towards the unit circle as a result of increasing the AR order (Kay, 1988), only periodicities corresponding to roots larger than 0.95 in the case of AR order of N/3 and 0.99 in the case of AR order of N/2 are considered as potentially significant components. Whether a given component is spurious or actually significant can then be confirmed through the comparison of the different AR order results as well as the comparison of the results from the MTM analysis. Also, the existence of a component at a given period in a large number of records indicates that such periodicity is not likely to be spurious. In the following figures (Figs. 4.1 - 4.12) the components are indicated at their corresponding frequencies with circles. The size of each circle is made proportional to the closeness of its respective root to the unit circle, scaled between 0.95 and 1.0 for the *N*/3 case, and between 0.98 and 1.0 for the *N*/2 case. The following results were obtained for each of data sets:

a) USGS Flow data

The results from the MEM analysis of the USGS flow data using AR orders of *N*/2 and *N*/3 are shown in Figs. 4.1 and 4.2, respectively. A number of periodicities in Figs. 4.1 and 4.2 can be identified which are common to several stations. In Fig. 4.1 the longest periodicity is 106.7 years which is present in only one station, followed by 73.7 and 67.6 years in another two stations. No periodicities in that range were found in Fig. 4.2. Groups of periodicities found in the USGS flow data are summarized in Table IV.1. In Table IV.1 the first column is the approximate range of the group, the second and third columns are for AR order of *N*/2 and give the number of stations *n* which have significant components in that range and the mean period in that group. The fourth and fifth columns give the same information for AR order of *N*/3. According to Table IV.1, the most common periods for the USGS flow data are 2.6, 3.2, 4.4, 5.7, 7.6, and 10.9 years.

b) HCN Rainfall Data

The results from the MEM analysis of the HCN rainfall data are shown in Figs. 4.3 and 4.4. Only one station with periodicity of 100.4 years is observed in Figs. 4.3 and 4.4, followed by two or three periods of around 80 years. In the range from 18 to 40 years there is no clear grouping of frequencies, but this range can be roughly divided into groups of 17-20, 20-25, and 25-40 year periodicities. Groups of periodicities found in the HCN rainfall data are summarized in Table IV.2. According to the results in Table IV.2, the most common periods in the HCN rainfall data are 2.3, 2.7, 3.2, 3.8, and 5.7 years.

c) NCDC Divisional Average Rainfall Data

The results from the MEM analysis of the NCDC divisional rainfall data are shown in Figs. 4.5 and 4.6. It can be observed from Figs. 4.5 and 4.6 that these groups of periods are more consistent than in the case of HCN rainfall data. This is due to the averaging of the NCDC divisional data, which has the effect of reducing noise in the data. Groups of periodicities found in the NCDC divisional rainfall data are summarized in Table IV.3. The most common periods in Table IV.3 are 2.5, 2.7, 3.2, 4.8, 5.6, and 11.1 years.

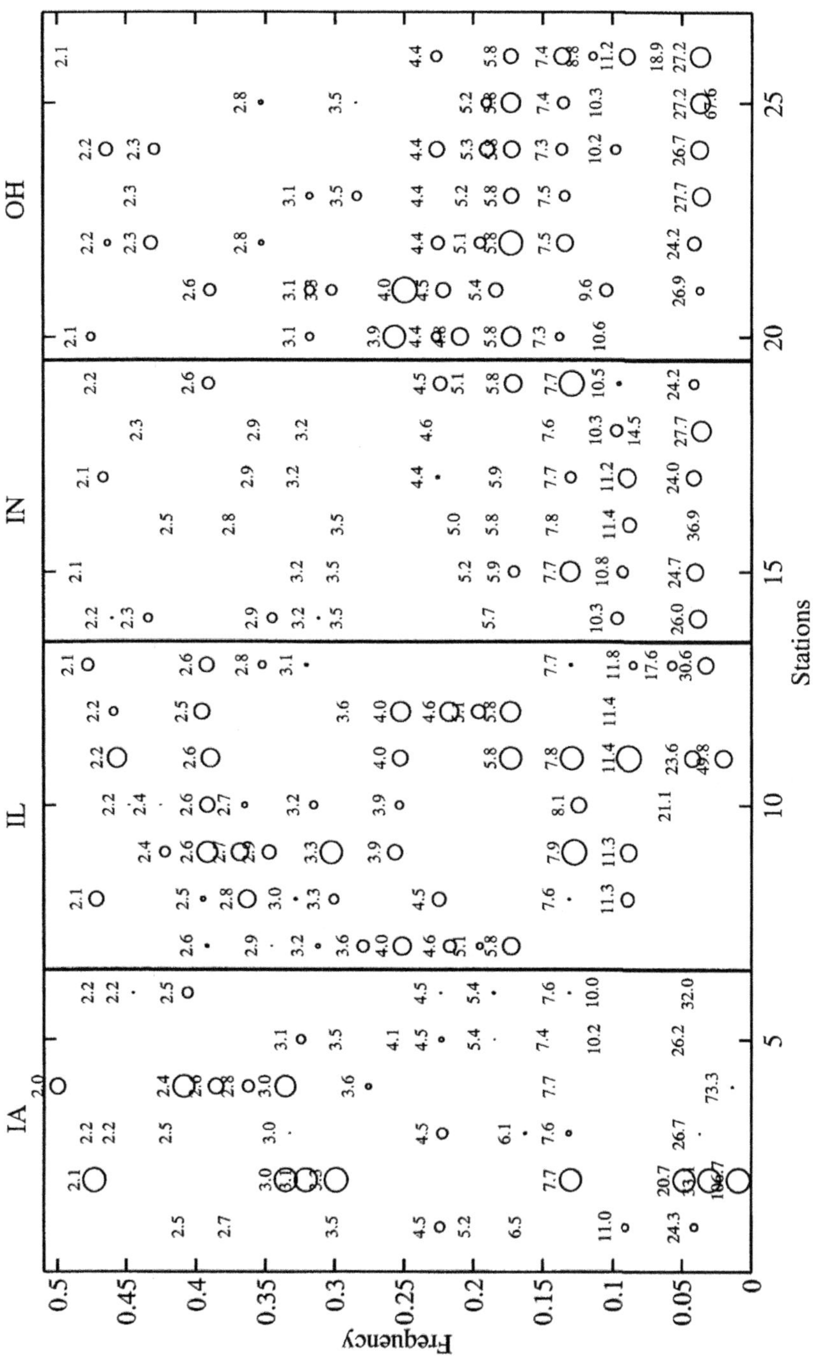

Figure 4.1. Periodicities in the USGS flow data using MEM analysis with N/2 AR order.

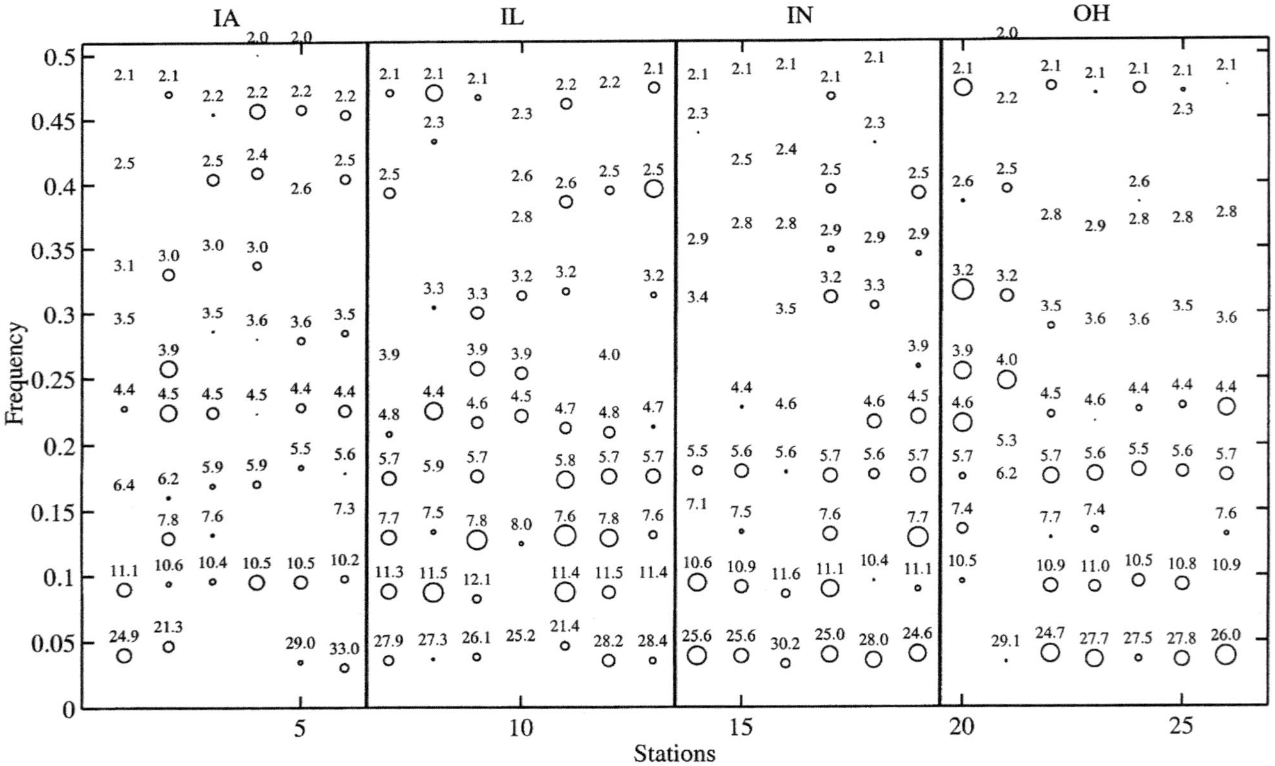

Figure 4.2. Periodicites in the USGS flow data using MEM analysis with N/3 AR order.

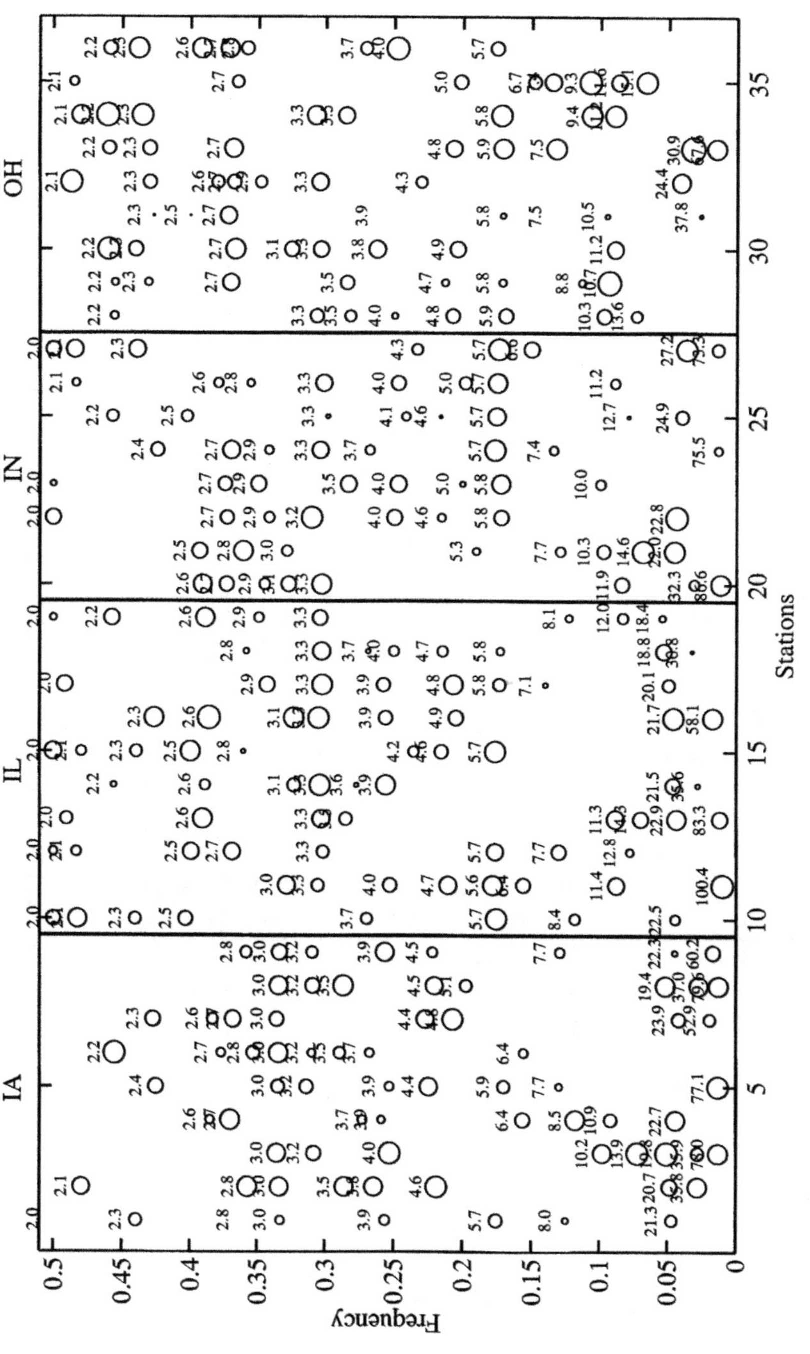

Figure 4.3. Periodicities in the HCN rainfall data using MEM analysis with N/2 AR order.

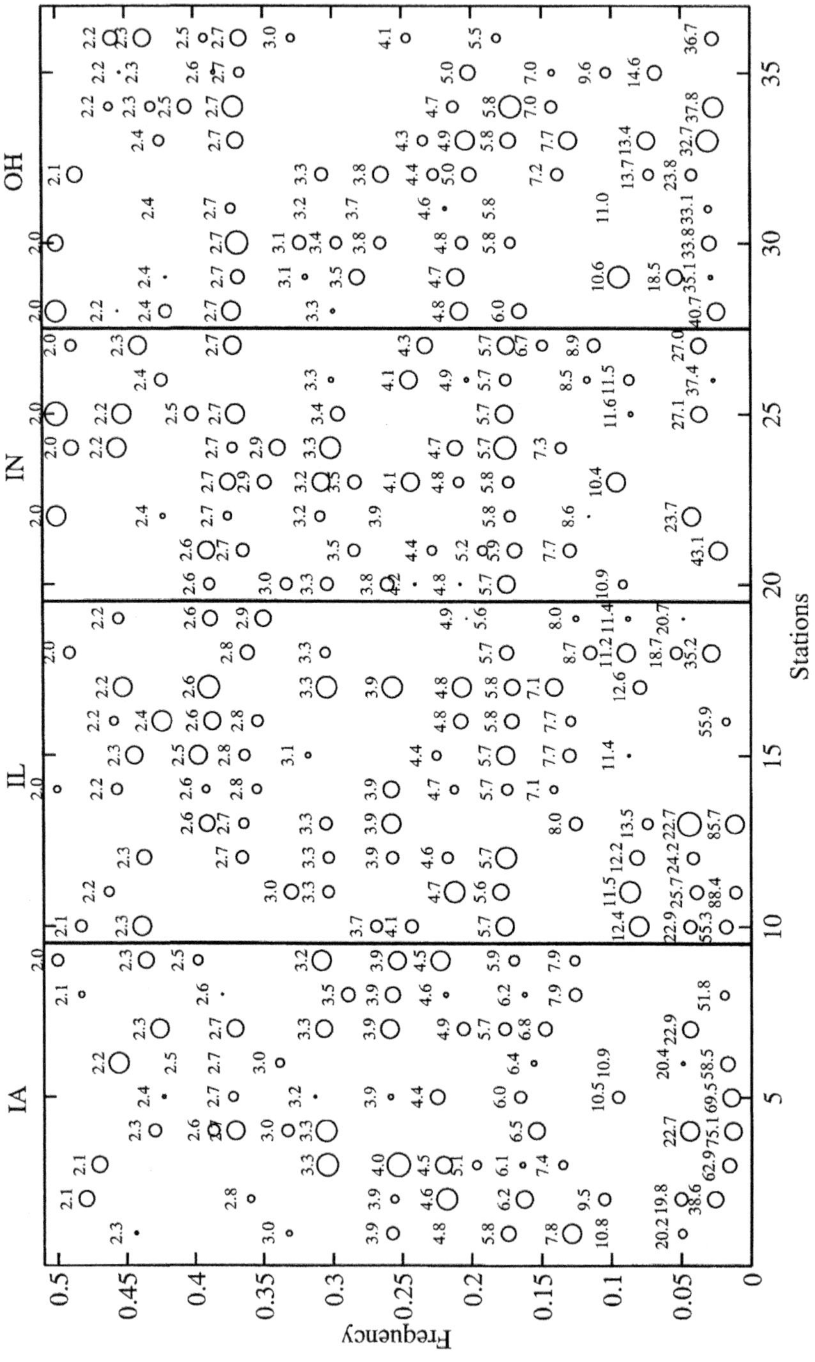

Figure 4.4. Periodicities in the HCN rainfall data using MEM analysis with N/3 AR order.

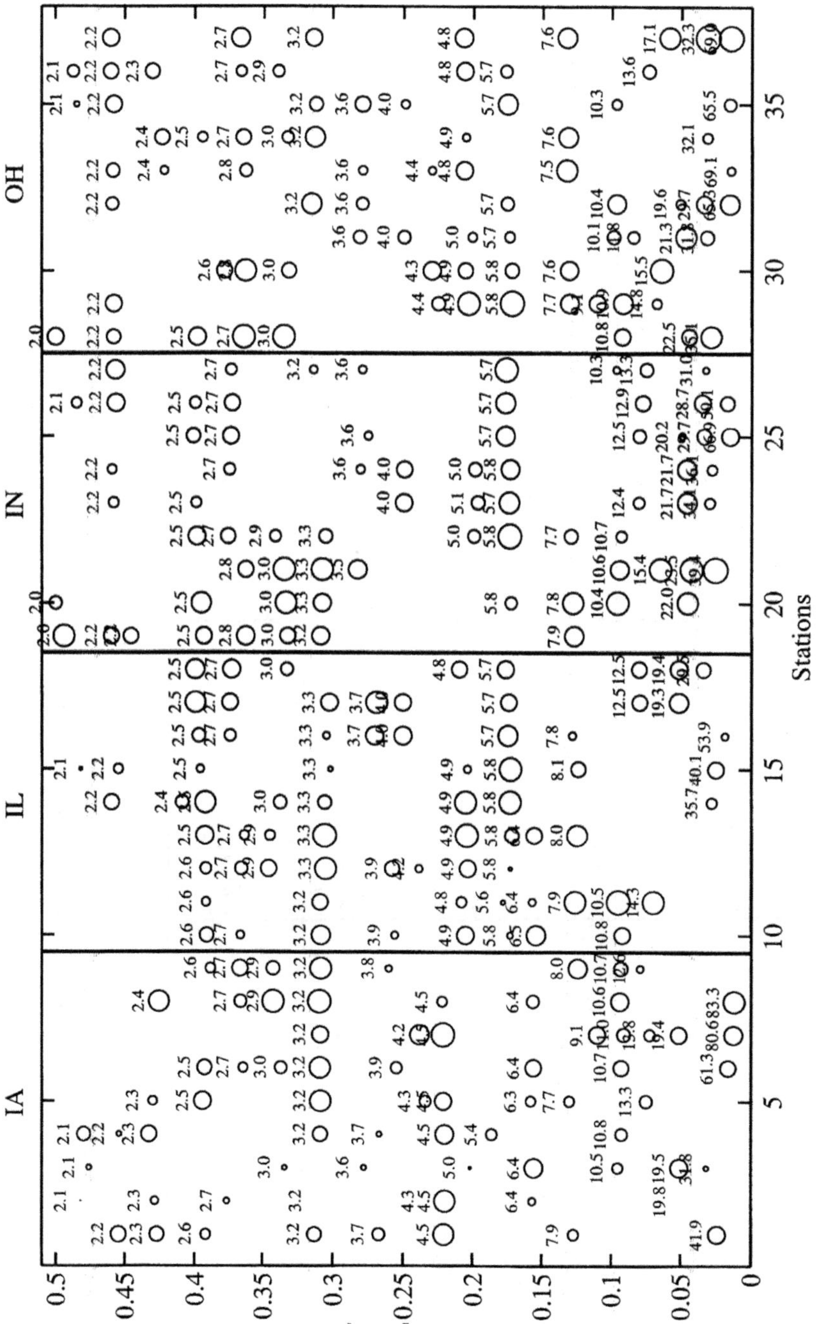

Figure 4.5. Periodicities in the NCDC divisional rainfall data using MEM analysis with N/2 AR order.

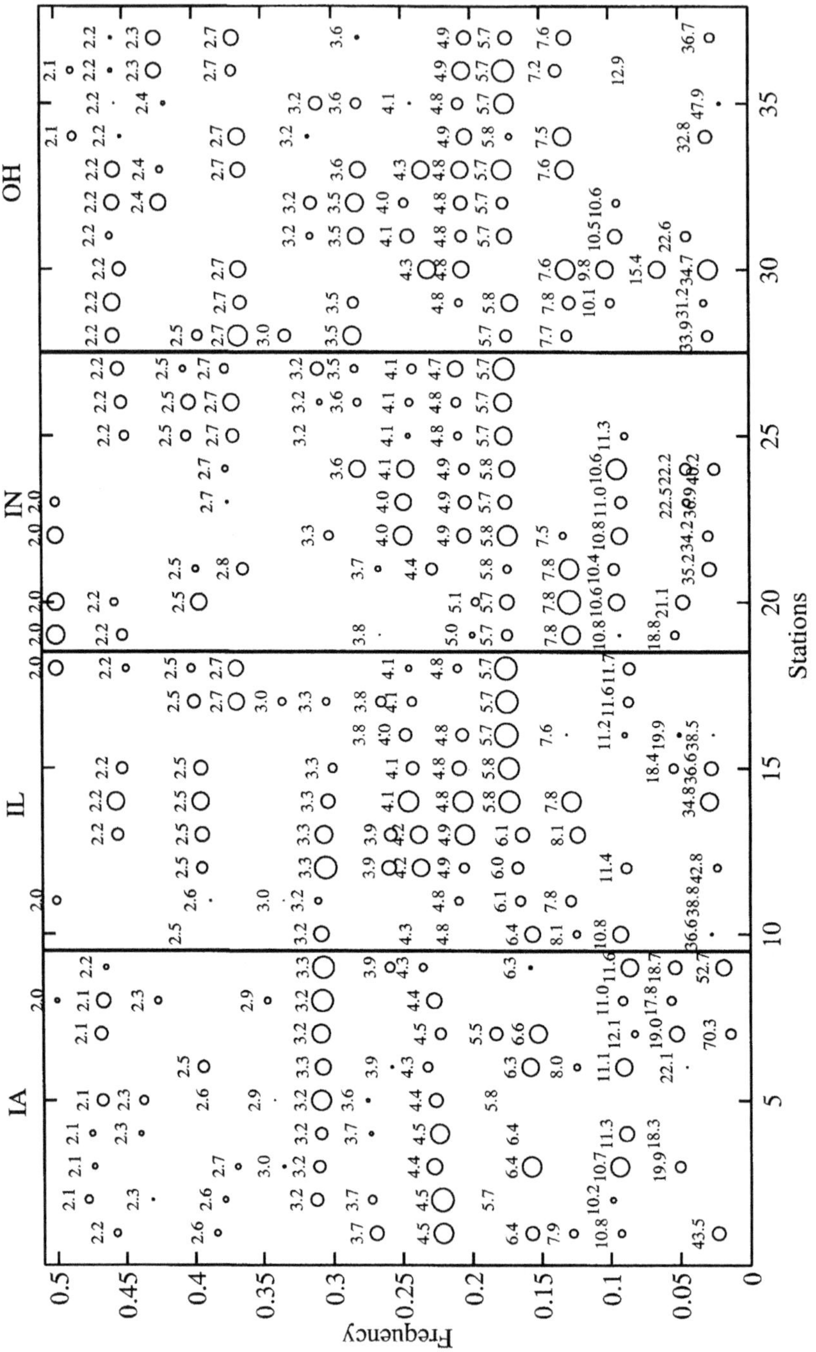

Figure 4.6. Periodicities in the NCDC divisional rainfall data using MEM analysis with N/3 AR order.

Table IV.1. Periodicities in the USGS flow data, MEM analysis.

Periodicity Group (Years)	$p = N/2$		$p = N/3$	
	No. of Stations	Mean Period	No. Of Stations	Mean Period
30-40	4	33.1±2.7	2	31.6±1.9
20-30	17	25.2±2.2	21	26.3±2.2
10-12	18	10.9±0.6	23	10.9±0.4
7-8	20	7.6±0.2	17	7.6±0.2
5.3-6	17	5.7±0.2	23	5.7±0.1
4.9-5.3	10	5.2±.08	0	---
4-5	19	4.4±0.2	24	4.5±0.2
3.5-4	14	3.7±0.2	14	3.7±0.2
3-3.5	20	3.2±0.1	16	3.3±0.1
2.8-3	10	2.9±0.05	11	2.9±0.06
2.4-2.8	21	2.6±0.2	20	2.6±0.1
2.25-2.4	7	2.3±0.02	5	2.3±0.02
2.15-2.25	11	2.2±0.04	7	2.2±0.05
2-2.15	7	2.1±0.02	17	2.1±0.02

Table IV.2. Periodicities in the HCN rainfall data, MEM analysis.

Periodicity Group (Years)	$p = N/2$		$p = N/3$	
	No. Of Stations	Mean Period	No. Of Stations	Mean Period
50-80	9	69.1±10	7	61.3±8
25-40	9	33.7±4	12	33.3±5
20-25	14	22.4±1.4	10	22.4±1.5
17-20	4	19.1±0.6	3	19.6±0.7
10-12	14	10.9±0.6	13	11.1±0.4
7-8	9	7.5±0.2	16	7.5±0.4
6-7	5	6.5±0.17	9	6.3±0.3
5-6	23	5.7±0.2	27	5.7±0.2
4.5-5	17	4.7±0.2	23	4.8±0.1
4-4.5	11	4.2±.2	12	4.3±0.2
3.5-4	25	3.8±0.2	20	3.8±0.1
3-3.5	32	3.2±0.1	28	3.2±0.1
2.6-2.8	25	2.7±0.05	24	2.7±0.04
2.4-2.6	13	2.5±0.04	15	2.5±0.04
2.25-2.4	15	2.3±0.03	19	2.3±0.04
2.15-2.25	10	2.2±0.01	12	2.2±0.02
2-2.15	11	2.1±0.02	8	2.1±0.03

Table IV.3. Periodicities in the NCDC divisional rainfall data, MEM analysis.

Periodicity Group (years)	$p = N/2$		$p = N/3$	
	No. Of Stations	Mean Period	No. of Stations	Mean Period
50-80	8	63.8±5	2	61.5±13
25-40	14	32.6±3	13	35.4±2.2
20-25	7	21.8±1	5	22.1±0.6
17-20	7	19.2±1	8	18.9±0.75
10-13	24	11.1±0.9	24	11.0±0.6
7-8	12	7.7±0.1	15	7.7±0.2
6-7	8	6.4±0.05	10	6.3±0.2
5-6	26	5.6±0.2	28	5.7±0.2
4.5-5	21	4.8±0.2	25	4.8±0.1
4-4.5	11	4.2±0.2	27	4.2±0.2
3.5-4	19	3.7±0.2	22	3.7±0.1
3.1-3.4	25	3.2±0.04	23	3.2±0.04
2.6-2.8	24	2.7±0.04	19	2.7±0.04
2.4-2.6	22	2.5±0.03	15	2.5±0.03
2.25-2.4	8	2.3±0.02	9	2.3±0.04
2.15-2.25	17	2.2±0.02	21	2.2±0.02
2-2.15	8	2.1±0.02	8	2.1±0.04

d) HCN Temperature Data

The results from the MEM analysis of the HCN temperature data are shown in Figs. 4.7 and 4.8. The groups of periodicities are more consistent than in the case of flow or rainfall data. Groups of periods found in the HCN temperature data are summarized in Table IV.4. The most common periods in Table IV.4 are 2.8, 3.1, 5.4, 8.3, 10.9, and 50-80 years.

e) NCDC Divisional Average Temperature Data

The results from the MEM analysis of the NCDC divisional temperature data are shown in Figs. 4.9 and 4.10. The groups of periodicities are the most consistent of all data sets analyzed. Groups of periodicities found in the NCDC divisional temperature data are summarized in Table IV.5. The most common periodicities in Table IV.5 are 2.1, 2.5, 5.4, 8.4, 10.9, 17.9 and 50-80 years.

f) NOAA Tree-ring Data

The results from the MEM analysis of the NOAA tree-ring data are shown in Figures 4.11 and 4.12. In these figures, no clear grouping of periodicities is observed. In an attempt to compare periodicities in this group of data with other data, MEM analysis was repeated for the last 100 years of each time series, which are concurrent with the available flow, rainfall, and temperature data. The results from the MEM analysis of the last 100 years of tree-ring data are shown in Figures 4.13 and 4.14. Much better grouping of periodicities is seen in these figures, and periodicities in the last 100 years of tree-ring data are comparable to those found in flow, rainfall, and temperature data. Groups of periodicities found in the last 100 years of NOAA tree-ring data are summarized in Table IV.6. The most common periods in Table IV.6 are 2.1, 2.3, 2.5, 3.3, and 20-40 years.

4.5.2. Results From MTM Analysis. One advantage of using MTM is that in addition to identifying periodic components, a significance level can also be estimated for each component. When a large number of regional data are available, additional inference can be made about such periodicities. Generally speaking, a periodicity which is found to be significant in a large number of stations in a given region suggests that such component is evidence of a regional periodicity. In statistical terms, if the number of stations with such periodicity is significantly larger than the expected number under the null hypothesis of no periodicity, then this periodicity is statistically significant; otherwise it is spurious. The expected number of stations under the null hypothesis is simply equal to the significance level α times total number of stations. A 95% confidence interval can then be constructed for the average number by considering the binomial distribution with a rate of failure of α. The use of a significance level of $1/N$ recommended by Thomson (1990a) seems to be too strict for such an analysis, since the goal is not to assess significance for a single station but rather to compare the actual and expected number of occurrences of significant components in the studied region. Therefore a statistical significance level of α of 0.1 (probability of obtained F-value is 0.90 or more) was used for the significance of line components in addition to the $1/N$ level (probability equal to or exceeding $1-1/N$), where N is the number of data points. In the following figures (Figs. 4.15 - 4.28), components which are significant at the 0.1 level are shown as circles which are scaled between 0.9 and 1.0 according to the probability of their F-test. Components which are significant at the $1/N$ level are further identified by a plus sign within their respective circles. The following results were obtained from MTM analysis:

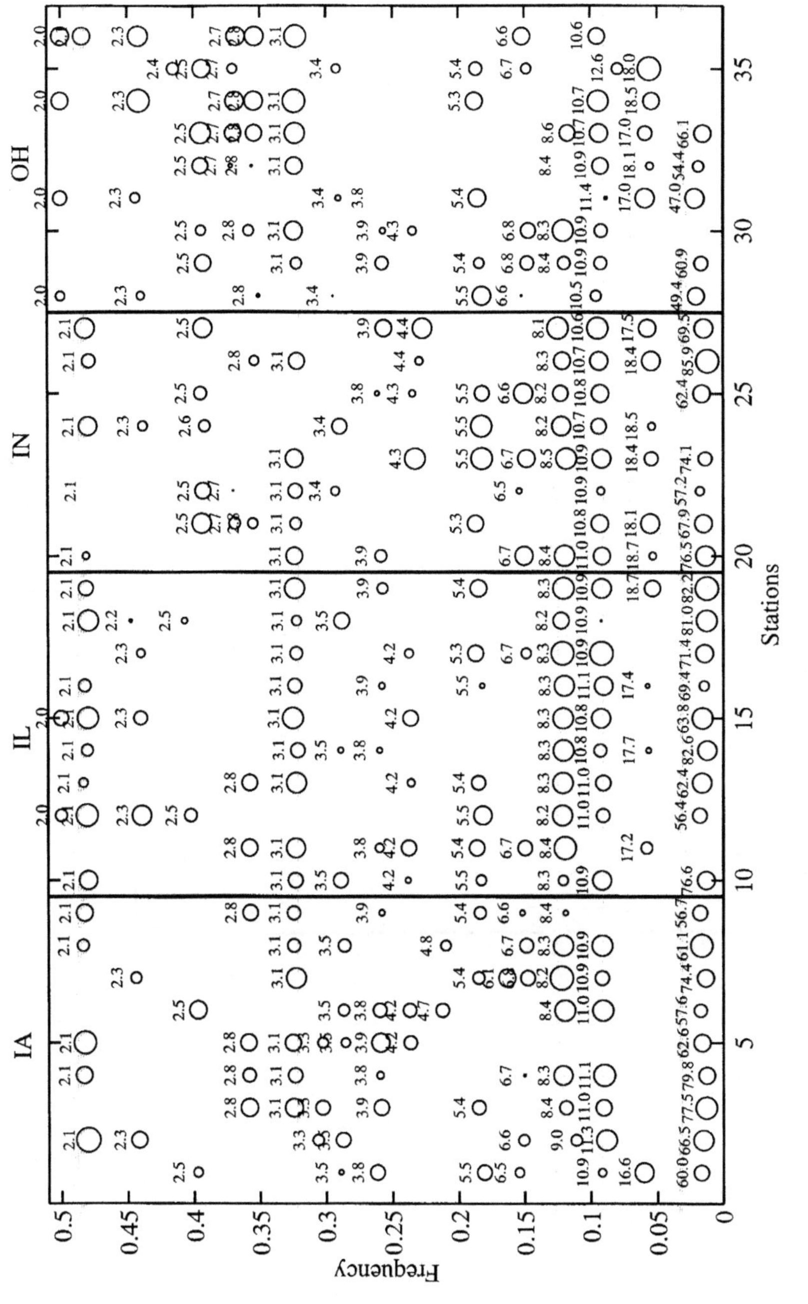

Figure 4.7. Periodicities in the HCN temperature data using MEM analysis with N/2 AR order.

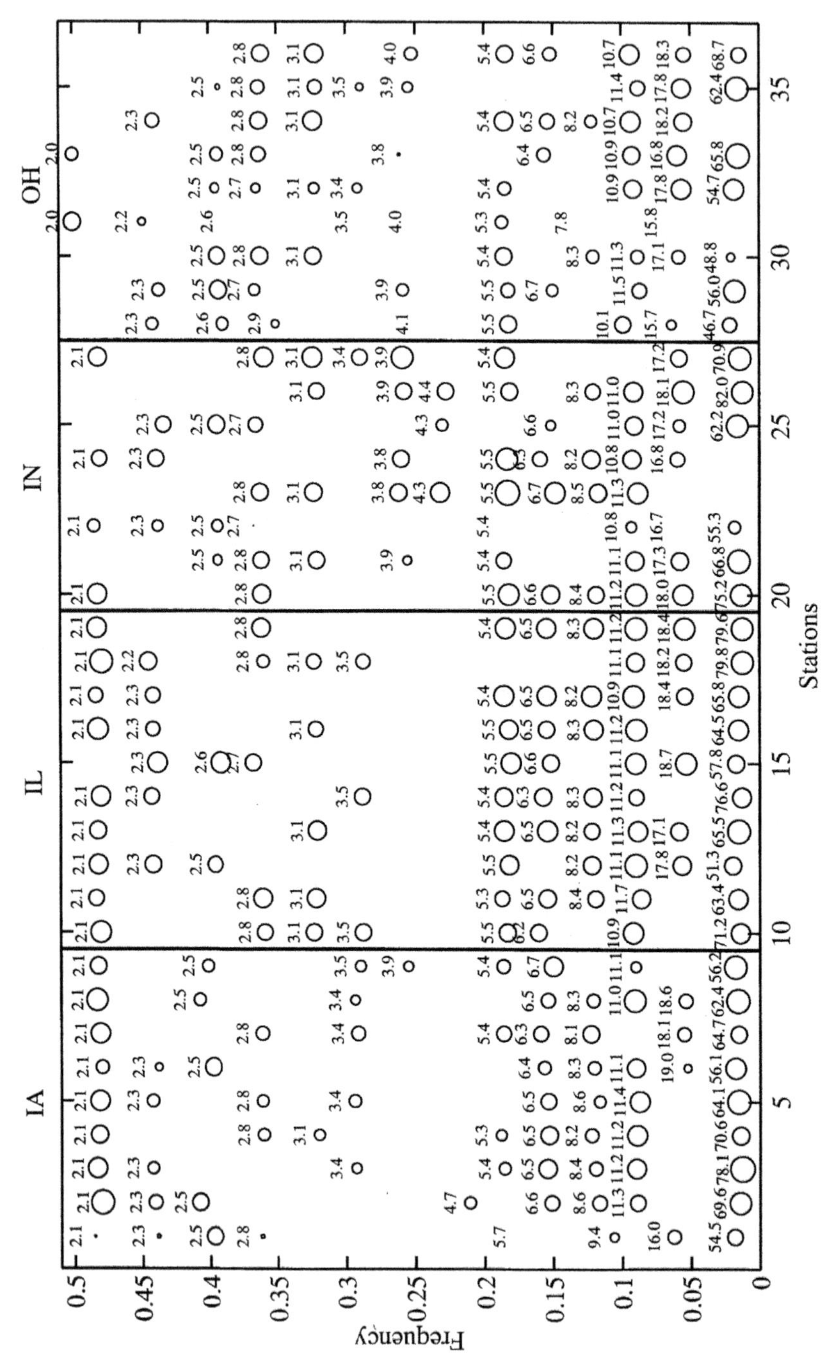

Figure 4.8. Periodicities in the HCN temperature data using MEM analysis with N/3 AR order.

Table IV.4. Periodicities in the HCN temperature data, MEM analysis.

Periodicity Group (years)	$p = N/2$		$p = N/3$	
	No. Of Stations	Mean Period	No. Of Stations	Mean Period
50-80	24	66.1±8	29	65.2±8
17-20	15	18±0.6	19	18±0.6
10-12	32	10.9±0.2	32	11.1±0.3
8-9	26	8.3±0.1	20	8.3±0.1
6-7	18	6.6±0.2	24	6.5±0.1
5-6	20	5.4±0.06	28	5.4±0.08
4-4.5	12	4.3±0.06	5	4.2±0.2
3.5-4	16	3.9±0.02	10	3.9±0.05
2.9-3.2	26	3.1±0.01	15	3.1±0.01
2.6-2.9	21	2.8±0.05	23	2.8±0.03
2.4-2.6	15	2.5±0.04	17	2.5±0.04
2.25-2.4	10	2.3±0.01	16	2.3±0.02
2-2.15	19	2.1±0.01	22	2.1±0.01

Table IV.5. Periodicities in the NCDC divisional temperature data, MEM analysis.

Periodicity Group (years)	$p = N/2$		$p = N/3$	
	No. of Stations	Mean Period	No. of Stations	Mean Period
50-80	36	68.2±5	37	64.8±6
17-20	17	17.9±0.3	30	17.8±0.3
10-12	33	10.9±0.1	37	11.1±0.1
8-9	28	8.4±0.1	30	8.3±0.1
6-7	2	6.4±0.03	30	6.5±0.1
5-6	25	5.4±0.05	32	5.4±0.05
4.5-5	2	4.7±0.01	2	4.7±0.02
4-4.5	7	4.2±0.01	19	4.3±0.03
3.6-4	33	3.8±0.01	5	3.8±0.01
2.9-3.2	23	3.1±0.02	0	---
2.6-2.9	14	2.8±0.04	14	2.8±0.01
2.4-2.6	17	2.5±0.01	28	2.5±0.02
2.25-2.4	6	2.3±0.02	7	2.3±0.01
2-2.2	23	2.1±0.01	19	2.1±0.01

Table IV.6. Periodicities in the last 100 years NOAA tree-ring data, MEM analysis.

Periodicity Group (years)	$p = N/2$		$p = N/3$	
	No. of Stations	Mean Period	No. of Stations	Mean Period
50-80	4	62.5±9	3	6078±5.4
20-40	11	25.5±4.2	19	29.1±6.8
17-20	2	17.8±0.9	4	18.2±0.58
10-12	9	11.2±0.26	10	11.1±0.31
9-10	2	9.2±0.17	4	9.4±0.38
7-8	6	7.6±0.13	3	7.4±0.16
6-7	5	6.5±0.14	7	6.4±0.26
5-6	8	5.5±0.45	9	5.3±0.38
4.5-5	4	4.7±0.22	4	4.9±0.04
4-4.5	3	4.3±0.23	6	4.4±0.09
3.6-4	11	3.8±0.11	16	3.8±0.07
3.2-3.6	16	3.3±0.05	16	3.3±0.06
2.9-3.2	8	3.0±0.04	4	3.0±0.04
2.6-2.9	15	2.8±0.06	9	2.8±0.06
2.4-2.6	17	2.5±0.04	17	2.5±0.04
2.25-2.4	20	2.3±0.06	12	2.3±0.05
2-2.2	10	2.1±0.05	13	2.1±0.03

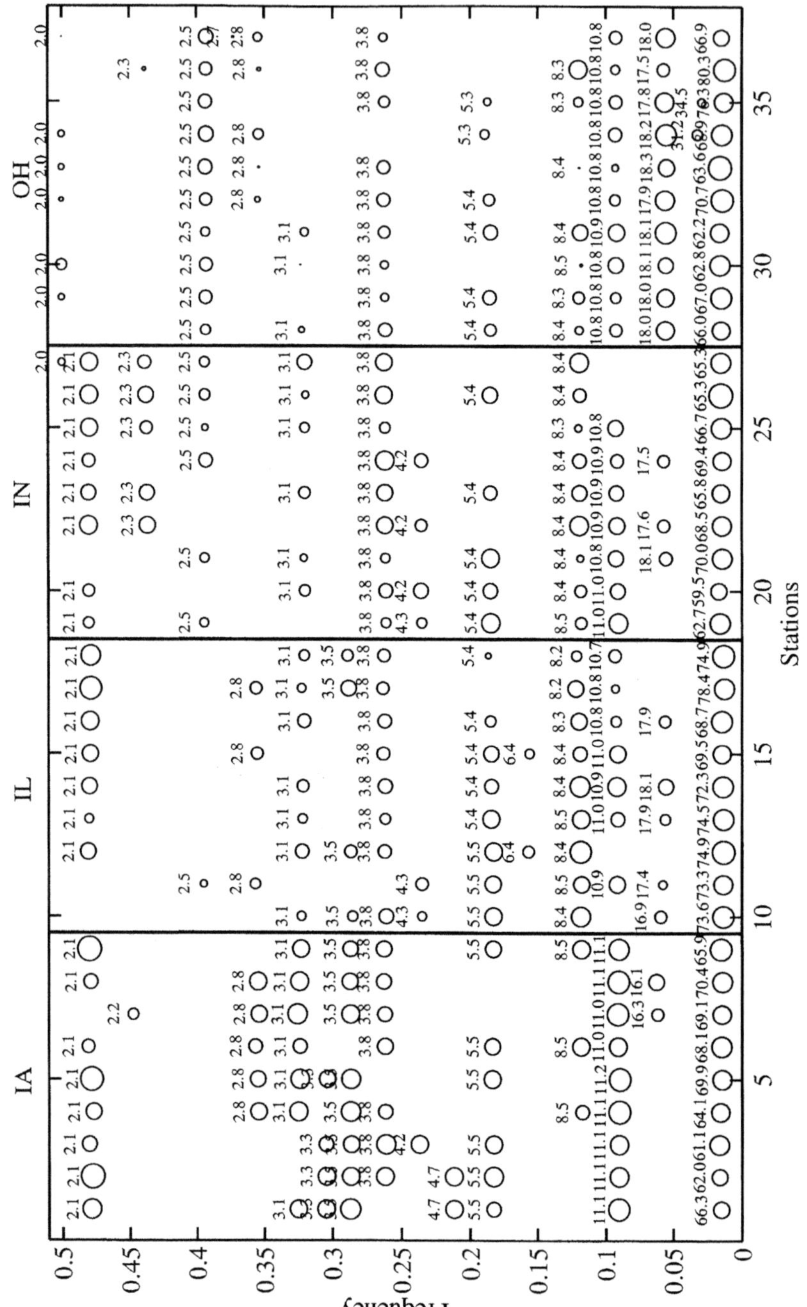

Figure 4.9. Periodicities in the NCDC divisional temperature data using MEM analysis with N/2 AR order.

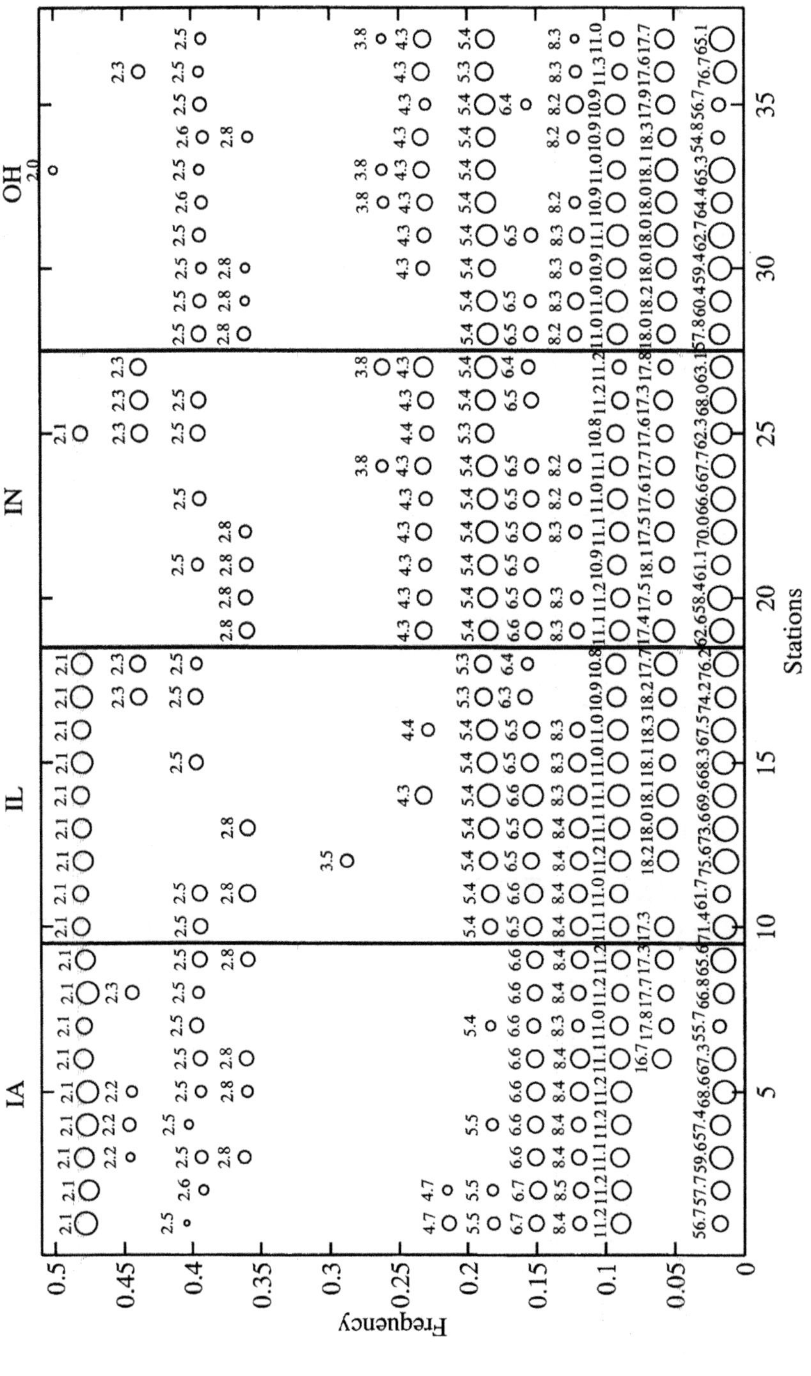

Figure 4.10. Periodicities in the NCDC divisional temperature data using MEM analysis with AR order of N/3.

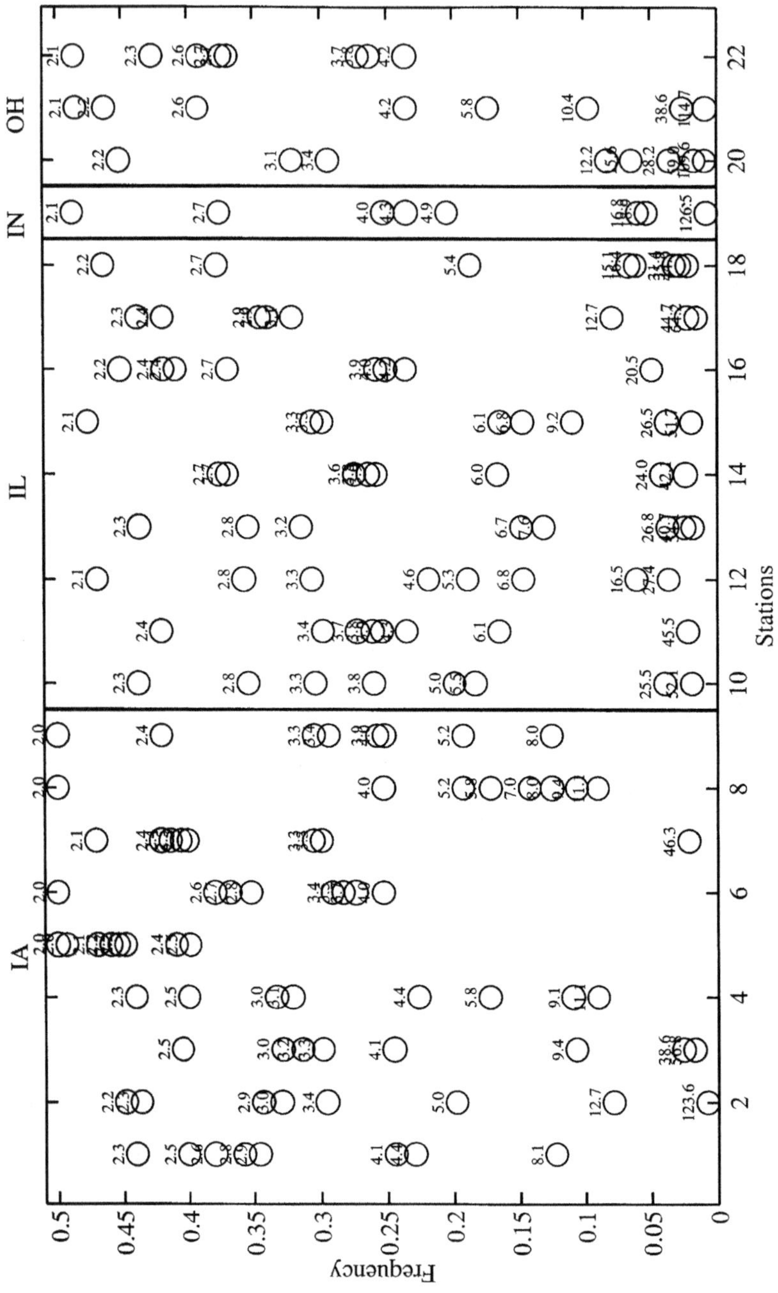

Figure 4.11. Periodicities in the NOAA Treering data using MEM analysis with AR order of N/2.

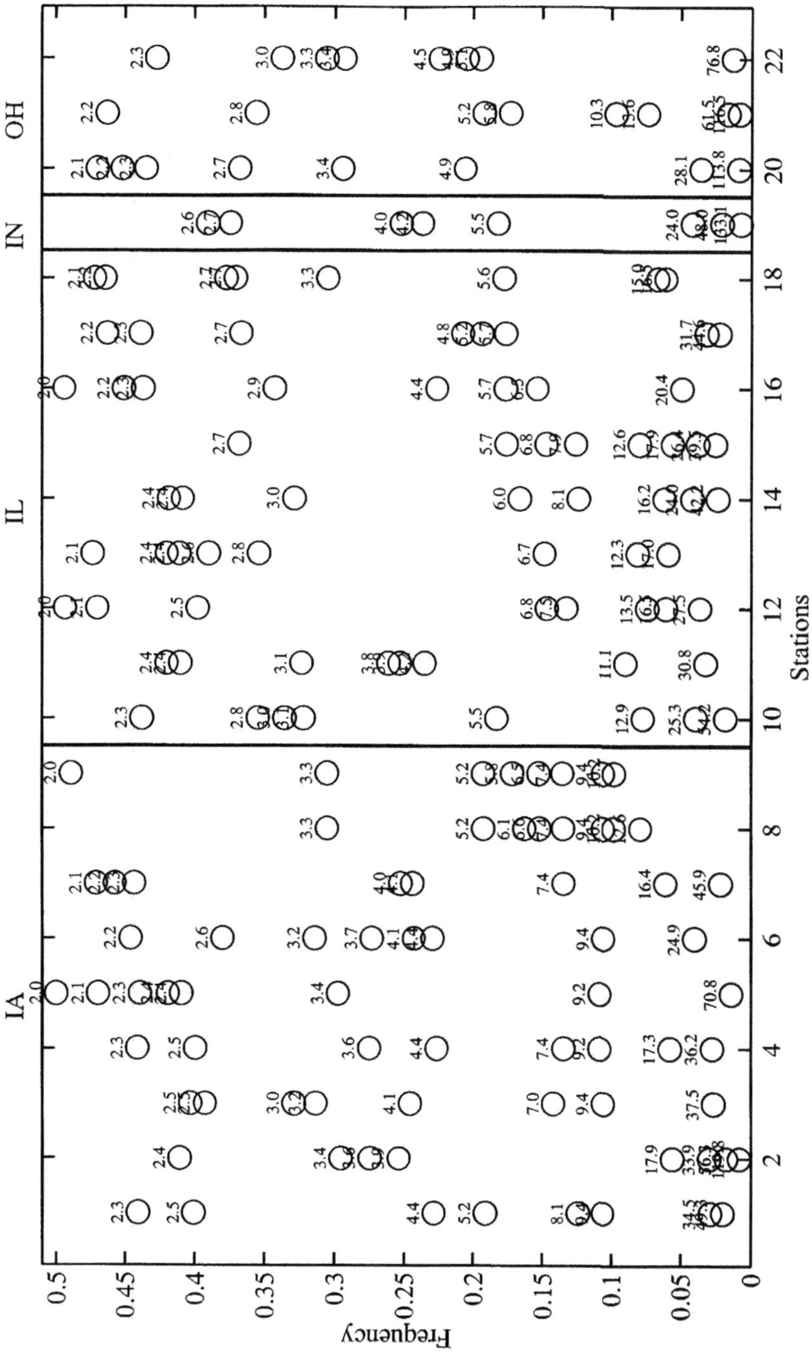

Figure 4.12. Periodicities in the NOAA Treering data using MEM analysis with AR order of N/3.

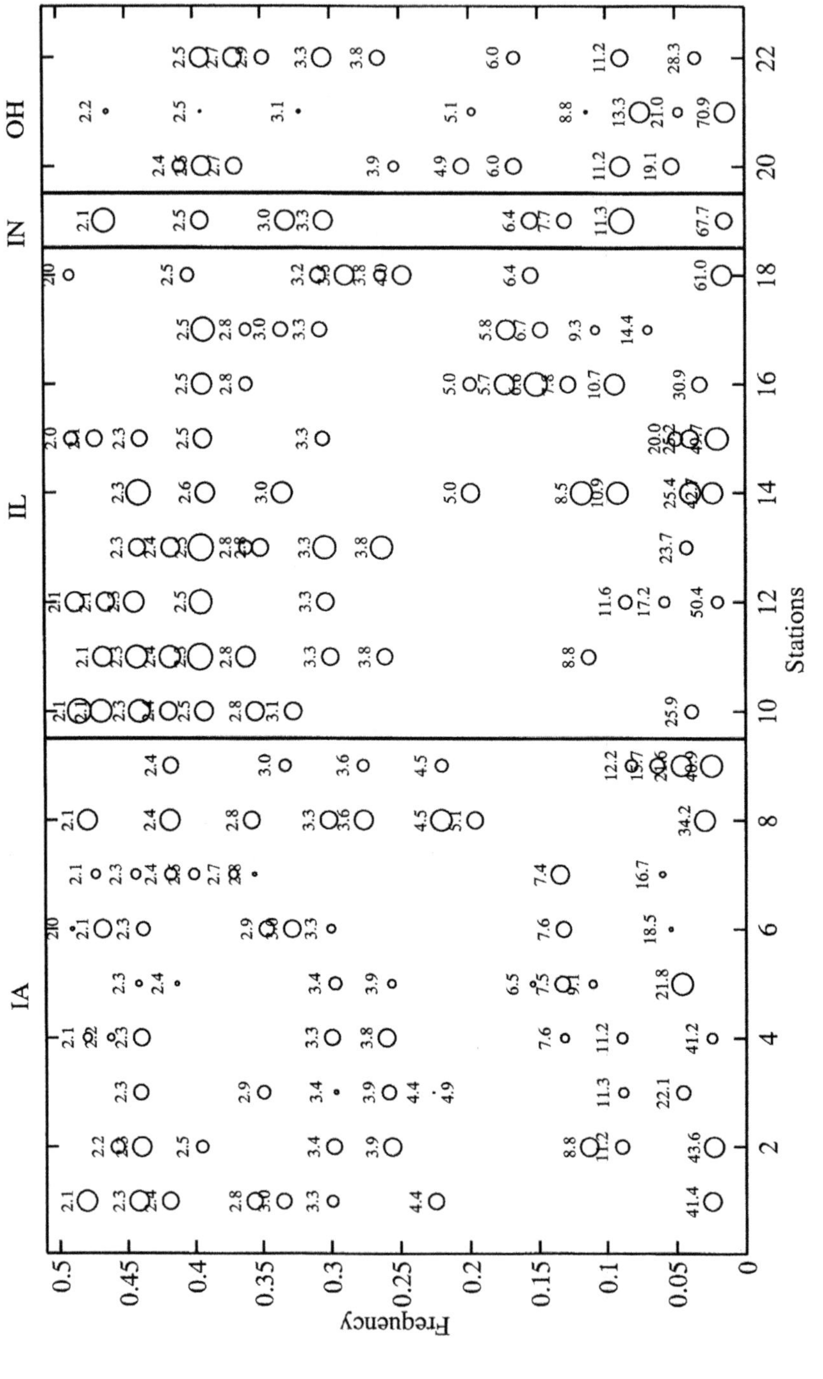

Figure 4.13. Periodicities in the last 100 years of NOAA treering data using MEM analysis with AR order of N/2.

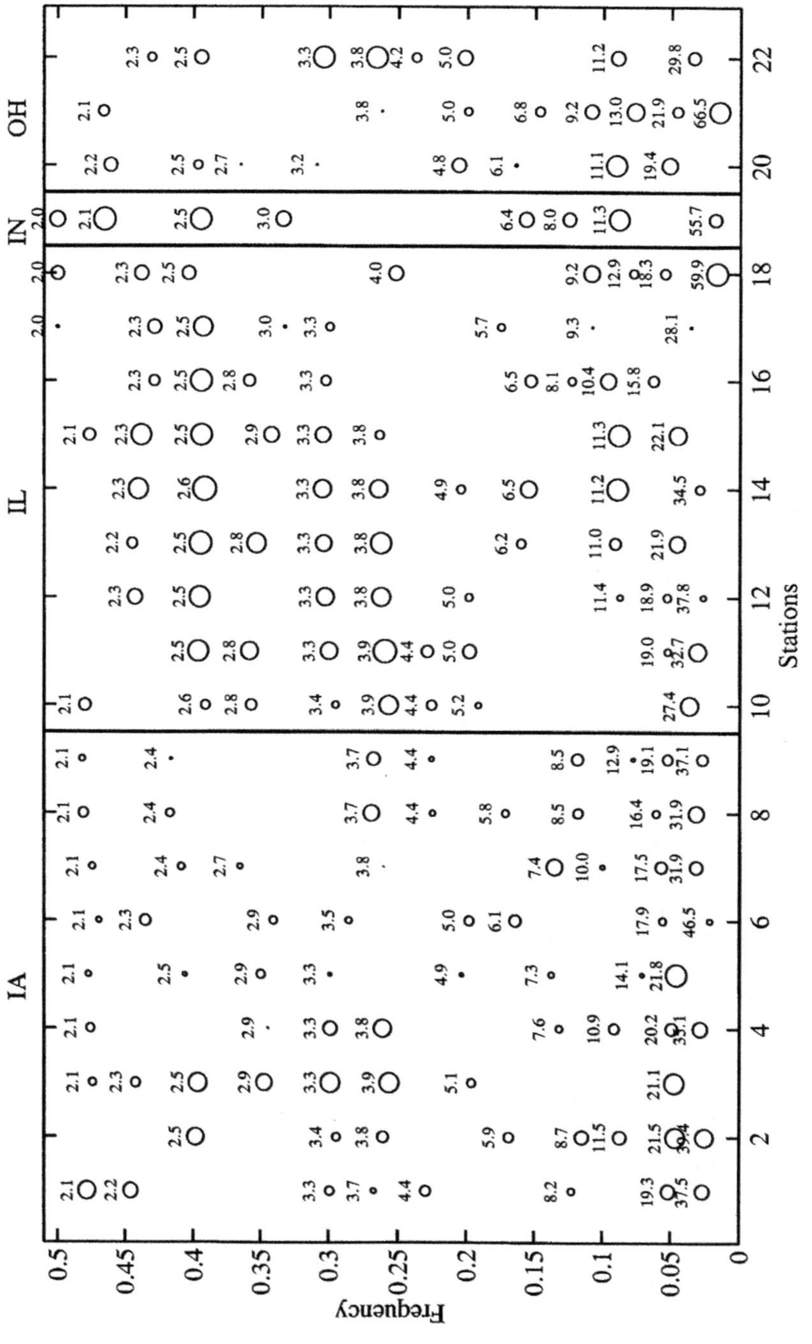

Figure 4.14. Periodicities in the last 100 years of NOAA treering data using MEM analysis with AR order of N/3.

a) USGS Flow Data

The results from the MTM analysis of the USGS flow data are shown in Figs. 4.15 and 4.16 for band width values of 2.5/*N* and 4/*N*, respectively. Periodicities found in the USGS flow data are summarized in Table IV.7. The first column in Table IV.7 gives the approximate range of each group of periodicities. Columns 2 to 5 pertain to the case of *W*=2.5/*N* and give the number of stations with significant components at the 0.1 level, the number of stations with significant components at the *1/N* level, the mean of the period in the respective group, and the average percentage of the total variance explained by components in that group. Columns 6 to 9 give the same information but with regard to the case of *W*=4/*N*. For this group of 26 stations, considering a 95% confidence interval, no more than five stations should have significant components at a given frequency at the 0.1 level under the null hypothesis of no periodicity. Furthermore, no more than one station should have significant components at a given frequency at the 1/*N* level under the null hypothesis of no periodicity. Based on the numbers in Table IV.7, sufficient evidence does not exist for periodicities in the range from 89 to 121 years, although a component with such periodicity would explain 8 to 15% of the total variance in about seven stations. However, it is seen from Figures 4.15 and 4.16 that most of these stations are found in Iowa, which raises the question of a possible spatial variation of periodicities. The group of 24 to 30 years with an average period of 25.7 years is found in more stations than would be only due to chance. Periodicities of 10-12, 7-8, and 4-5 years also seem to be evident. Periodicities of 2-3 years seem to be less evident at the 1/*N* level, with the groups centered at 2.1 and 2.6 being the most common at the 0.1 level. The percentage of total variance explained by the identified periodicities range from 4 to 14%. The largest contribution comes from the 5.7 years component at 13-14%, followed by the 11.1 years component at about 12% of the total variance.

b) HCN Rainfall Data

The results from the MTM analysis of the HCN rainfall data are shown in Figs. 4.17 and 4.18. Groups of periodicities found in the HCN rainfall data are summarized in Table IV.8. For this group of 36 stations, considering a 95% confidence interval, no more than seven stations should have significant components at a given frequency at the 0.1 level under the null hypothesis of no periodicity. Furthermore, no more than one station should have significant components at a given frequency at the 1/*N* level under the null hypothesis of no periodicity. Based on the results in Table IV.8 there is little evidence of periodicities in the range of 100-150, 30-40, 20-25 years. Periodicities of 10-12, 5-6, 4-5, and 2-4 years are the most common in terms of the number of stations having significant components at both the 0.1 and 1/*N* levels. The percentage of total variance explained by these periodicities range from 5 to 15% with a maximum of 15-16% for the group of 100-150 years followed by the 30-40 years group at 7-8% of the total variance.

c) NCDC Divisional Average Rainfall Data

The results from the MTM analysis of NCDC divisional rainfall data are shown in Figs. 4.19 and 4.20. It can be noticed from these figures that the results are more uniform than were those for the HCN rainfall data for the same reasons discussed before. Groups of periodicities found in the NCDC divisional rainfall data are summarized in Table IV.9. For this group of 37 divisions, considering a 95% confidence interval, no more than seven divisions should have significant components at a given frequency at the 0.1 level under the null hypothesis of no periodicity. Furthermore, no more than one division should have significant components at a given frequency at the $1/N$ level under the null hypothesis of no periodicity. Based on the results in Table IV.9 the number of stations with periodicities of 90 to 140 years is too small to support the existence of such periods in the data. The same is more or less true with periods of 30-40, 20-25, and 17-20 years. The most significant groups are 2.6-2.8, 3-3.5, 4-5, and 5.4-6 years. The percentage of total variance explained by these periodicities range from 5 to 8%, with the largest contribution from components 5.4-6 years at 7 to 8% of the total variance.

An observation in both Figs. 4.19 and 4.20 is the switch of periodicity from around 5.8 years in Ohio, Indiana, and most of Illinois to 6.4 years in Iowa and the first four divisions in Illinois. This, again, raises the question of spatial variability of periodicities. Another example is periods of 30 to 33 years in Fig. 4.20 which seem to be found only in Indiana and Ohio and almost disappear in Iowa and Illinois in exchange for periodicities of 19 to 21 years.

d) HCN Temperature Data

The results from the MTM analysis of the HCN temperature data are shown in Figs. 4.21 and 4.22. Groups of periodicities found in the HCN temperature data are summarized in Table IV.10. Based on the results in Table IV.10, there is not enough evidence to establish the existence of periods of 90 to 120 years. The most significant groups are 50-80, 17-20, 10-12, 5-6, and 2-2.2 years. Other periods seem to be much more significant for the case of $W=2.5/N$ than for the case of $W=4/N$. This may be attributed to a narrower bandwidth of the temperature data for which $W=4/N$ is large enough to smooth these periodicities. The percentage of the total variance explained by these periodicities ranges from 3 to 14%. The largest contribution comes from periods of 90-120 years at 14%, followed by periods of 50-80 years at 12-14%, then 10-12 years at 11% of the total variance.

e) NCDC Divisional Average Temperature Data

The results from the MTM analysis of the NCDC divisional temperature data are shown in Figs. 4.23 and 4.24. Groups of frequencies found in the NCDC divisional temperature data are summarized in Table IV.11. Periods from 50 to 80 years are strongly significant for $W=2.5/N$ at both the 0.1 and $1/N$ levels. For $W=4/N$ only one station is significant at the $1/N$ level. Again, this may be attributed to the narrowness of the actual bandwidth of periodicities in the temperature data. Periods of 10-12 years and 2-2.15 years are also highly significant. Other common periods are 2.2-2.4, 2.9-3.2, and 5-6 years. The percentage of total variance explained by these periodicities range from 2 to 14%. The largest contribution is from periods of 50 to 80 years at 12-14%, followed by periods 10 to 12 years at 9-11%, then 2.4 to 2.6 years at 8-9% of the total variance.

As was the case with the NCDC divisional rainfall data, some periodicities seem to vary spatially. In Figs. 4.23 and 4.24 periodicities of 18 to 19 years seem to disappear in Iowa and the first two divisions in Illinois. Also, there seems to be an east-to-west drift in the 10 to 11 years periodicities from 10.6 in Ohio to 10.9 in Iowa. The same is observed for periods of 6 to 7 years which consistently change from 6.5 to 6.7 years from east to west.

f) NOAA Tree-ring Data

The results from the MTM analysis of the NOAA tree-ring data are shown in Figs. 4.25 and 4.26. Since the length of the times series in this group is more than 200 years, a significance level of 0.95 instead of 0.90 was considered as well as the $1/N$ significance

Table IV.7. Periodicities in the USGS flow data, MTM analysis.

Periodicity Group (years)	$W = 2.5/N$				$W = 4/N$			
	n^1 at 0.1	n^1 at $1/N$	Mean Period	% Var.	n^1 at 0.10	n^1 at $1/N$	Mean Period	% Var.
89-121	2	0	93.3±6	8	7	0	109.2±	15
24-30	17	7	25.7±1	10	13	8	27.3±1.	9
10-12	18	5	11.1±0	12	18	1	10.9±0.	12
7-8	19	6	7.6±0.	10	4	2	7.8±0.1	13
5-6	21	3	5.7±0.	13	14	4	5.7±0.2	14
4-5	24	6	4.5±0.	9	14	7	4.5±0.2	10
3.5-4	13	1	3.7±0.	4	12	1	3.7±0.2	5
3-3.5	14	2	3.2±0.	5	16	4	3.3±0.1	5
2.8-3	9	2	2.9±0.	5	10	0	2.9±0.0	6
2.4-2.8	22	0	2.6±0.	4	22	3	2.6±0.0	4
2.25-2.4	6	0	2.3±0.	4	6	0	2.3±0.0	4
2.15-2.25	8	0	2.2±0.	4	11	1	2.2±0.0	4
2-2.15	16	1	2.1±0.	4	20	4	2.1±0.0	4

[1] n is the number of stations with components that are significant in a given periodicity group at the specified significance level.

Table IV.8. Periodicities in the HCN rainfall data, MTM analysis.

Periodicity Group (years)	$W = 2.5/N$				$W = 4/N$			
	n^1 at 0.10	n^1 at $1/N$	Mean Period	% Var.	n^1 at 0.10	n^1 at $1/N$	Mean period	% Var.
100-150	4	0	132±18.	16	12	2	134±12.	15
30-40	5	0	33.8±2	8	10	4	34.2±2.5	7
20-25	11	1	22.6±1.2	5	9	1	22.5±1	5
17-20	2	0	18.7±1.7	5	2	0	19.1±0.6	6
10-12	15	1	10.7±0.6	6	17	2	10.4±0.8	6
7-8	18	2	7.6±0.2	5	14	2	7.6±0.2	6
6-7	9	0	6.4±0.2	5	10	0	6.5±0.3	4
5-6	26	8	5.7±0.2	6	26	10	5.7±0.2	6
4-5	33	7	4.6±0.3	5	33	3	4.5±0.3	5
3.5-4	30	6	3.8±0.2	6	19	3	3.8±0.2	6
3-3.5	31	7	3.2±0.1	6	31	6	3.3±0.1	5
2.6-2.8	20	4	2.7±0.04	8	23	6	2.7±0.05	7
2.4-2.6	13	1	2.5±0.06	4	10	0	2.6±0.04	5
2.25-2.4	20	3	2.3±0.04	6	24	4	2.3±0.04	5
2.15-2.25	14	2	2.2±0.01	6	9	2	2.2±0.03	6
2-2.15	6	0	2.1±0.03	5	14	1	2.1±0.03	6

[1] n is the number of stations with components that are significant in a given periodicity group at the specified significance level.

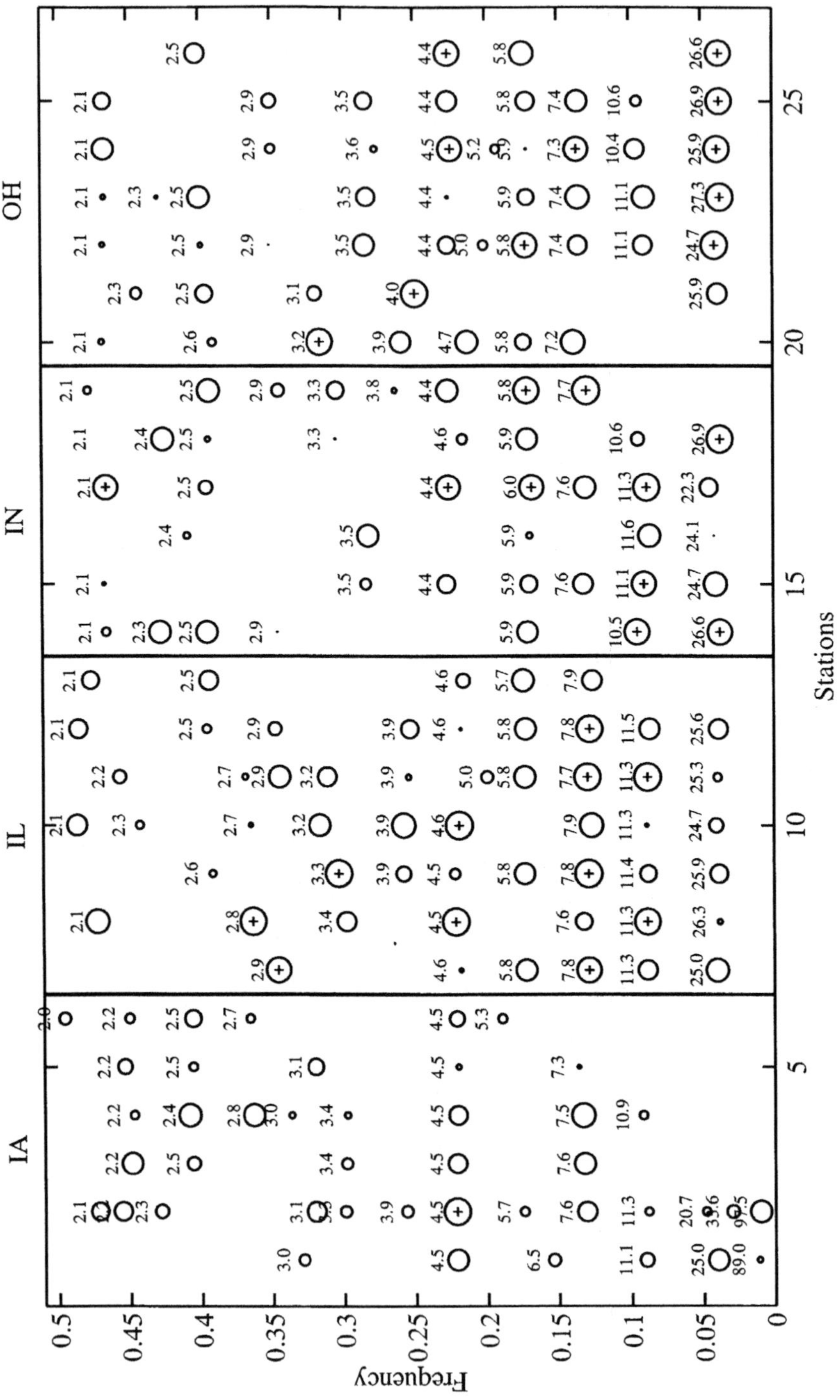

Figure 4.15. Periodicities in the USGS flow data using MTM analysis with band width W=2.5/N.

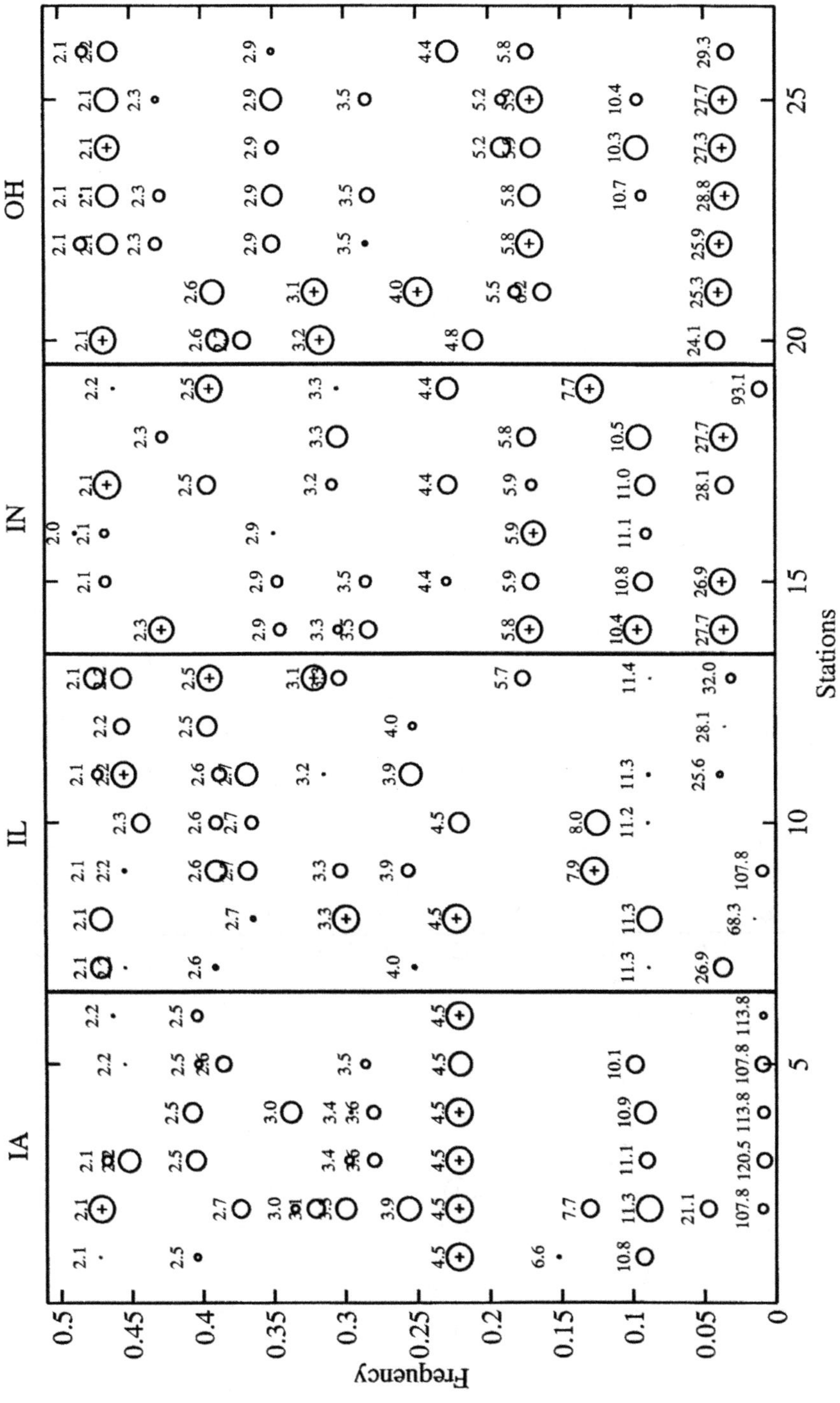

Figure 4.16. Periodicities in the USGS flow data using MTM analysis with band width W=4/N.

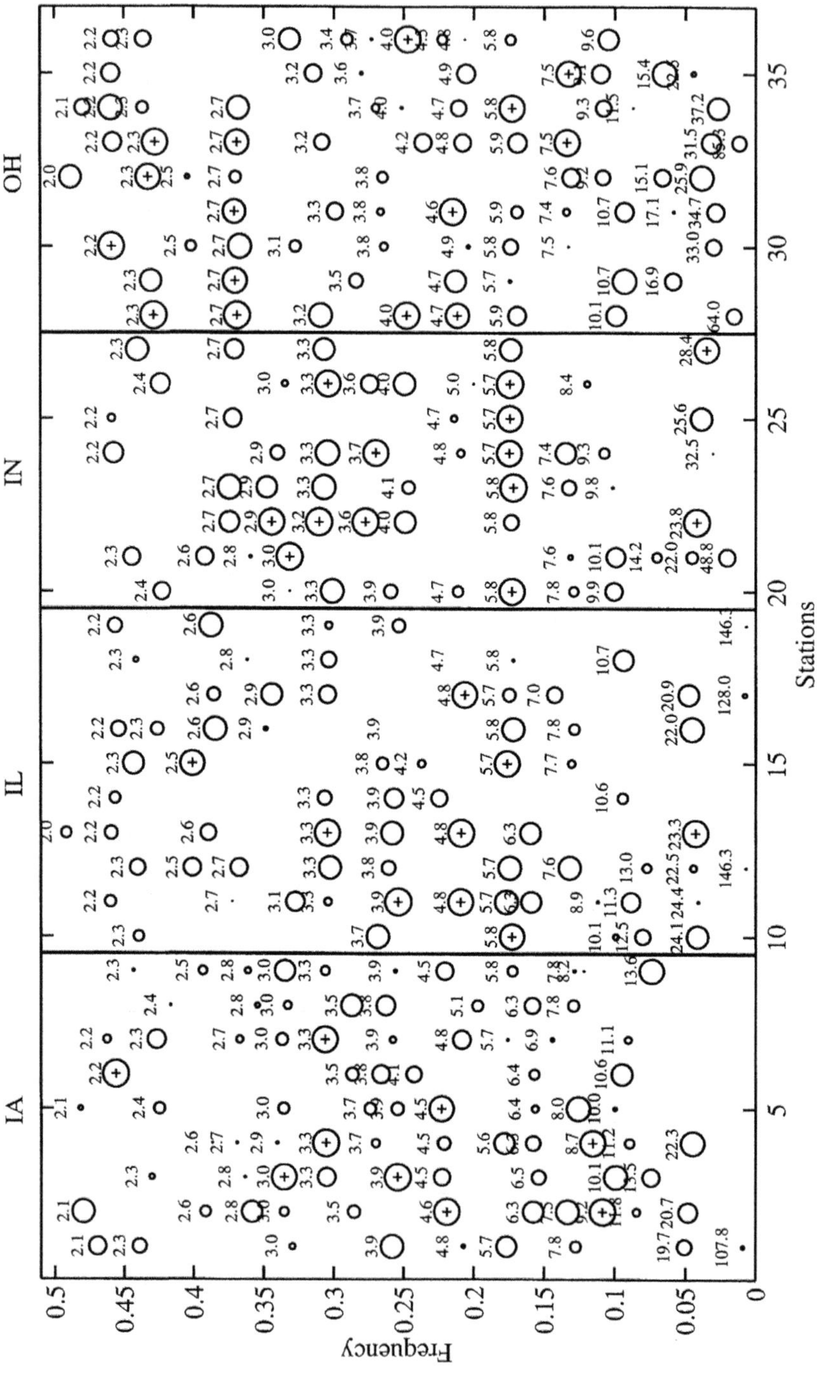

Figure 4.17. Periodicities in the HCN rainfall data using MTM analysis with band width W=2.5/N.

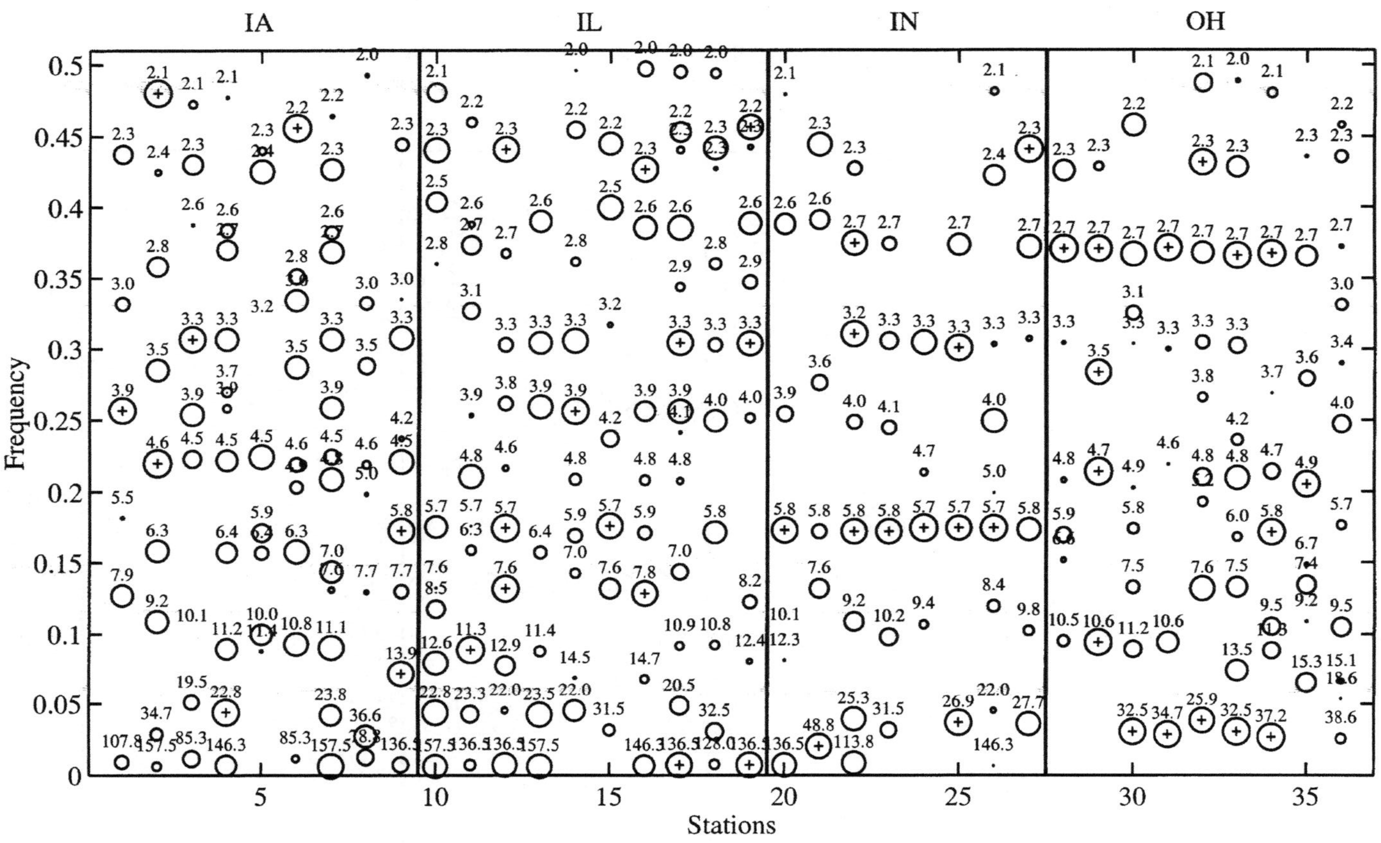

Figure 4.18. Periodicities in the HCN rainfall data using MTM analysis with band width W=4/N.

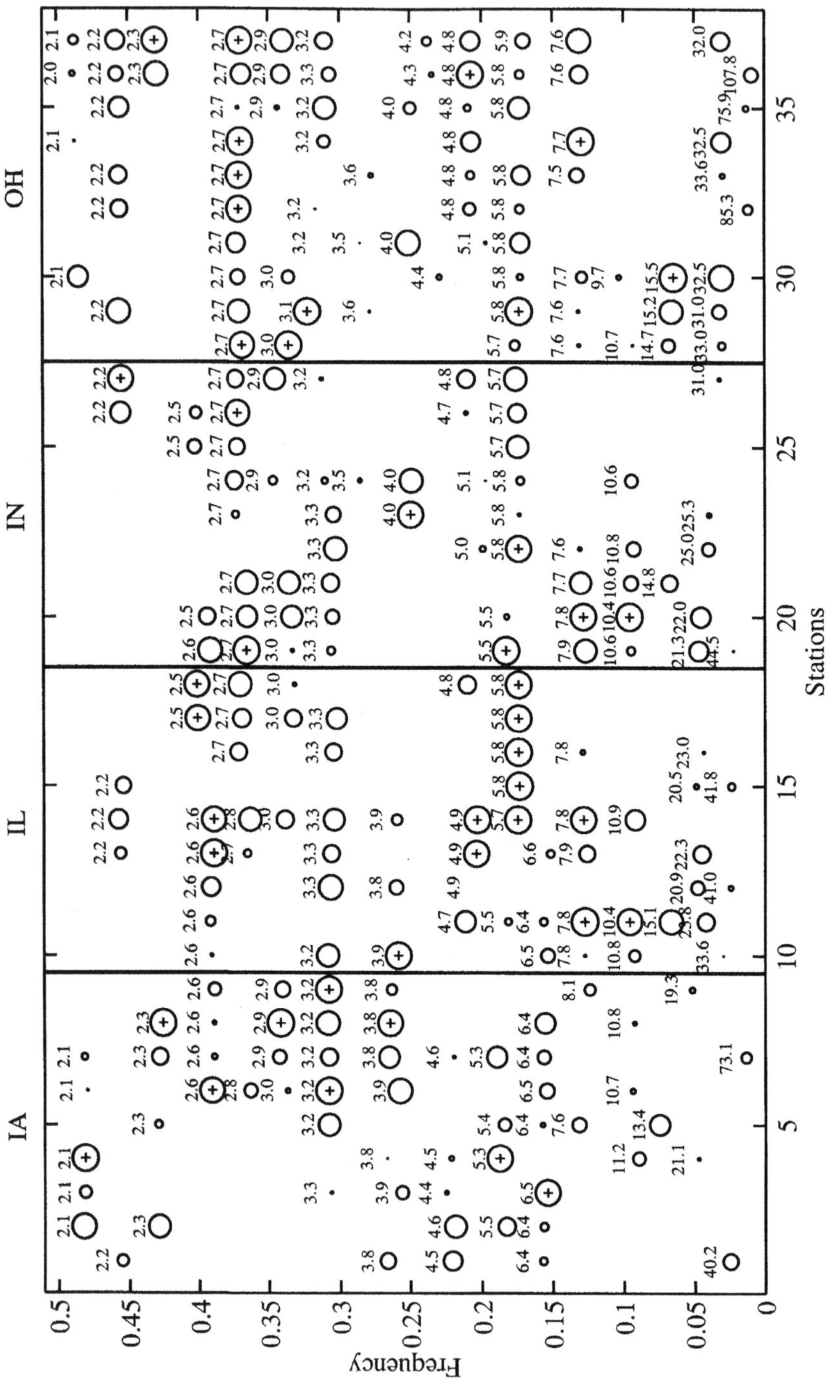

Figure 4.19. Periodicities in the NCDC divisional rainfall data using MTM analysis with band width W=2.5/N.

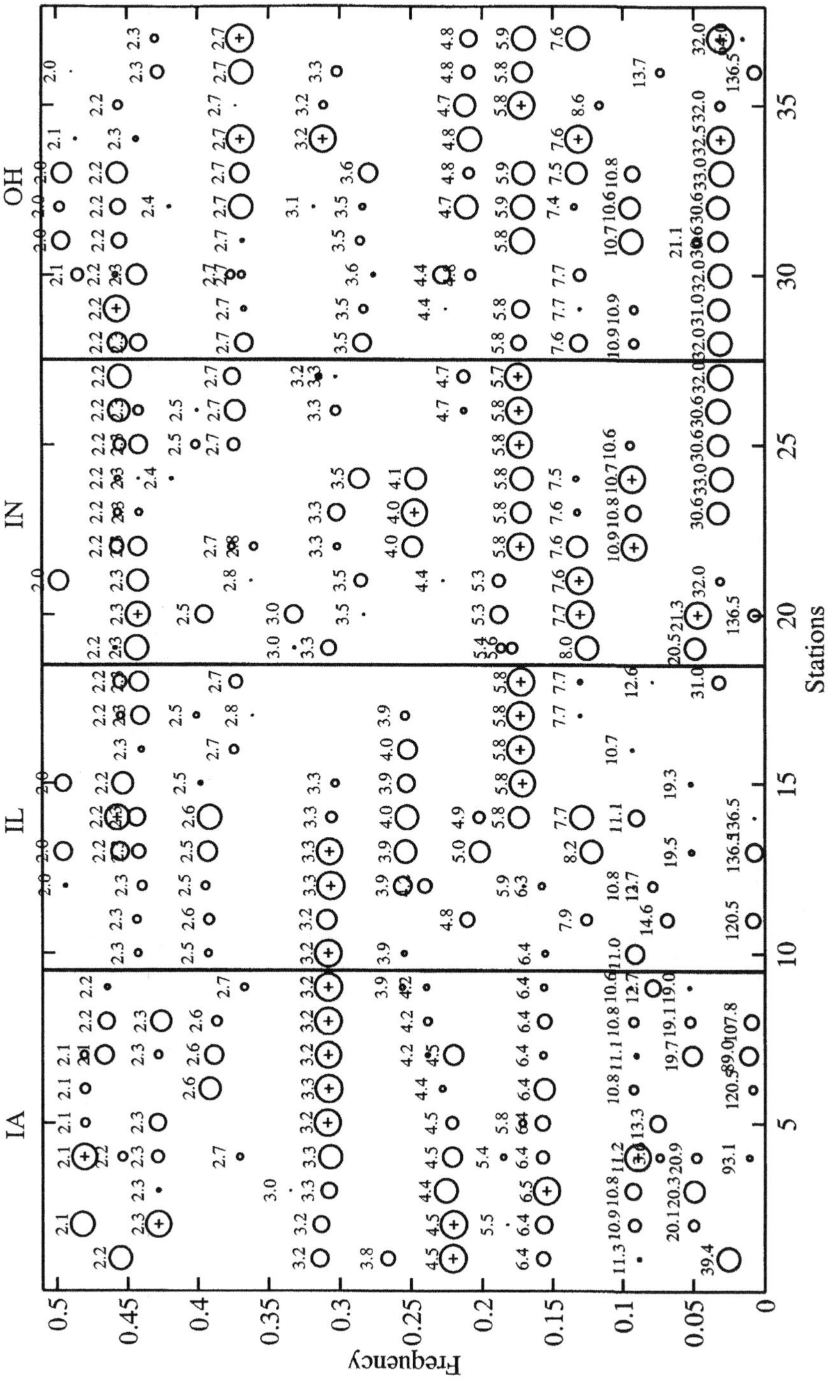

Figure 4.20. Periodicities in the NCDC divisional rainfall data using MTM analysis with band width W=4N.

Table IV.9. Periodicities in the NCDC divisional rainfall data, MTM analysis.

Periodicity Group (years)	$W = 2.5/N$				$W = 4/N$			
	n^1 at 0.10	n^1 at $1/N$	Mean Period	% Var.	n^1 at 0.10	n^1 at $1/N$	Mean Period	% Var.
90-140	1	0	107.8	6	8	0	123±16.0	5
30-40	8	0	32.4±1.0	7	17	2	32.1±2.0	6
20-25	9	0	22.2±1.5	4	6	1	20.7±0.5	4
17-20	1	0	20.2±0.8	4	5	0	19.3±0.3	5
10-12	12	2	10.7±0.2	5	21	2	10.9±0.2	6
7-8	17	4	7.7±0.12	5	17	3	7.6±0.14	5
6-7	10	1	6.4±0.1	4	11	1	6.4±0.04	5
5.4-6	25	8	5.7±0.1	7	24	9	5.8±0.1	8
4-5	24	3	4.6±0.3	4	29	3	4.5±0.3	6
3.5-4	15	3	3.8±0.2	5	17	1	3.7±0.2	6
3-3.5	29	1	3.2±0.1	5	28	9	3.2±0.1	7
2.6-2.8	24	12	2.7±0.02	7	22	2	2.7±0.03	6
2.4-2.6	15	5	2.5±0.03	6	13	0	2.5±0.03	5
2.25-2.4	6	2	2.3±0.01	4	29	2	2.3±0.04	5
2.15-2.25	12	1	2.2±0.01	6	23	2	2.2±0.01	7
2-2.15	9	1	2.1±0.01	7	16	1	2.1±0.04	6

[1] n is the number of stations with components that are significant in a given periodicity group at the specified significance level.

level. As can be seen from Figs. 4.25 and 4.26, groups of periodicities can hardly be identified. Also, very few of the observed periodicities are significant at the 1/*N* level. The MTM analysis was repeated for the last 100 years of the NOAA tree-ring data, and the results are shown in Figs. 4.27 and 4.28. A much clearer grouping of periodicities can be seen in these figures, and periodicities that are significant at the 1/*N* level are more common than in the case when the whole record is analyzed.

In general, one would expect periodicities to be more well-defined as the length of the records increases. The fact that periodicities are not well grouped when the entire record is considered for MEM or MTM analysis is counterintuitive. One reason for such behavior may have to do with the quality of the earlier part of the tree-ring data. Another concern is whether a different type of nonstationarity is involved, in which not only periodic components are added to a stationary process, but also the amplitude and frequency of such components change with time. In this case the structure of the process may vary in a periodic manner (Thomson, 1990a, p. 213). This type of nonstationarity can be studied by means of evolutionary spectral analysis (Priestley, 1965) as well as other time-frequency techniques. This aspect of the analysis will be discussed in Chapter 5.

Groups of periodicities found in the last 100 years of the NOAA tree-ring data are summarized in Table IV.12. The most common periods in Table IV.12 are 2.5, 3.3, 3.8, and 10-12 years. The percentage of total variance explained by these periodicities range from 5 to 18%.

4.6. DISCUSSION OF RESULTS

Based on the results of harmonic analysis of a number of hydrologic as well as climatic time series in the study region which were presented in the previous sections, it is evident that periodicities do exist in flow, rainfall, temperature, and tree-ring data. These periodicities are consistent within different time series of the same data type. Furthermore, some of these periodicities are detected in all four data categories analyzed. The results of the harmonic analysis are summarized in Table IV.13, in which the number of stations with significant periodic components at different average periods identified by both the MEM and MTM methods are given as a percentage of the total number of stations for each data category. Based on a 95% confidence interval of the number of stations with significant components at the 0.1 level in the absence of real periodicities, the upper limit of the percentage of stations is about 20%. Accordingly, percentages which are less than 20% in Table IV.13 are insignificant and are therefore not considered.

Previous research on climatic as well as hydrologic time series suggests that some of these periodicities are of natural origin. A number of natural phenomena are believed to have an effect on hydrologic as well as climatic data. These periodicities include the well known Sunspot cycle with a period of about 11 years, the sun's magnetic (Hale) cycle with a period of 22 years, the Luni-Solar cycle with a period of about 18.6 years, and the El-Niño-Southern Oscillations (ENSO) with periods of 2 to 3 and 4 to 6 years. Periods of 2 to 3 years are also known as the Quasi-Biennial Oscillations (QBO). It is also believed that climatic systems are nonlinear systems rather than linear ones (Burroughs, 1992). The nonlinearity of the weather results in a number of effects on the periodicities exhibited by the system under the influence of external driving forces. The first of these effects is called harmonic generation. When a nonlinear system is forced to oscillate at

Table IV.10. Periodicities in the HCN temperature data, MTM analysis.

Periodicity Group (years)	$W = 2.5/N$				$W = 4/N$			
	n^1 at 0.10	n^1 at $1/N$	Mean Period	% Var.	n^1 at 0.10	n^1 at $1/N$	Mean Period	% Var.
90-120	1	0	113.8	25	11	1	103±8.0	14
50-80	28	6	66.6±7	12	15	1	67.5±8.0	14
17-20	22	6	18.6±0.7	5	18	3	18.8±0.4	7
10-12	34	17	10.8±0.2	11	35	11	10.7±0.14	11
8-9	30	16	8.3±0.1	7	8	0	8.5±0.2	7
6-7	31	3	6.7±0.1	5	13	0	6.7±0.2	7
5-6	30	12	5.4±0.06	4	22	4	5.4±0.07	4
4-4.5	11	1	4.3±0.15	2	4	0	4.2±0.04	3
3.8-4	14	0	3.9±0.03	4	17	0	3.9±0.04	4
3.2-3.8	10	0	3.5±0.02	3	18	0	3.5±0.07	4
2.9-3.2	33	12	3.1±0.02	4	22	1	3.1±0.01	5
2.6-2.9	5	0	2.8±0.1	6	2	0	2.7+0.07	6
2.4-2.6	26	1	2.5±0.02	9	22	0	2.5±0.02	8
2.25-2.4	27	4	2.3±0.03	3	23	1	2.3±0.01	3
2 -2.15	25	12	2.1±0.01	3	30	13	2.1±0.02	3

[1] n is the number of stations with components that are significant in a given periodicity group at the specified significance level.

Table IV.11. Periodicities in the NCDC divisional temperature data, MTM analysis.

Periodicity Group (years)	$W = 2.5/N$				$W = 4/N$			
	n^1 at 0.10	n^1 at $1/N$	Mean Period	% Var.	n^1 at 0.10	n^1 at $1/N$	Mean Period	% Var.
50-80	37	26	65.2±6.0	12	37	1	68.2±8.0	14
18-19	24	1	18.6±0.2	5	21	0	18.6±0.2	9
10-12	37	19	10.7±0.1	11	37	9	10.8±0.1	9
8-9	33	3	8.4±0.05	6	0	0	---	---
6-7	35	0	6.6±0.08	5	12	0	6.7±0.05	7
5-6	35	6	5.5±0.03	4	20	0	5.4±0.02	4
4.5-5	9	0	4.7±0.03	1	2	0	4.8±0.01	5
4-4.5	25	0	4.3±0.02	3	9	0	4.3±0.01	4
3.6-4	30	0	3.8±0.02	4	32	0	3.8±0.02	5
3.2-3.6	5	0	3.5±0.1	4	25	0	3.4±0.1	6
2.9-3.2	36	15	3.1±0.02	4	23	0	3.1±0.01	4
2.6-2.9	12	0	2.9±0.02	4	0	0	---	---
2.4-2.6	11	0	2.5±0.02	9	36	2	2.5±0.02	8
2.2-2.4	37	22	2.3±0.01	3	16	2	2.3±0.01	2
2 -2.2	26	15	2.1±0.01	3	32	15	2.1±0.03	3

[1] n is the number of stations with components that are significant in a given periodicity group at the specified significance level.

Table IV.12. Periodicities in the last 100 years of the NOAA tree-ring data, MTM analysis.

Group	$W = 2.5/N$				$W = 4/N$			
	n at 0.90	n at 1-1/N	Mean	% Var.	n at 0.95	n at 1-1/N	Mean	% Var.
50-80	3	2	67.0±4.4	18	3	0	66.9±8.8	18
20-40	13	0	26.4±5.0	7	16	3	24.8±4.2	7
17-19	3	1	18.0±0.56	10	2	0	18.5±0.47	9
10-12	8	3	11.0±0.31	9	7	4	11.1±0.16	8
9-10	7	0	9.4±0.33	4	4	0	9.3±0.20	5
7-8	4	0	7.6±0.12	6	8	0	7.6±0.09	7
6-7	7	0	6.5±0.27	6	7	1	6.6±0.15	5
5-6	10	1	5.4±0.41	6	10	0	5.6±0.37	6
4.5-5	7	1	4.7±0.19	6	6	1	4.7±0.13	6
4-4.5	4	0	4.3±0.16	5	6	1	4.3±0.23	7
3.6-4	14	1	3.8±0.10	9	22	7	3.8±0.07	7
3.2-	18	4	3.3±0.06	5	17	5	3.3±0.05	5
2.9-	10	1	3.0±0.06	5	6	0	3.0±0.09	4
2.6-	7	0	2.8±0.05	4	8	0	2.8±0.03	3
2.4-	15	9	2.5±0.02	8	14	9	2.5±0.02	7
2.2-	7	2	2.3±0.03	3	13	1	2.3±0.06	4
2 -2.2	15	2	2.1±0.03	4	11	1	2.1±0.05	4

[1] n is the number of stations with components that are significant in a given periodicity group specified significance level.

Table IV.13. Summary of the periodicities in the data, MEM and MTM analyses. Given are the percentages of the total number of stations with significant periodicities.

Periods	Maximum Entropy Method (MEM)						Multi-Taper Method (MTM)					
	USGS Flow	Rainfall		Temperature		Tree-ring	USGS Flow	Rainfall		Temperature		Tree-ring
		HCN	DIV	HCN	DIV			HCN	DIV	HCN	DIV	
2.1	65%	31%	22%	61%	62%	59%	77%	39%	43%	86%	86%	68%
2.2	42%	33%	57%	---	---	---	42%	39%	62%	---	---	---
2.3	27%	53%	24%	50%	27%	90%	23%	67%	78%	78%	100%	59%
2.5-2.6	100%	42%	59%	47%	76%	77%	100%	36%	41%	72%	97%	68%
2.7-3	42%	69%	65%	64%	38%	68%	38%	64%	65%	---	32%	45%
3-3.5	77%	89%	68%	72%	62%	73%	62%	86%	78%	92%	97%	82%
3.5-4	54%	69%	59%	44%	89%	73%	50%	83%	46%	47%	86%	100%
4-5	92%	64%	68%	33%	51%	27%	92%	92%	78%	31%	68%	32%
5-6	88%	75%	76%	78%	86%	41%	81%	72%	68%	83%	95%	45%
6-7	---	25%	27%	67%	81%	32%	---	28%	30%	86%	95%	32%
7-9	77%	44%	41%	72%	81%	27%	73%	50%	46%	83%	89%	36%
10.7-11.1	88%	39%	65%	89%	89%	45%	69%	47%	57%	97%	100%	36%
17.9-19.6	----	---	22%	53%	81%	---	---	---	---	61%	65%	---
21-26	81%	39%	---	---	---	50%	65%	31%	24%	---	---	73%
32-34	---	33%	38%	---	---	---	---	22%	46%	---	---	---
65-69	---	25%	22%	81%	100%	---	---	33%	---	78%	100%	---
89-150	---	---	---	---	---	---	27%	33%	22%	31%	----	---

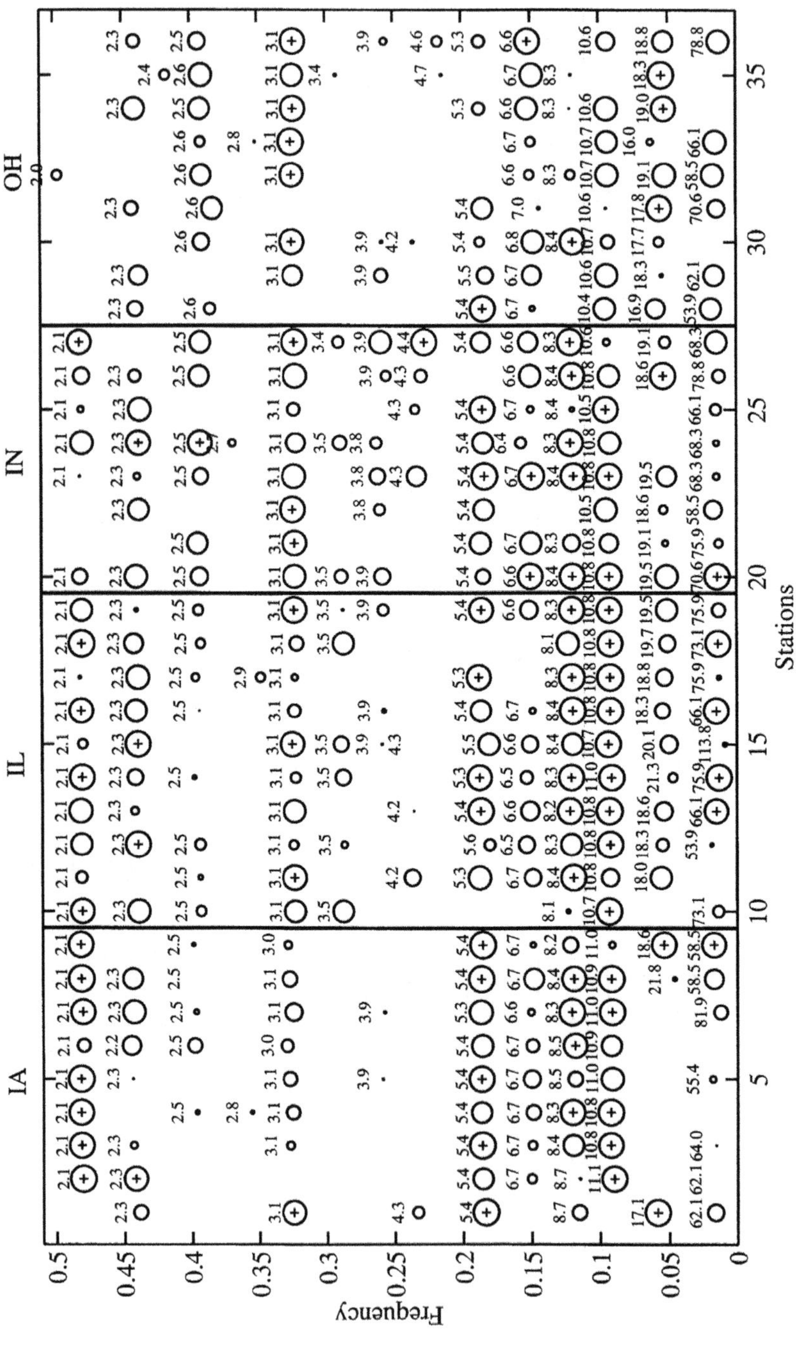

Figure 4.21. Periodicities in the HCN temperature data using MTM analysis with band width W=2.5/N.

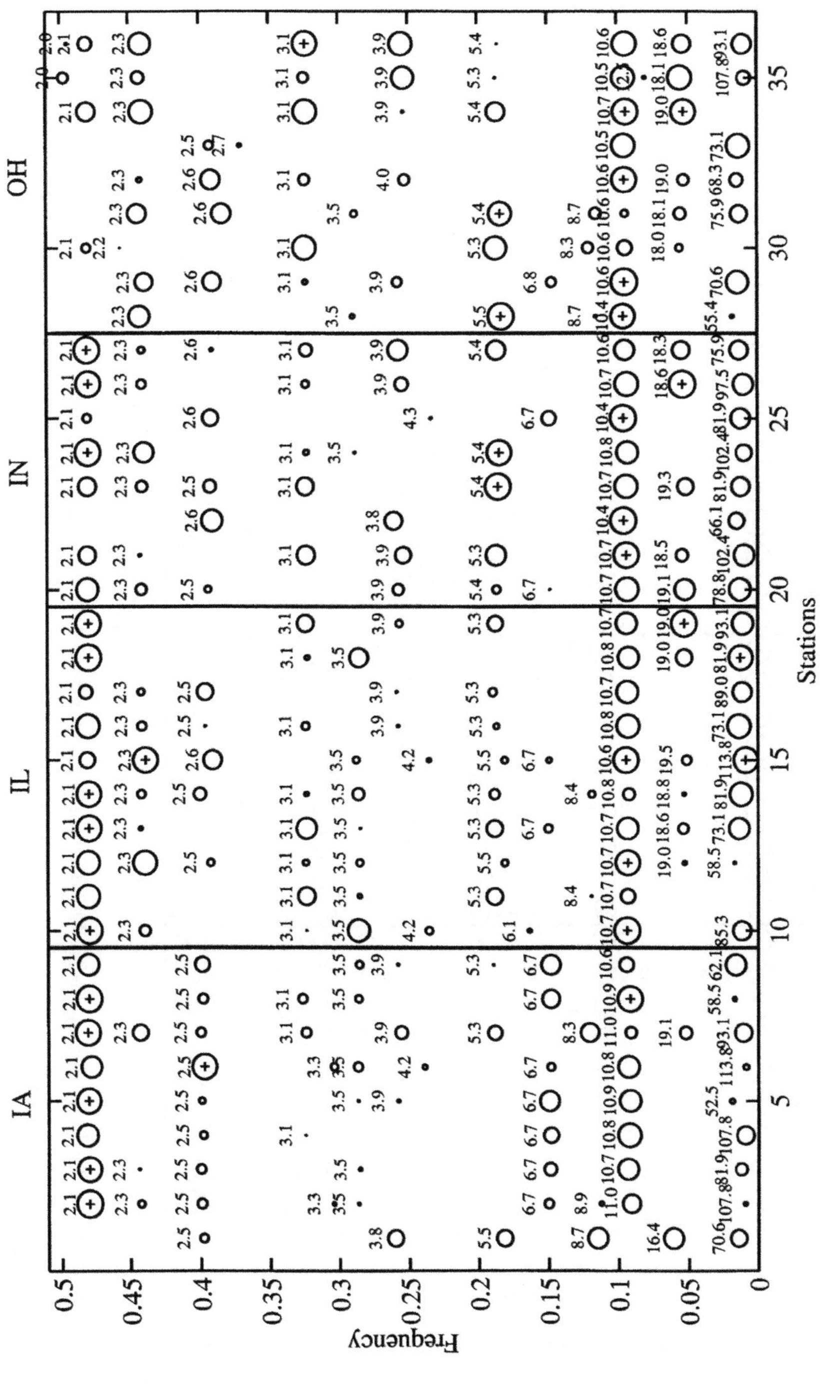

Figure 4.22. Periodicities in the HCN temperature data using MTM analysis with band width W=4/N.

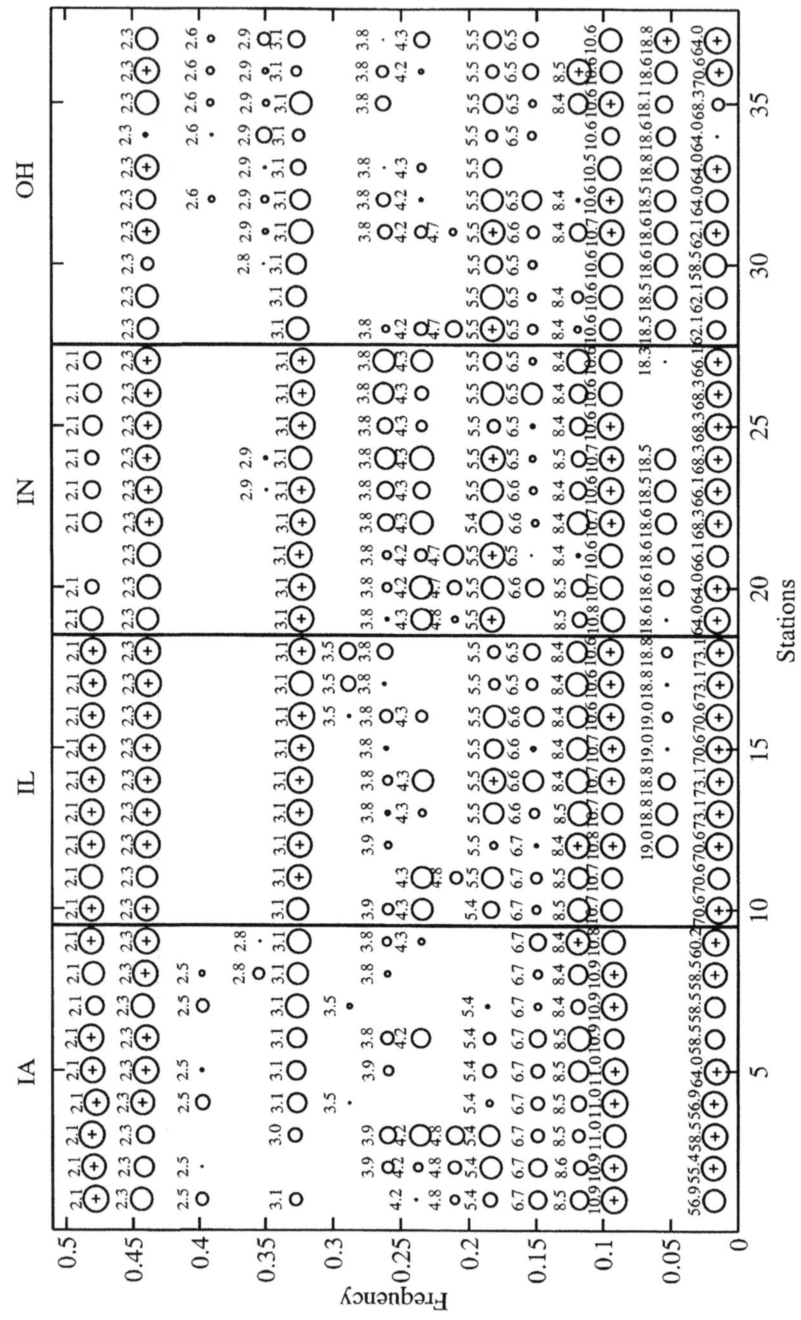

Figure 4.23. Periodicities in the NCDC divisional temperature data using MTM analysis with band width W=2.5/N.

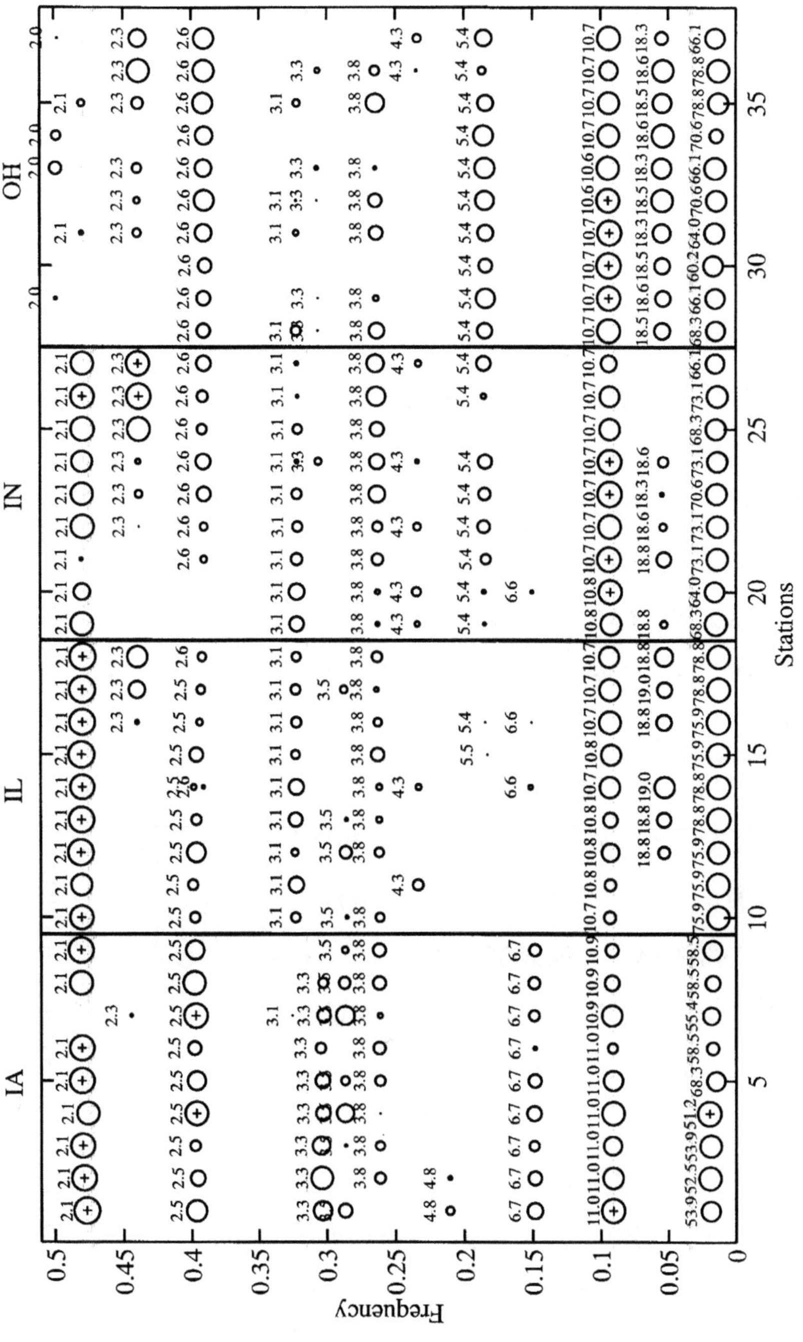

Figure 4.24. Periodicities in the NCDC divisional temperature data using MTM analysis with band width W=4/N.

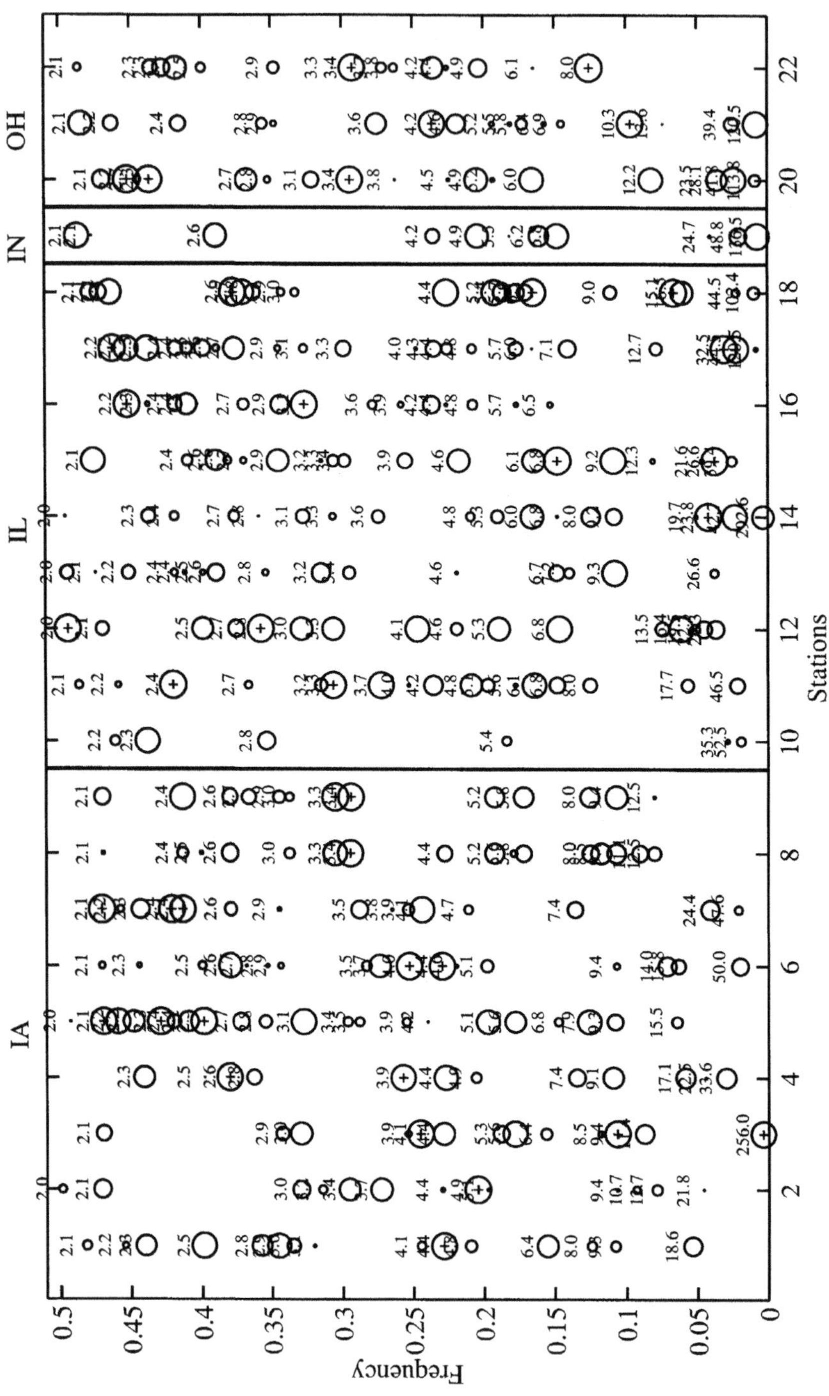

Figure 4.25. Periodicities in the NOAA treering data using MTM analysis with band width W=2.5/N.

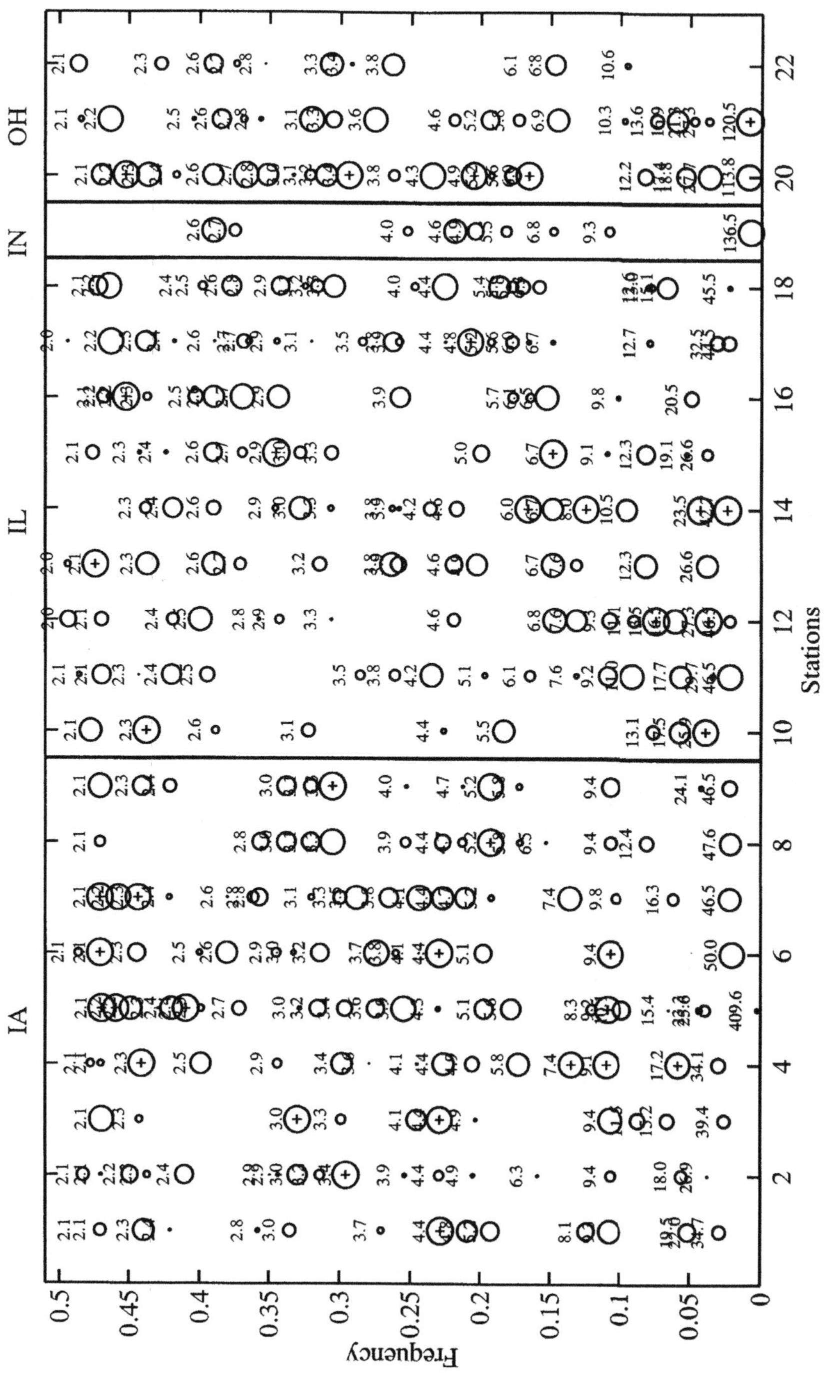

Figure 4.26. Periodicities in the NOAA treering data using MTM analysis with band width W=4/N.

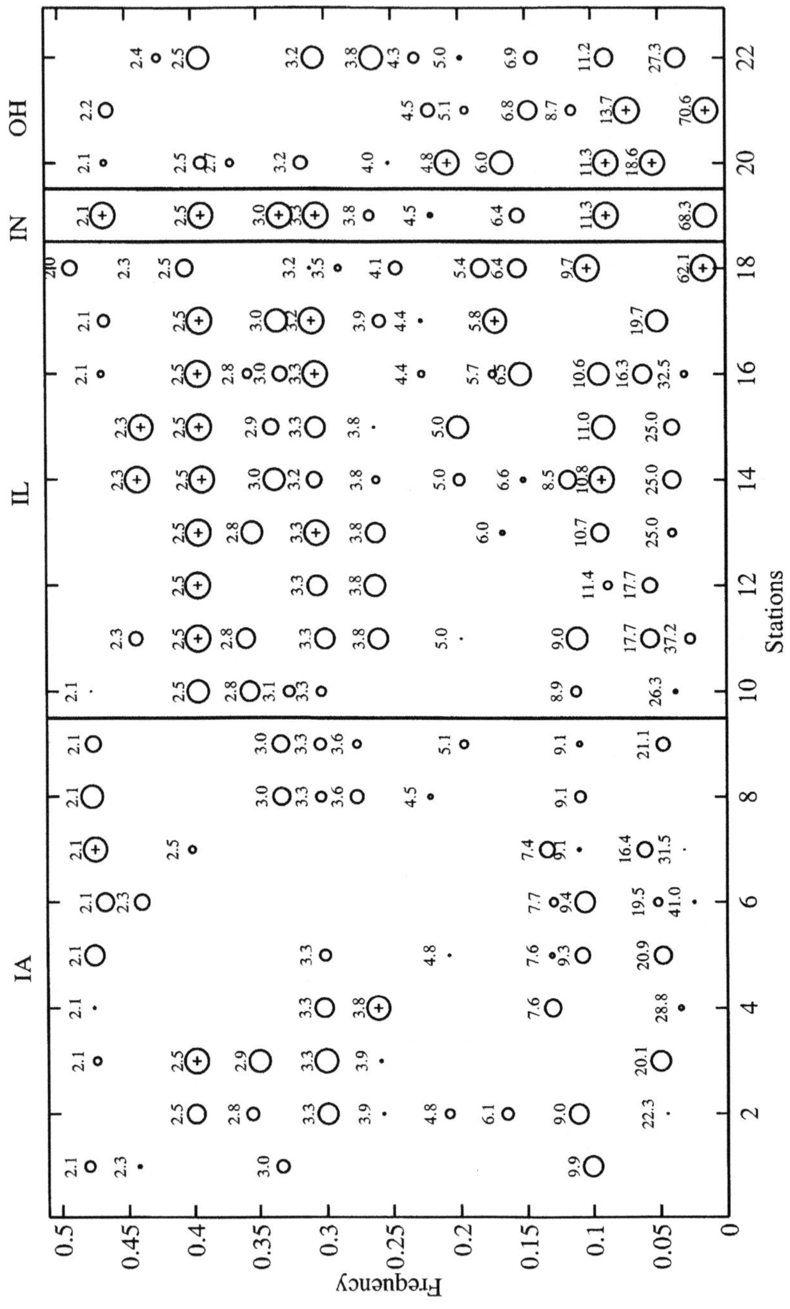

Figure 4.27. Periodicities in the last 100 years of NOAA treering data using MTM analysis with band width W=2.5/N.

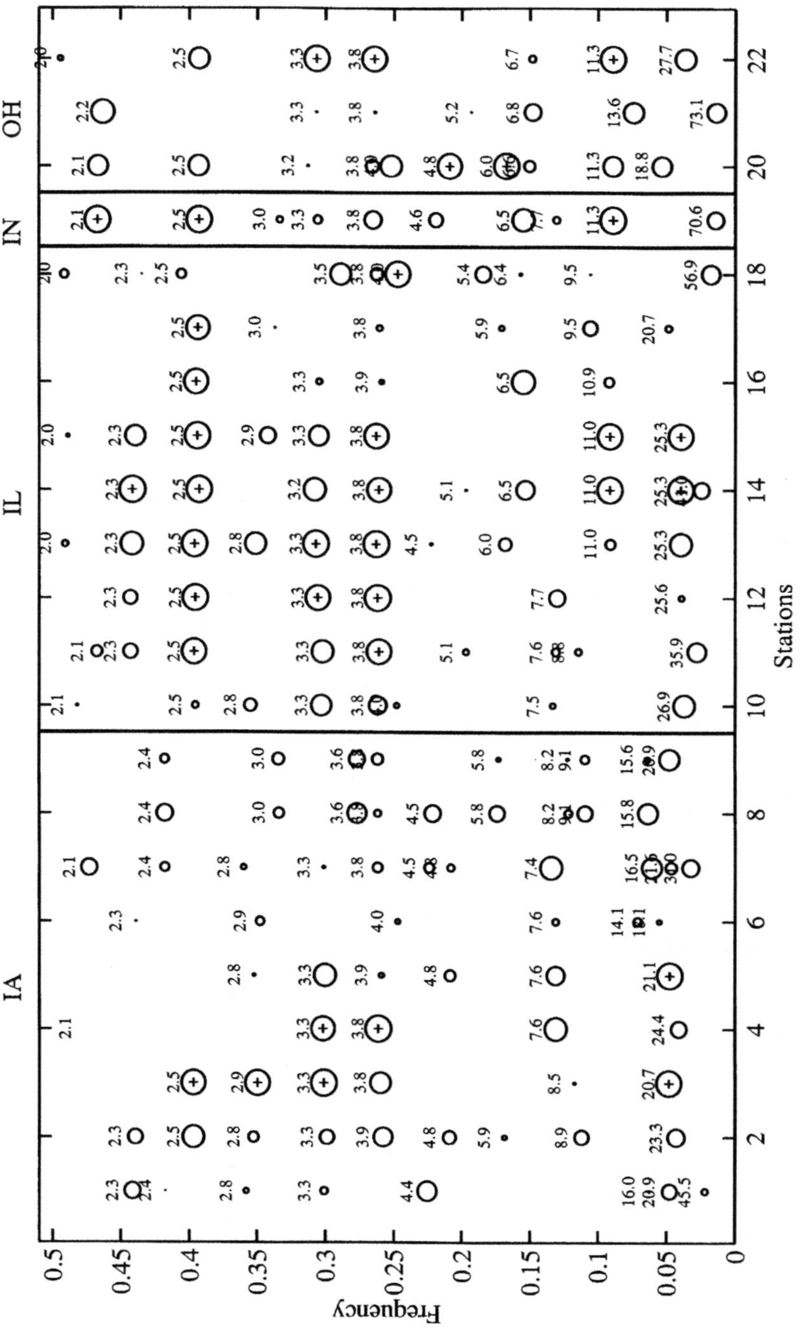

Figure 4.28. Periodicities in the last 100 years of NOAA treering data using MTM analysis with band width W=4/N.

a given frequency, the system also produces higher harmonics of this frequency. In addition, when the system is excited at two or more frequencies, the sums and differences of these frequencies are produced in the form $mf_1 \pm nf_2$ where f_1 and f_2 are the excitation frequencies, and m and n are integers. This would produce subharmonics such as $f_1 - f_2$, $2f_1 - f_2$, $f_1 - 2f_2$ and so on (Burroughs, 1992). As an example, the 11 year sunspot cycle mentioned above will produce periods of 5.5 , 3.7, 2.75, and 2.2 years. Similarly, the 11-year sunspot cycle combined with the 22-year magnetic cycle will produce a period of 7.4 years. The sums and differences of the 11-year , 22-year, and the 18.6-year cycles produce periodicities of 6.9, 10.1, 26.9, and 120.3 years.

In light of this information, the following can be deduced from the results in Tables IV.1 through IV.12 as well as the summary given in Table IV.13:

- Periodicities in the 2 to 3-year range seem to be concentrated at 2.1, 2.2, 2.3, 2.5, and 2.7 to 3 years. These periodicities conform with the high frequency variability in QBO of order 2 to 3 years. Each of these periodicities accounts for 3 to 7% of the total variability in the data. Together, periods in this range account for 9 to 20% of the total variability in the data. Periodicities of 2.2 years seems to be less common in temperature as well as tree-ring data than in flow and rainfall data. Periodicities of 2.1 years, on the other hand seem to be less common in rainfall data than flow, temperature, and tree-ring data. Periodicities of 2.5 or 2.6 years are the most common in this group appearing in all flow data and a large percentage of other data as well.
- Periodicities in the range 3 to 3.5 years are the most common group in all data sets, with an average of 78% of the stations. This group, along with the group of 3.5 to 4 years may be related to either QBO or the higher harmonic of the 11-year Sunspot cycle at 3.7 years. Each of these groups contributes about 4 to 7% of the total variability in the data.
- Periodicities of 4 to 5 years are present in a large portion of flow data (92% of all stations) and rainfall data (64% to 92% of all stations), and are less common in temperature data (31% to 68% of all stations) and tree-ring data (27% to 32% of all stations). These periodicities may be related to the 4 to 6-year ENSO periodicities. These periodicities contribute 9 to 10% of the total variability in flow data, and 2 to 6% of the total variability in other data.
- Periodicities of 5 to 6 years are the second most common among all periodicities with an average of 74% of all stations. The most probable origin of these periodicities is the 5.5 year harmonic of the Sunspot cycle, in addition to the 4 to 6-year ENSO periodicities. These periodicities contribute 13 to 14% of the total variability in flow data, and 4 to 8% of the total variability in other data.
- Periodicities of 6 to 7 years are absent in flow data. They are also less common in rainfall data (25% to 30% of all stations) and tree-ring data (32% of all stations) than in temperature data (67% to 95% of all stations). The period of 6.9 years resulting from the interaction between the 11-year sunspot cycle and the 18.6 Luni-Solar cycle may be related to this group of periodicities. These periodicities account for 4 to 7% of the total variance in rainfall, temperature, and tree-ring data.

- Periodicities of the 7 to 9 years are less prominent in rainfall data (41% to 50% of all stations) and tree-ring data (27% to 36% of all stations) than in flow data (73% and 77% of all stations) and temperature data (72% to 89% of all stations). Periodicities in this range may be related to the 7.4-year harmonic resulting from the combination of the sun's 11 and 22-year cycles. Periodicities in this range account for 10 to 13% of the total variability in flow data and 5 to 7% of the total variability in other data.
- Periodicities in the range from 10.7 to 11.1 years are the third most common group in all data categories. Periodicities in this range are often attributed to the sunspot cycle of 11 years. It can be noted that these periods are less common in rainfall data (39% to 65% of all stations) and tree-ring data (36% to 45% of all stations) than in flow data (69% to 88% of all stations) or temperature data (89% to 100% of all stations). Periodicities in this range account for about 12% of the total variability in flow data, 5 to 6% of the total variability in rainfall data, 9 to 11% of the total variability in temperature data, and 8 to 9% of the total variability in tree-ring data.
- Periodicities of 17.9 to 19.6 years are mostly common only to temperature data. Periodicities in this range are often attributed to the 18.6-year Luni-Solar cycle. These periodicities account for 5 to 9% of the total variability in temperature data.
- Periodicities of 21 to 26 years, on the other hand, seem to be common to flow data (65% to 81% of all stations) and much less common to rainfall data (24% to 39% of all stations). Periodicities in this range are often attributed to the 22-year Hale cycle. These periodicities contribute 9 to 10% of the total variability in flow data and 4 to 5% of the total variability in rainfall data.
- Periods from 32 to 34 years seem to be common only to rainfall data (22% to 46% of all stations). These periodicities contribute 6 to 8% of the total variability in rainfall data. Periodicities from 60 to 69 years, on the other hand, seem to be common only to temperature data (78% to 100% of all stations). These periodicities account for 12 to 14% of the total variability in temperature data. Neither of the 32 to 34-year or 60 to 69-year periodicities have been related to physical phenomena in the literature.
- Periods greater than 89 years are less common in all data, and they appear only when MTM analysis is used. This is attributed to the relatively short length of the records available for analysis which is seldom more than 100 years. The detection of periods of 89 years or more using 100 years or less of data is questionable, since at most one cycle would be present in the data. Taking into account the effect of noise in the data and the possibility of superimposed trends, such cycles cannot be reliably detected using such short records. The few detected periodicities, however, indicate that these periodicities may be responsible for 8 to 15% of the total variability in flow data, 5 to 15% of the total variability in rainfall data, and 14% of the total variability in temperature data.

4.7. CONCLUSIONS

Based on the results from the harmonic analysis of the yearly flow, rainfall, temperature, and tree-ring data discussed in this chapter the following conclusions may be presented.

1. It is evident that significant components of different periods are present in the data. As such, these processes cannot be considered stationary with continuous spectra, but

rather centered stationary with mixed spectra. In that respect, the results of MEM and MTM analysis are in good agreement.

2. Most of the periodicities found in the data can be attributed to natural phenomena such as the sunspot cycle, the Hale cycle, ENSO, and QBO. However, longer periodicities (32 to 34 and 60 to 69 years) do not seem to have a known origin. Periods of more than 89 years are hard to establish due to the shortness of the available records. The most common four periodicities in all data are 2.5 to 2.6 years, 3 to 3.5 years, 5 to 6 years, and 10.7 to 11.1 years. Periodicities of 17.9 to 19.6 years and 60 to 69 years are found mostly in temperature data, while periodicities of 21 to 26 years are found mostly in flow data, and periodicities of 32 to 34 years are found mostly in rainfall data.

3. The percentage of the total variance explained by different periodicities varies from 3 to 18%. High frequency components contribute 3 to 6% of the total variability, while there seems to be an increase of percentage for lower frequency components up to 18% of the total variability.

4. It was observed in some data sets, especially divisional average data, that some type of frequency drift occurs in the study region from east to west. In some cases frequencies are discontinued and other frequencies evolve also moving from east to west. This would suggest some type of spatial distribution of frequencies. However, a much larger database would be required to further investigate this behavior.

5. Based on the results of the harmonic analysis of tree-ring data, there is an indication of a different type of nonstationarity. This type of nonstationarity involves changes in the structure of the process where both the frequency and amplitude of cyclic components are time dependent. This type of nonstationarity needs further investigation through time-frequency analysis.

5. Time-Frequency Analysis

5.1. INTRODUCTION

Conventionally, time series have been studied either in the time domain or the frequency domain. The representation of a signal in the time domain is localized in time, i.e. the value of the signal at each instant in time is well defined. However, the time representation of a signal is poorly localized in frequency, i.e. little information about the frequency content of the signal at a certain frequency can be known by looking at the signal in the time domain. On the other hand, the representation of a signal in the frequency domain is well localized in frequency, but is poorly localized in time, and as a consequence it is impossible to tell when certain events occurred in time.

In studying stationary or conditionally stationary processes with mixed spectra, the separate use of time domain and frequency domain analyses is sufficient to reveal the structure of the process. Results discussed in the previous chapters suggest that the time series analyzed in this book are conditionally stationary processes with mixed spectra. Additionally, there is some indication of nonstationarity, especially in longer time series. In this case, a different type of nonstationarity must be considered in which the spectral properties of the process change with time. This change may be in the form of either a change in the shape of the spectral density function of the background continuum or a change in either the amplitudes or frequencies of the line components, or both. Since the detection of such type of nonstationarity requires both time and frequency information neither the time nor the frequency representation of the signal is sufficient for the analysis. A more suitable approach in this case is to use a different type of representation which is localized in both time and frequency.

One suitable technique for the analysis of nonstationary time series is the Short Time Fourier Transform (STFT) first described by Gabor (1946). STFT is achieved by analyzing successive segments of the signal by using a sliding window of fixed length. This technique represents a compromise between the time and frequency representations of a signal. Information about the change in the signal regarding both the time and frequencies at which such changes occur can thus be obtained by using the STFT technique. However, the compromise between time and frequency results in reduction in precision which depends on the size of the sliding window used for the analysis.

The basic theory of spectral analysis of nonstationary time series was developed by Priestley (1965) under the name of "Evolutionary Spectral Analysis". According to Priestley (1984), evolutionary spectra of a nonstationary time series have essentially the same physical interpretation as the spectra of stationary processes. However, while the spectrum of a stationary process describes the power-frequency distribution for the whole process, the evolutionary spectrum is time-dependent and describes the local power-frequency distribution at different instants in time.

In this chapter, flow, rainfall, and temperature as well as tree-ring data are analyzed for detecting and evaluating nonstationary behavior of these time series. The evolutionary spectra for each of the time series is obtained and the evolution of spectral properties of both the line components and the background are studied.

5.2. EVOLUTIONARY SPECTRAL ANALYSIS

Classical spectral estimation methods are based on the assumption that time series are either stationary or can be rendered stationary by using simple transformations (Priestley, 1984). Therefore, classical analysis when applied to data from nonstationary processes may give misleading results, depending on the degree of nonstationarity of the process. Priestley (1965) developed an approach to study the spectra of nonstationary processes which has the same physical interpretation as the spectra of stationary processes. The main difference between classical and evolutionary spectra is that the evolutionary spectrum of a time series is time dependent and describes the local power-frequency distribution at each instant in time (Priestley, 1984). In addition to having a physical interpretation similar to the spectra of stationary processes, evolutionary spectra provide information about the stochastic structure of the process, which is useful in modeling, and can be estimated by using simple techniques which do not require prior information about the structure of the process. The numerical computation of evolutionary spectra is roughly equivalent to splitting up a realization in several segments and then estimating individual spectra for each of the segments. Priestley (1965) suggests a method for calculating the evolutionary spectra $S_t(f)$ of a nonstationary time series based on a double windowing technique. First, a raw estimate of $S_t(f)$ in the form of Eq. 5.1 is obtained by using a data window $g(u)$.

$$U_t(f) = \sum_{u=-\infty}^{\infty} g(u)\, X_{t-u}\, e^{-if(t-u)} \tag{5.1}$$

The estimate in Eq. 5.1 is similar to the conventional periodogram, and suffers from large variance independent of the length of the time series. To reduce the variance, a second averaging window $W_T(t)$ is applied to $U_t(f)$ to obtain a better estimate of $S_t(f)$ in the form of Eq. 5.2,

$$V_t(f) = \sum_{v=-\infty}^{\infty} W_T(v)\, |U_{t-v}(f)|^2 \tag{5.2}$$

where T is the width of the averaging window. In this case, the variance of the evolutionary spectral estimate $V_t(f)$ in Eq. 5.2 is given by Eq. 5.3,

$$var\{V_t(f)\} = C\, S_t^2(f) \tag{5.3}$$

where C is a constant dependent on the forms of the two windows $g(u)$ and $W_T(t)$. For example, if $g(u)$ and $W_T(t)$ are given by,

$$g(u) = \begin{cases} \dfrac{1}{2\sqrt{h\pi}}, & |u| \leq h \\ 0, & |u| > h \end{cases} \tag{5.4}$$

$$W_T(t) = \begin{cases} \dfrac{1}{T}, & -T/2 \leq t \leq T/2 \\ 0, & otherwise \end{cases} \tag{5.5}$$

then the value of C in Eq. 4.3 will be given by Eq. 5.6,

$$C = \frac{4h}{3T} \tag{5.6}$$

where h and T are the widths of the windows $g(u)$ and $W_T(t)$, respectively. Priestley (1965) gives relationships for choosing the values of h and T. However, prior information about the shape of the spectrum is required for the optimal choice of these two parameters.

The reason for using a double window technique, as seen above, is to reduce the variance of the estimate of the evolutionary spectrum. Data windows applied to segments of the data with width h result in good time resolution, but produce a large variance estimator that must be smoothed by averaging a number T of spectral estimates. The averaging process reduces the variance of the estimator while sacrificing some resolvability in the time domain (Priestley, 1984). This technique is similar to that applied in the Multi Taper Method (MTM) for spectral analysis, with the difference being that in MTM the variance is reduced by averaging the spectra from the same data segment using multiple data tapers, thus preserving, to some extent, time resolvability. For the analysis in this chapter the MTM is used as an alternative to the double windowing technique for evaluating evolutionary spectra of the hydrologic and climatic data.

In applying the MTM to study evolutionary spectra, the signal is divided into a number of (possibly overlapping) segments each of length T and MTM spectra are calculated for each segment. The variance of the MTM estimate of the spectral density $S_t(f)$ of a Gaussian white noise is given by Eq. 5.7,

$$var\{\hat{S}_t(f)\} = \frac{1}{K} S_t^2(f) \tag{5.7}$$

where K is the number of data tapers used, which is equal to $2NW-1$, N is the number of observations and W is the band width of the data tapers as discussed in the previous chapter. For colored Gaussian processes, this approximation is still good as long as S_t

(f) does not vary too rapidly over the interval $(f-W, f+W)$ (Percival and Walden, 1993).

Based on the harmonic analysis discussed in chapter four it is evident that the data analyzed in this study have mixed spectra, i.e. with both continuous and discrete components. Therefore, the evolutionary spectral analysis of the data will be divided in two parts. The first part of the analysis deals with the evolution of line components by investigating their significance at different points in time based on the F-test offered by the MTM method. The second part of the analysis deals with the continuous part of the spectrum. In this case the shapes of the spectral density function estimated at different points in time are compared and tested for stationarity.

Kottegoda (1985) suggested a technique for the assessment of stationarity in annual hydrologic series. However, his technique is based on fitting an ARMA model to the data and comparing the spectra at different points in time with the spectrum obtained from the fitted model. The technique is thus a parametric one, and depends largely on the choice of the appropriate model for the data. Furthermore, the theoretical distribution of the test statistic in this technique is not known and can only be obtained through simulation using realizations from the fitted model. With the uncertainties encountered in modeling hydrologic and climatic time series, especially with the existence of a number of cyclic components, it seems that a nonparametric test would be more appropriate for assessing the stationarity in such time series.

Priestley and Subba Rao (1969) suggest a test for nonstationarity in time series based on evolutionary spectra. The test has the form of the analysis of variance for a two-factor design, based on the logarithms of the evolutionary spectral estimates, Y_{ij}, given in Eq. 5.8.

$$Y_{ij} = \log \hat{S}_{t_i}(f_j) \tag{5.8}$$

The subscripts i and j in Eq. 5.8 refer to time segments and frequencies, respectively. The test is based on the sums of uncorrelated variates. Therefore, time segments t_i used in the test should have little or, preferably, no overlapping. The same condition applies for frequencies f_j also, which should have a spacing of at least $2W$, the band width of the data tapers. Statistics used in the test are given in Table V.1 below, where dot subscripts indicate averaging over the respective subscript.

The distribution of these statistics is given in Eqs. 5.9 to 5.11,

$$S_{I+R} / \sigma^2 \sim \chi^2_{(I-1)(J-1)} \tag{5.9}$$

$$S_T / \sigma^2 \sim \chi^2_{(I-1)} \tag{5.10}$$

$$S_F / \sigma^2 \sim \chi^2_{(J-1)} \tag{5.11}$$

Table V.1. Statistics for the Priestley and Subba Rao nonstationarity test.

Item	Degrees of Freedom	Sum of Squares
Between Times	$I - 1$	$S_T = J \sum_{i=1}^{I} (Y_{i\cdot} - Y_{\cdot\cdot})^2$
Between Frequencies	$J - 1$	$S_F = I \sum_{j=1}^{J} (Y_{\cdot j} - Y_{\cdot\cdot})^2$
Interaction + Residual	$(I - 1)(J - 1)$	$S_{I+R} = \sum_{i=1}^{I} \sum_{j=1}^{J} (Y_{ij} - Y_{i\cdot} - Y_{\cdot j} + Y_{\cdot\cdot})^2$
Total	$IJ - 1$	$S_o = \sum_{i=1}^{I} \sum_{j=1}^{J} (Y_{ij} - Y_{\cdot\cdot})^2$

where σ^2 is equal to the coefficient C in Eq. 5.3, which is equal to $1/K$ in Eq. 5.7 for the case of MTM estimates.
Based on the statistics calculated in Table V.1, the test procedure is as follows:

1. If the test based on S_{I+R} is significant, then the time series X_t is nonstationary and non-uniformly modulated.

2. If the test based on S_{I+R} is insignificant, proceed to test S_T. A significant S_T indicates that the process is nonstationary, while an insignificant S_T indicates the process is stationary.

3. Similarly, S_F can be used to test for complete randomness. Therefore, a significant S_F indicates a correlated process (colored noise), while an insignificant S_F indicates a random process (white noise).

4. The test can be extended to test stationarity at different frequencies f_j based on the statistic $S_{TF}(f_j)$ in Eq. 5.12, which is distributed as $\chi^2 (I\text{-}1)$.

$$S_{TF}(f_j) = \sum_{i=1}^{I} (Y_{ij} - Y_{\cdot j})^2 / \sigma^2 \qquad (5.12)$$

5. Similarly, the test can be extended to test for randomness at different times t_i based on the statistic $S_{FT}(t_i)$ in Eq. 5.13, which is distributed as $\chi^2 (J\text{-}1)$.

$$S_{FT}(t_i) = \sum_{j=1}^{J} (Y_{ij} - Y_{i\cdot})^2 / \sigma^2 \qquad (5.13)$$

5.3. EVOLUTION OF LINE COMPONENTS IN HYDROLOGIC AND CLIMATIC DATA

The study of the evolution of both the frequency and magnitude of spectral lines in hydrologic and climatic time series can give information about changes in the dynamics of the processes generating such time series. The study of the overall spectrum cannot reveal such information about the dynamics of the system. As discussed in Chapter Four, the weather system is believed to be a nonlinear one (Burroughs, 1992). The amplification and attenuation of internal mechanisms, which is typical of nonlinear systems (Lorenz, 1968; Ghil and Childers, 1987), induce the appearance and disappearance of different spectral peaks over some intervals in the observed time series. Evolutionary analysis of the spectral lines can thus be used to capture such variability in the systems producing such time series. The MTM is used in this section to investigate the evolution of line components in different time series. This part of the analysis is exploratory in nature since there are no statistical tests available for the assessment of the significance of changes in the line components. In the next section, however, statistical significance tests will be used to test for stationarity of the continuous spectrum.

One of the most common periodicities found in hydrologic and climatic data is the well known sunspot cycle at around 11 years. Berger et al. (1990) studied the evolution of the sunspot 11-year cycle as well as the sun's 22-year magnetic cycle by using both MEM and MTM spectral analysis. Their study showed that the 22-year cycle is much more stable than the 11-year cycle. Figure 5.1 shows the evolution of the frequencies as well as the significance of line components in the sunspot time series from 1868 to 1987 (120 years) using a sliding window of length 60 years applied at a time increment of two years. Line components which are significant at the 0.1 level are shown as circles sized proportional to their significance. A plus sign within a circle indicates that the line at that frequency is also significant at the $1/n$ level, where n is the length of the time segment. The rightmost values in the plot in Fig. 5.1 are those for the full time series. The dotted lines across the plot also correspond to the full time series. The same convention used in Fig. 5.1 is used to present the results from other data sets in the remainder of this section.

The evolution of amplitudes of line components for the sunspot time series is shown in Fig. 5.2. It can be observed from Figs. 5.1 and 5.2 that the cycle at ±11 years is the main feature in the sunspot time series. It can also be seen that this period is not constant throughout the time segments, but rather changes from 11.6 at the beginning of the record down to 10.2 towards the middle of the record then increases up to 10.6 towards the end of the record. The overall record indicates an average period of 10.7 years. This drift in periodicities is also observed at the 5.39-year period as well as in other periods in Fig. 5.1. It can also be noticed that the 5.39-year periodicity is not detected at the beginning of the record, then it is detected as two distinct periods of 5.3 and 4.8 years towards the middle of the record and then drifts from 5.1 to 5.4-year periods towards the end of the record. Similar observations can be made about other periodicities as well. The behavior of line components in the sunspot time series suggest the nonstationarity of the process. Frequency drifting and splitting also suggest the nonlinearity of the system producing this time series.

Another natural phenomenon that has an influence on hydrologic as well as climatic data is the El-Niño/Southern Oscillation phenomena (ENSO). El-Niño is used to describe the increase in the surface water temperature along the coasts of Peru and Ecuador, while the Southern Oscillation is a measure of the difference in atmospheric pressure between the pacific and Indo-Australian areas. The warm episodes of ENSO last for about two years and affect climate pattern over many areas in the world (Philander, 1990; Chiew et al. 1994). The evolution of the frequencies as well as the significance of line components in the Southern Oscillation Index time series from 1876 to 1995 (120 years) using a sliding window of length 60 years applied at time increments of two years is evident in Fig. 5.3. The evolution of the amplitudes of line components for the same time series is evident in Fig. 5.4. The main features in Figs. 5.3 and 5.4 are the periodicities at 6 to 7 years and about 3.6 years. Drifts in frequencies can be observed at 2.3 years, 6.1 to 7.3 years, and 3.4 to 3.6 periodicities. The behavior of line components in this time series is similar to that of the sunspot time series and suggests nonlinearity and nonstationarity of these processes.

The evolution of the frequencies as well as the significance of line components in the USGS flow data from station 2 are shown in Fig. 5.5. This time series is the longest of the available flow data in this region with a length of 119 observations (1874 to 1992). A sliding window of length 60 years is applied at 2-year increments to produce the evolutionary spectra. The evolution of the amplitudes of line components for the same time series is shown in Fig. 5.6. In Fig. 5.6, two main features can be identified. The first feature is the drift in the frequency of a number of components such as the ±4.51-year and the ±11.3-year components, which is similar to the behavior found in the sunspot time series. The second feature is the apparent aggregation and splitting of components. This is very clear in Fig. 5.6 for frequencies near 0.01 cpy (100 years period) and 0.04 cpy (25 years period) which aggregate at segment 5 and proceed to aggregate with the 0.046 cpy frequency (22 years period) at segment 15. These frequencies split back to three distinct frequencies at approximately their original positions at segments 20 and 27. A similar behavior can be observed at segment 20 for the component at 0.13 cpy (7.6 years period). The behavior observed in Figs. 5.5 and 5.6 is not typical of stationary linear systems for which line components should appear more or less as straight lines parallel to the time axis at their respective frequencies.

Figure 5.7 shows the evolution of the frequencies as well as the significance of line components in the HCN rainfall data from station 13 (N = 114 years, 1874-1987) computed by using a sliding window of length 60 at 2-year increments. The corresponding evolution of the magnitude of line components is clear in Fig. 5.8. According to Figs. 5.7 and 5.8, the most stable component is that around 3.3 years. Frequency drifts can be observed at low frequencies as well as at high frequencies. However, no clear frequency splitting or aggregation is observed in this time series. Similarly, Figs. 5.9 and 5.10 show the evolution of the frequencies and the amplitudes of line components, respectively, for the NCDC divisional average rainfall data from station 35 (N = 98 years, 1895-1992) . It would appear that for this data the drift in frequencies is less severe than in the case of the HCN rainfall data in Figs 5.7 and 5.8, which may be attributed to the effect of spatial averaging in obtaining the NCDC divisional average data

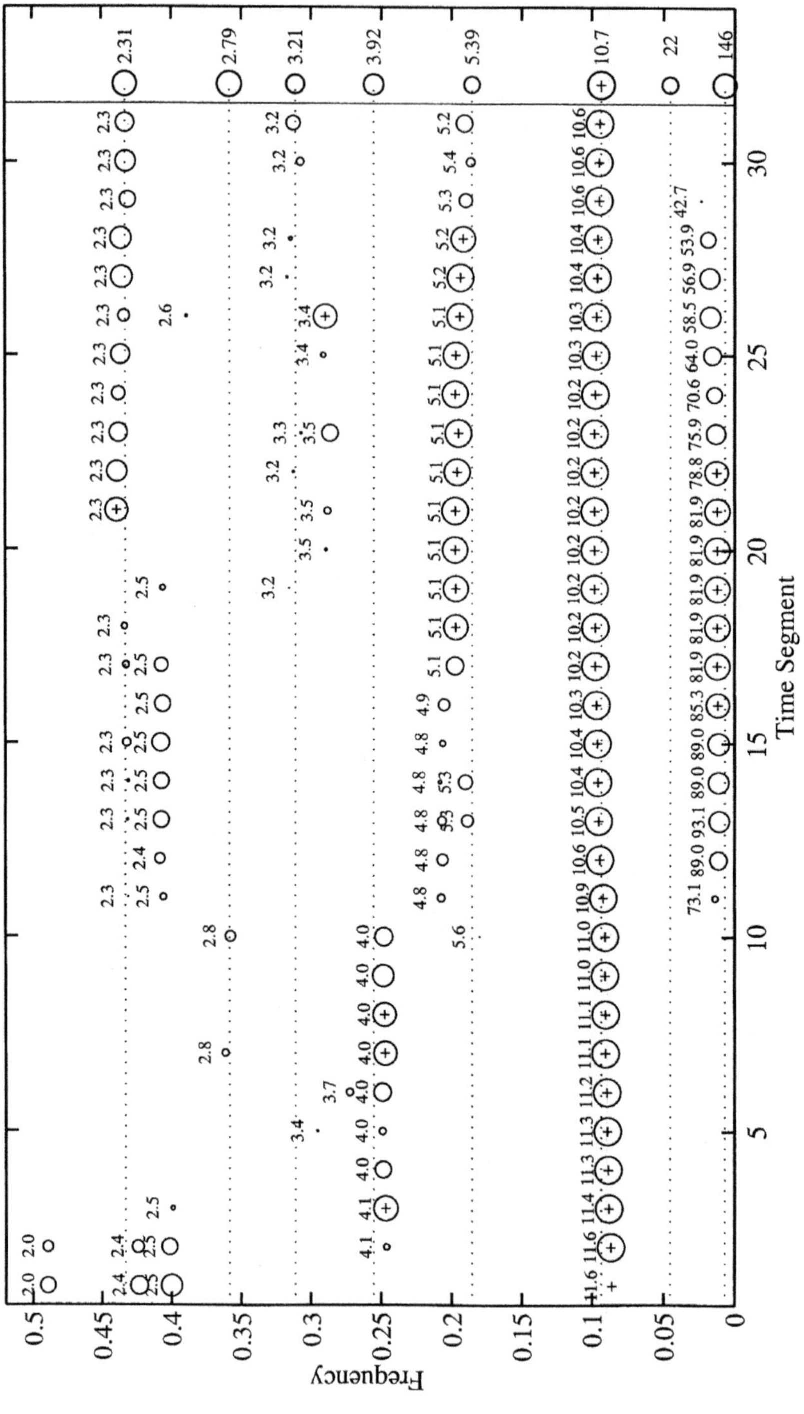

Figure 5.1. The evolution of line components in the sunspot time series from 1868 to 1987. A sliding window of length 60 is applied at 2-year increments.

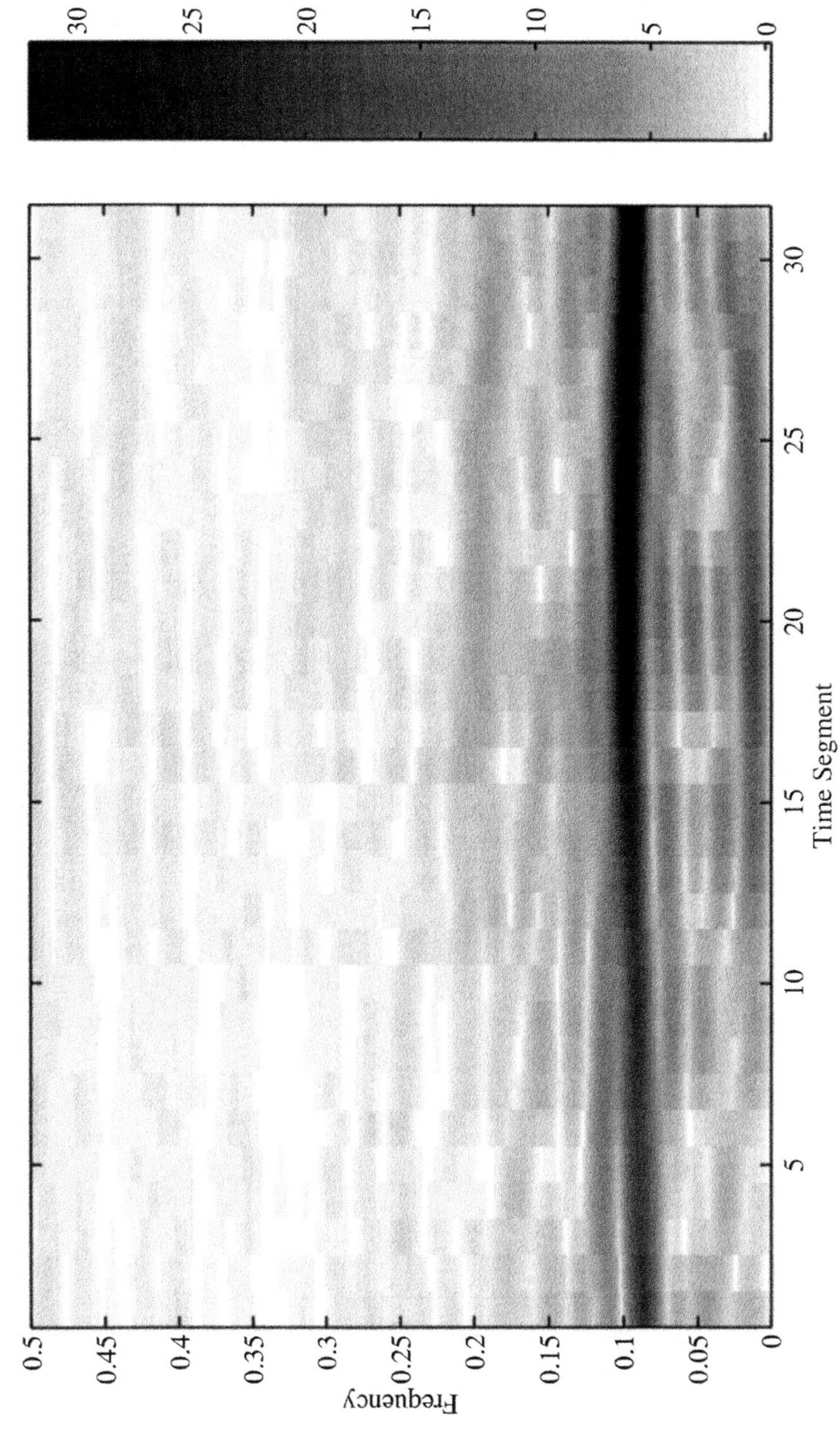

Figure 5.2. The evolution of the amplitude of line components in the sunspot time series from 1868 to 1987. A sliding window of length 60 is applied at 2-year increments.

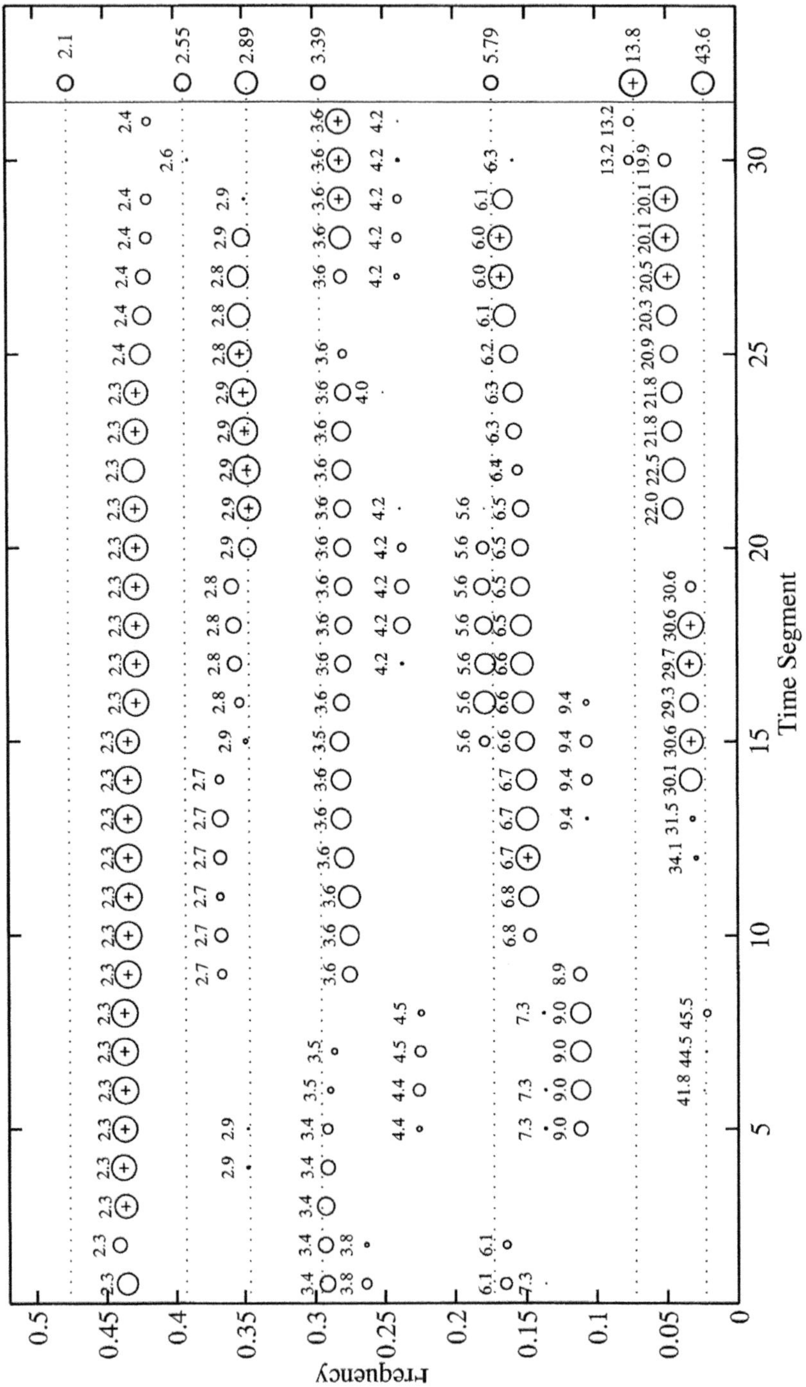

Figure 5.3. The evolution of line components in the SOI time series from 1876 to 1995. A sliding window of length 60 is applied at 2-year increments.

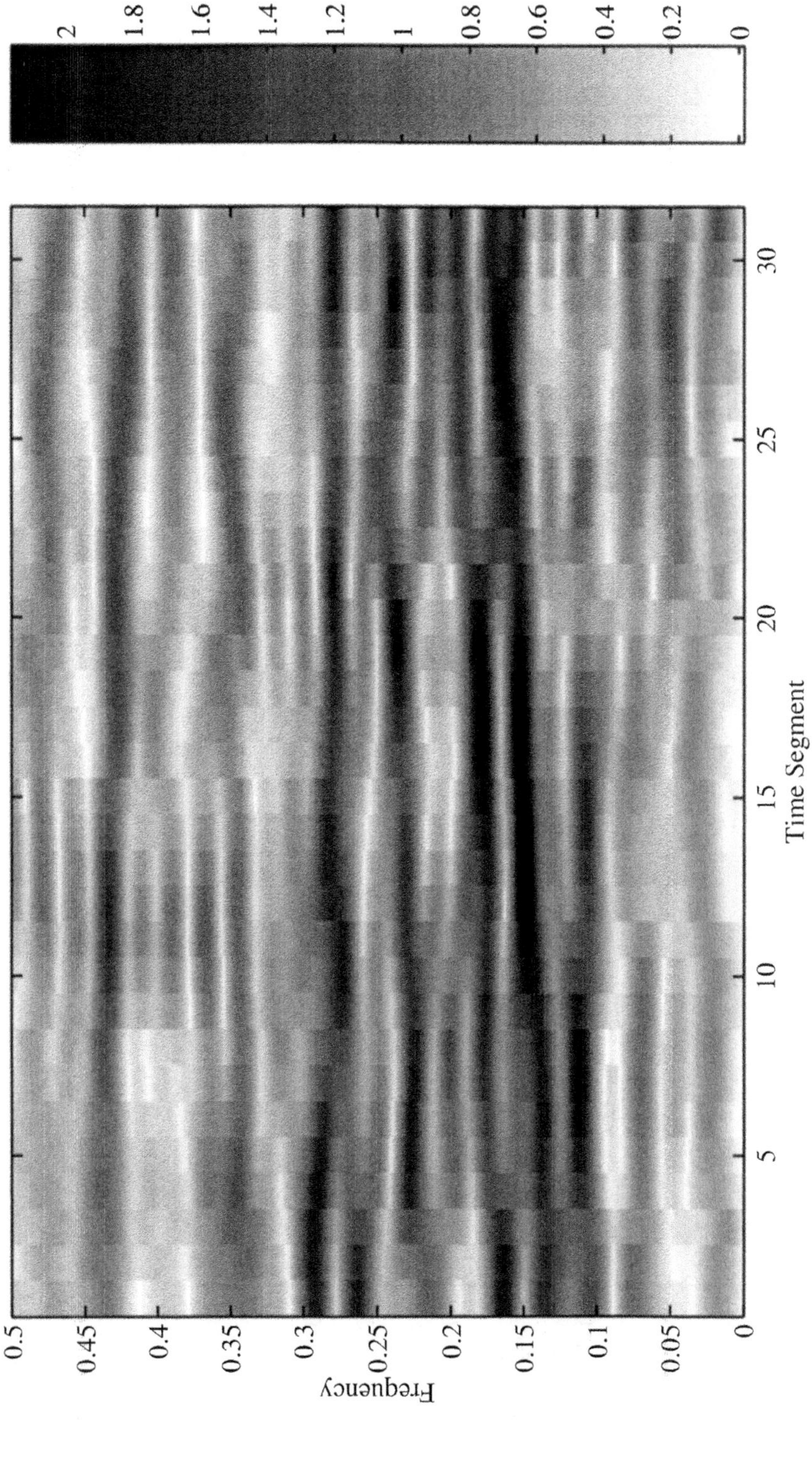

Figure 5.4. The evolution of the amplitude of line components in the SOI time series from 1876 to 1995. A sliding window of length 60 is applied at 2-year increments.

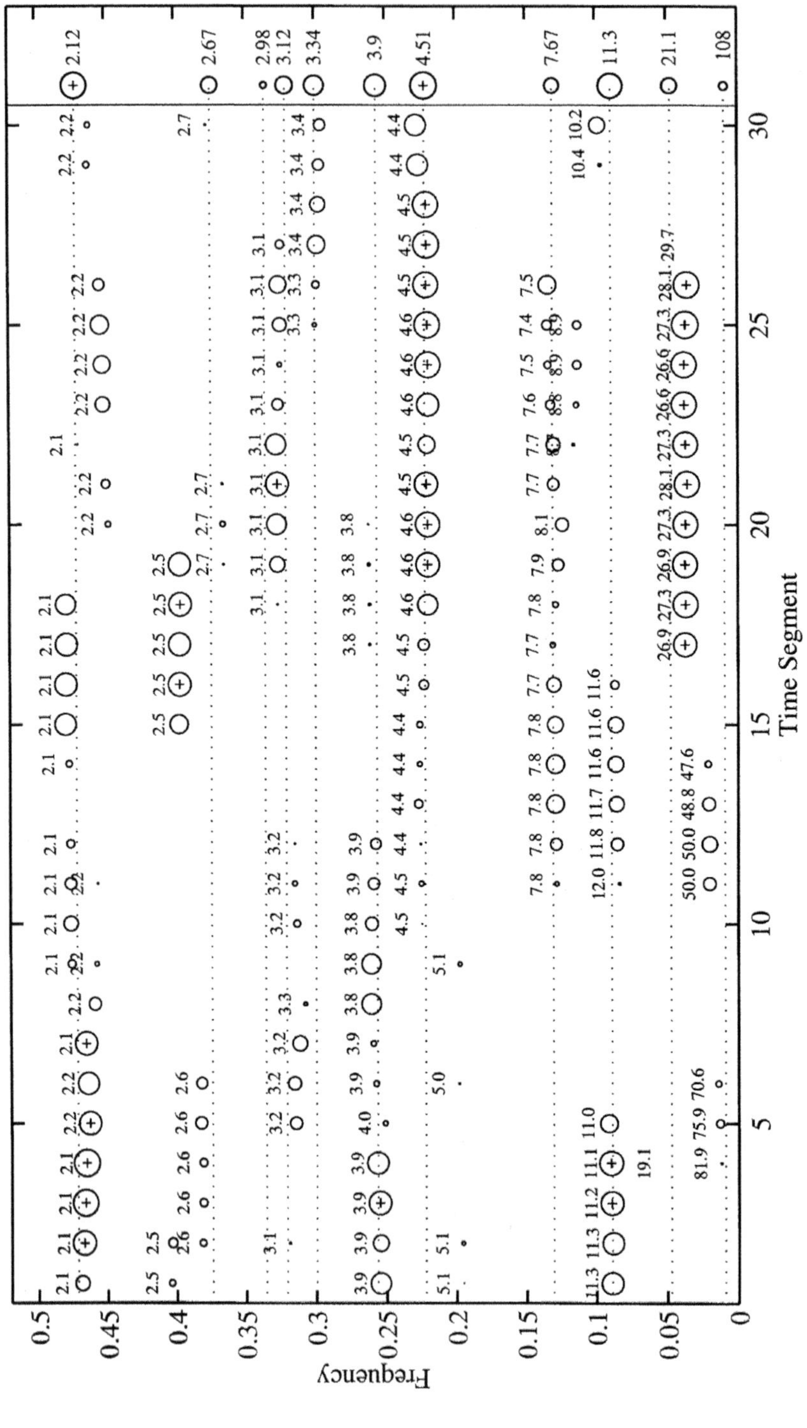

Figure 5.5. The evolution of line components in the USGS flow time series from station 2. A sliding window of length 60 is applied at 2-year increments.

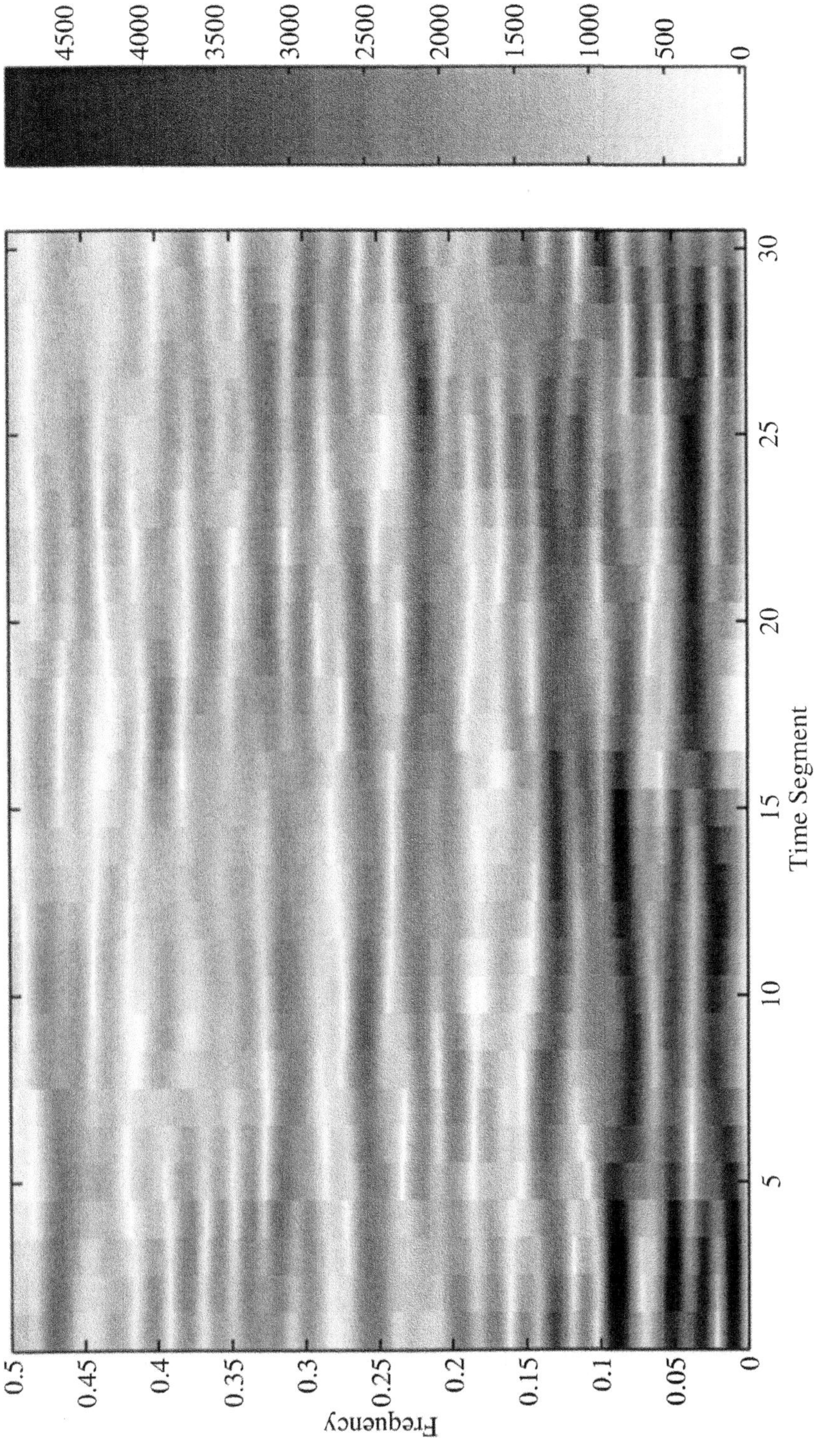

Figure 5.6. The evolution of the amplitude of line components in the USGS flow time series from station 2. A sliding window of length 60 is applied at 2-year increments.

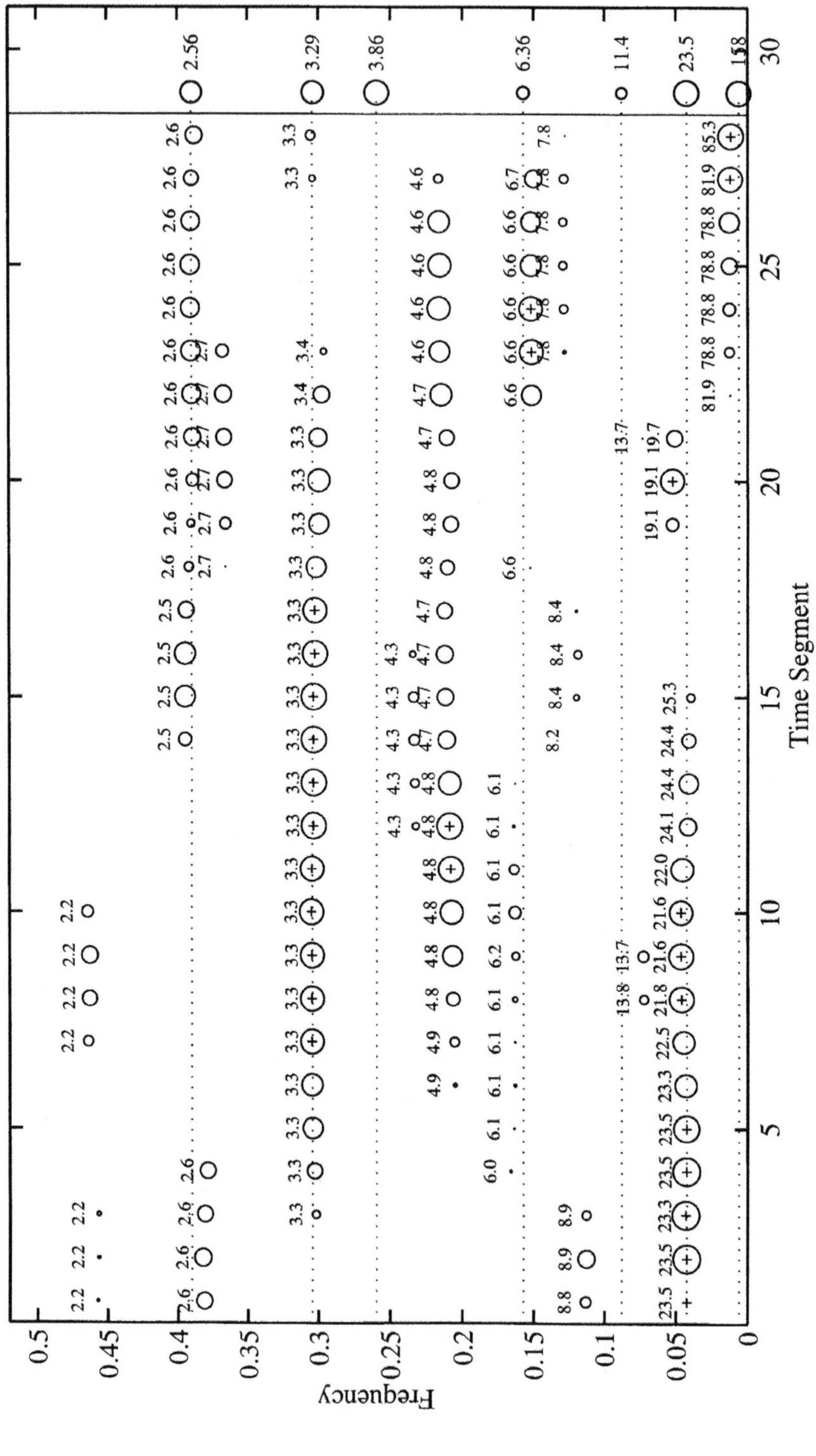

Figure 5.7. The evolution of line components in the HCN rainfall time series from station 13. A sliding window of length 60 is applied at 2-year increments.

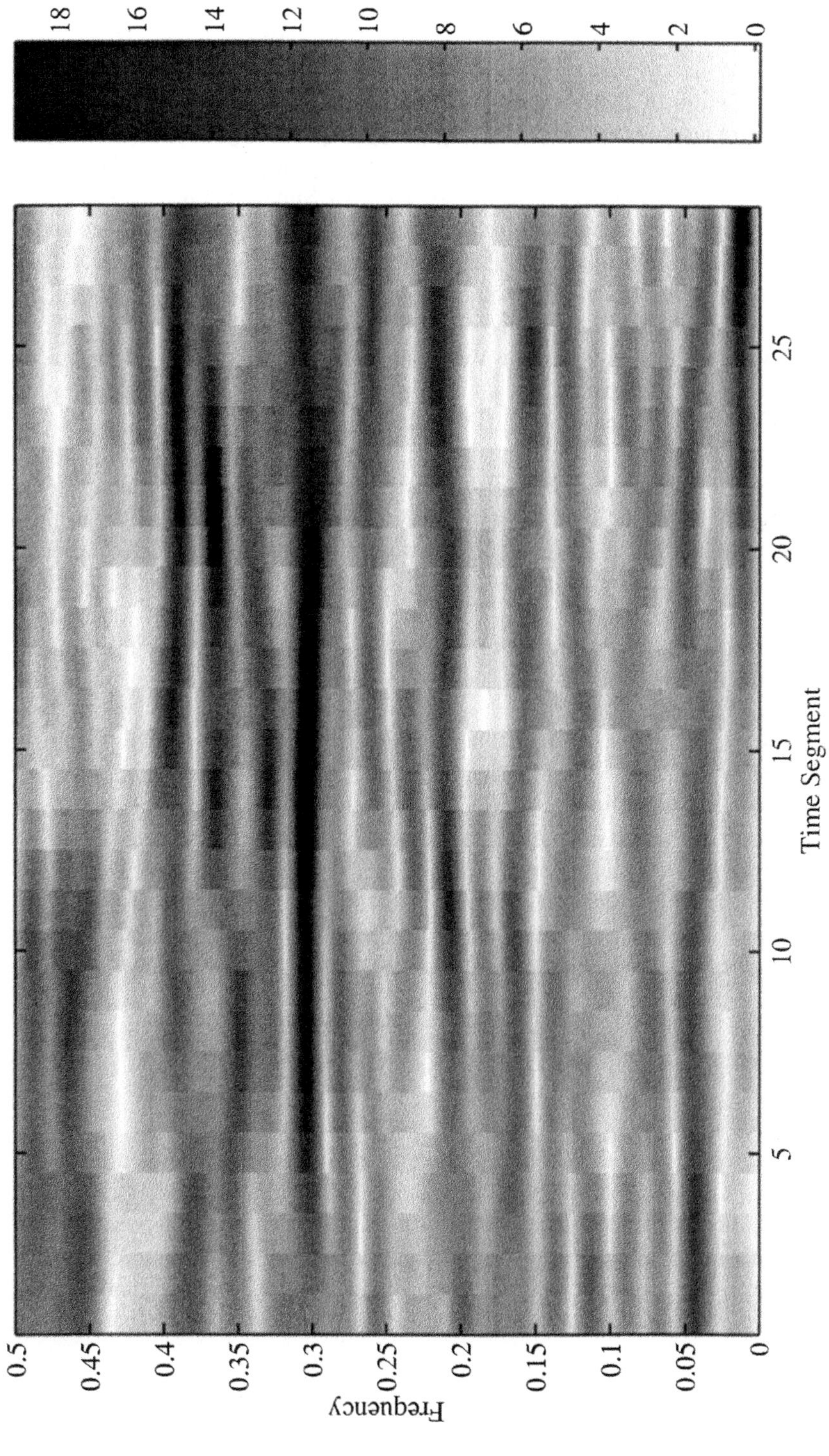

Figure 5.8. The evolution of the amplitude of line components in the HCN rainfall time series from station 13. A sliding window of length 60 is applied at 2-year increments.

The evolution of the frequencies as well as the significance of line components in the HCN temperature data from station 34 ($N = 107$ years, 1881-1987) computed by using a sliding window of length 60 at 2-year increments is shown in Fig. 5.11. The corresponding evolution of the magnitude of line components is shown in Fig. 5.12. Frequency drifts are also clear in Fig. 5.12 especially at the 3.95-year period. Also, an obvious feature is the aggregation of the 2.5-year and 2.8-year periods at the beginning of the record into a period of about 2.6 years at segment 18. Similarly, Figs. 5.13 and 5.14 show the evolution of the frequencies and the amplitudes of line components, respectively, for the NCDC divisional average temperature data from station 20 (N=101 years, 1895-1995). Periodicities of around 75 years at the beginning of the record drift to about 51 years at the end of the record, with a period of 64 years for the whole data set. Frequency drifts are observed at other frequencies as well. Similar to the HCN results in Figs. 5.11 and 5.12, periods of 2.5 years and 2.7 years in Figs. 5.13 and 5.14 are aggregated into a single period of 2.6 years at segment 12. (Due to the different starting dates of the two time series, segment 12 in Fig. 5.14 is actually concurrent with segment 18 in Fig. 5.12).

The evolution of the frequencies as well as the significance of line components in the NOAA tree-ring data from station 18 ($N = 513$ years, 1468-1980) computed by using a sliding window of length 100 at 14-year increments is shown in Fig. 5.15. This data series is the longest of all the time series in this study. Figure 5.16 shows the evolution of the amplitudes of line components for this time series. The first observation from these figures is that frequency drifts are more obvious than in the previous time series. Dominant periodicities are in the low frequency range, but no single frequency extends over a long period of time.

In general, the results in Figs. 5.1 to 5.16 show some similarities in terms of frequency drifts and the aggregation and splitting at a number of frequencies. This behavior was also observed in other time series in this study. This behavior, as mentioned before, is not typical of stationary linear systems. A limited number of simulations indicated that such behavior is sensitive to the noise level and the length of the time series. Due to the shortness of the available records and the lack of statistical tests for assessing the significance of such changes in the frequencies and/or amplitudes of line components, no final conclusions can be made about the nonlinearity of theses time series or the stationarity of line components.

5.4. EVOLUTION OF CONTINUOUS SPECTRA IN HYDROLOGIC AND CLIMATIC DATA

The study of the evolution of the continuous part of the spectrum provides information about the stochastic structure of the process, which is useful in model fitting (Priestley, 1984). The estimation of evolutionary spectra through MTM analysis is a powerful tool for assessing stationarity of time series. In this section, sample evolutionary spectra of different time series are presented and discussed. In addition, the results obtained from applying the test of stationarity proposed by Priestley and Subba Rao (1969) are reported for all the data.

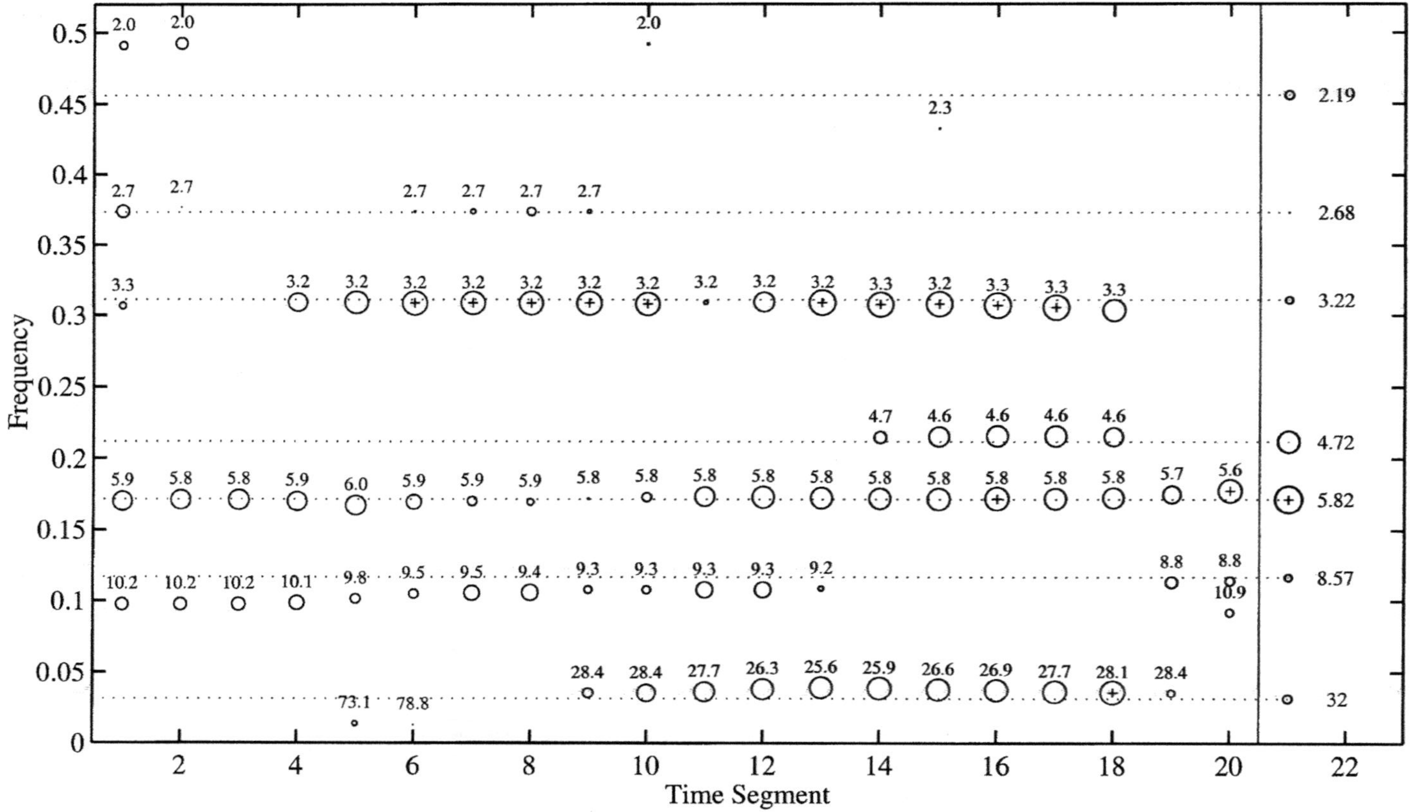

Figure 5.9. The evolution of line components in the NCDC rainfall time series from station 35. A sliding window of length 60 is applied at 2-year increments.

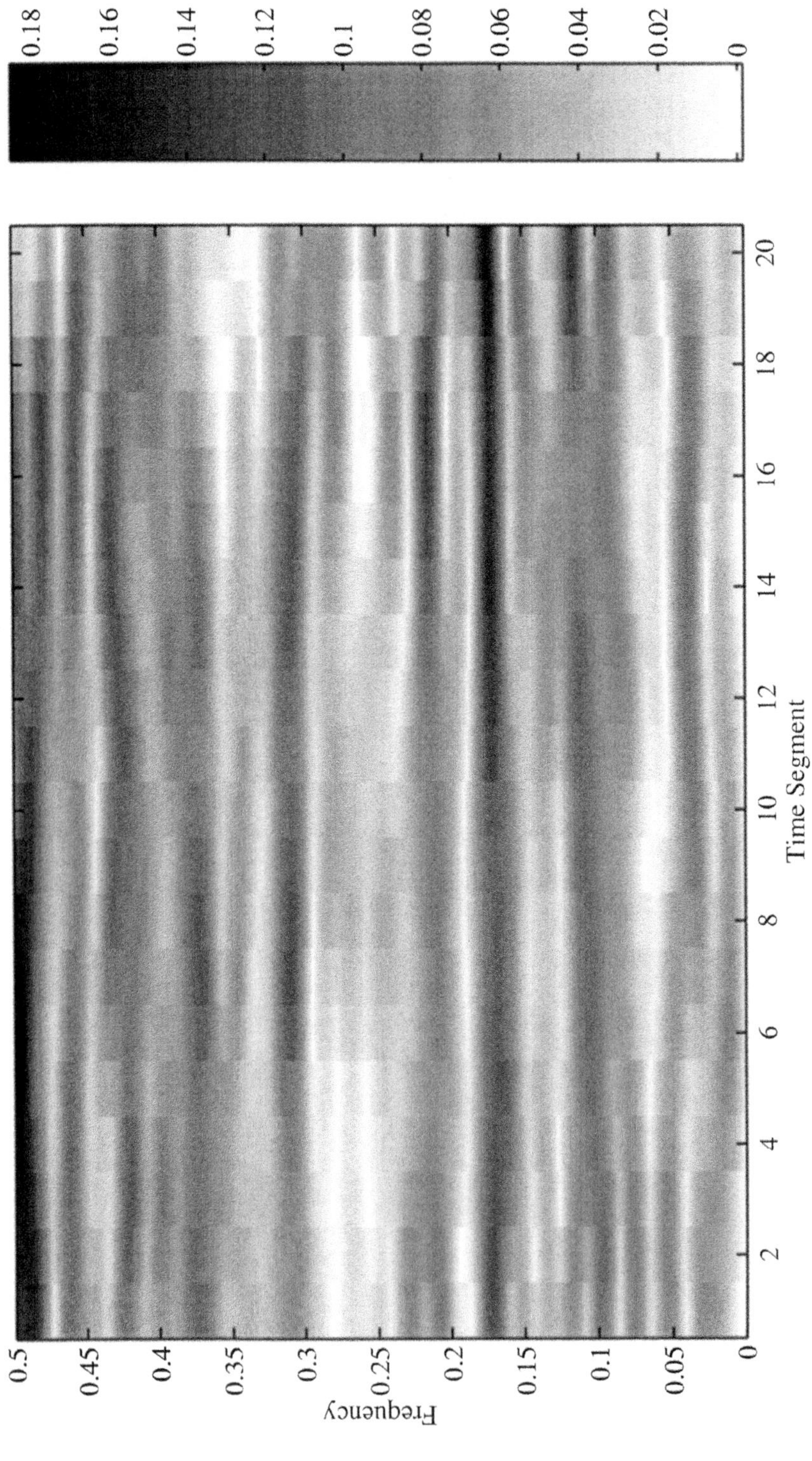

Figure 5.10. The evolution of the amplitude of line components in the NCDC rainfall time series from station 35. A sliding window of length 60 is applied at 2-year increments.

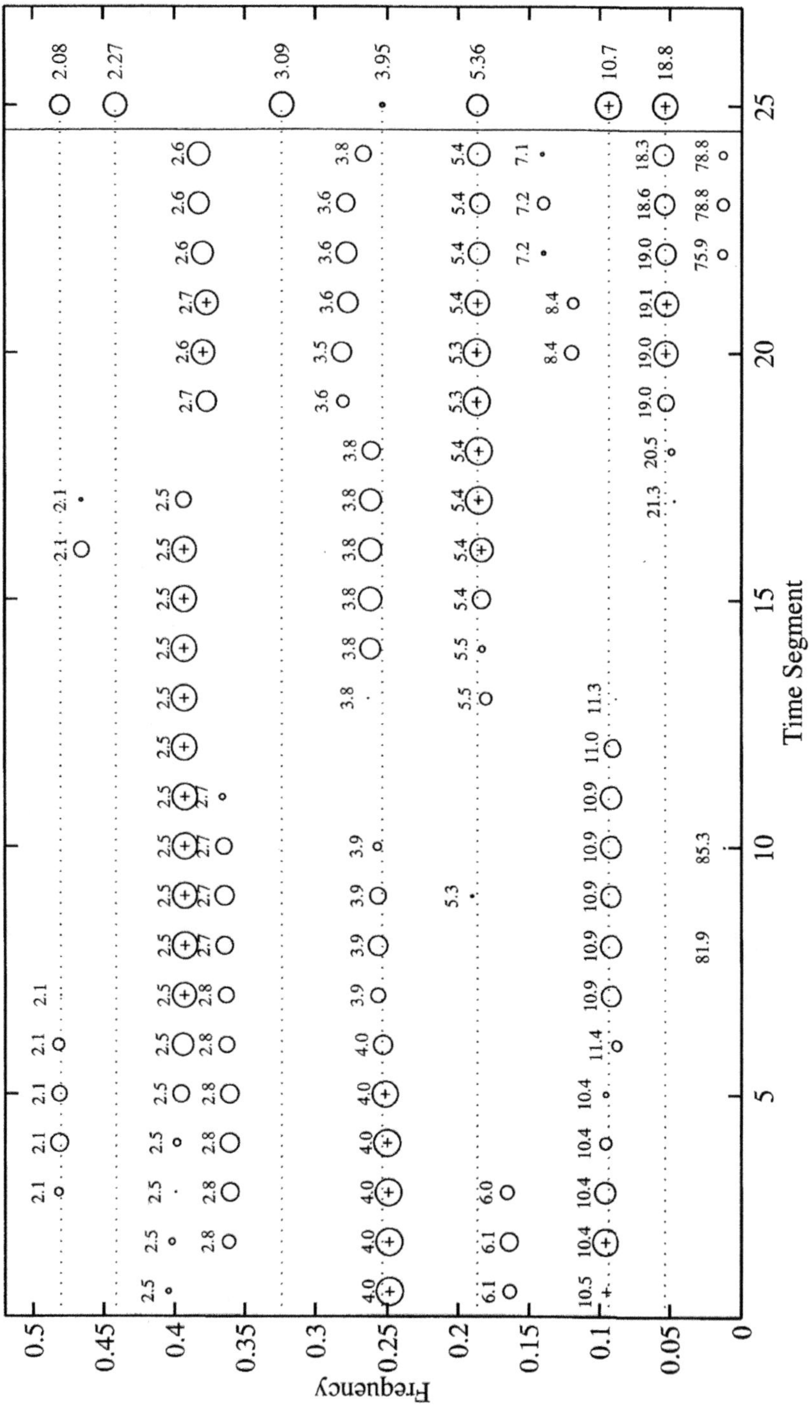

Figure 5.11: The Evolution of Line Components in the HCN Temperature Time Series from Station 34. A Sliding Window of Length 60 is Applied at 2-year Increments.

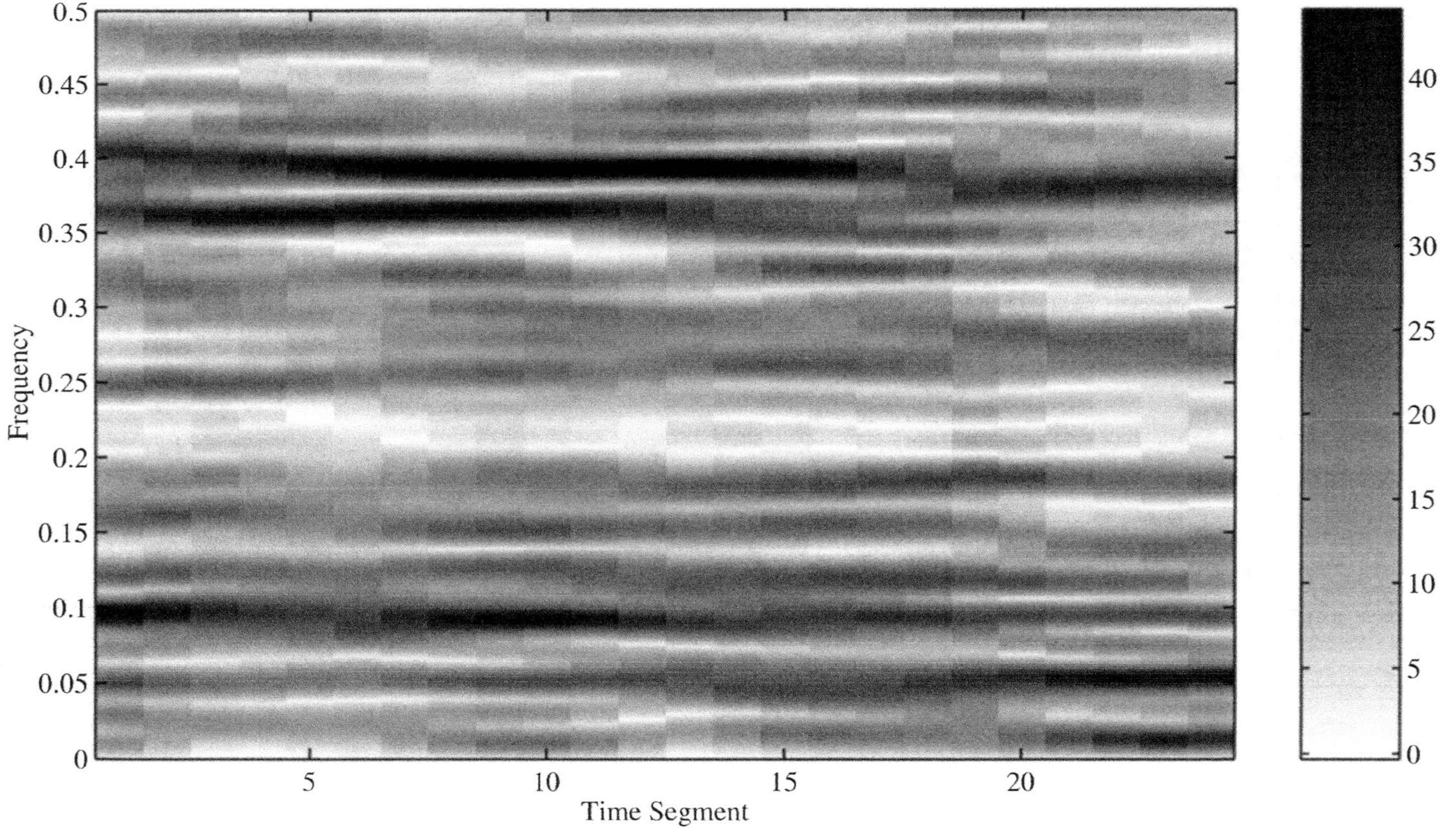

Figure 5.12. The evolution of the amplitude of line components in the HCN temperature time series from station 34. A sliding window of length 60 is applied at 20-year increments.

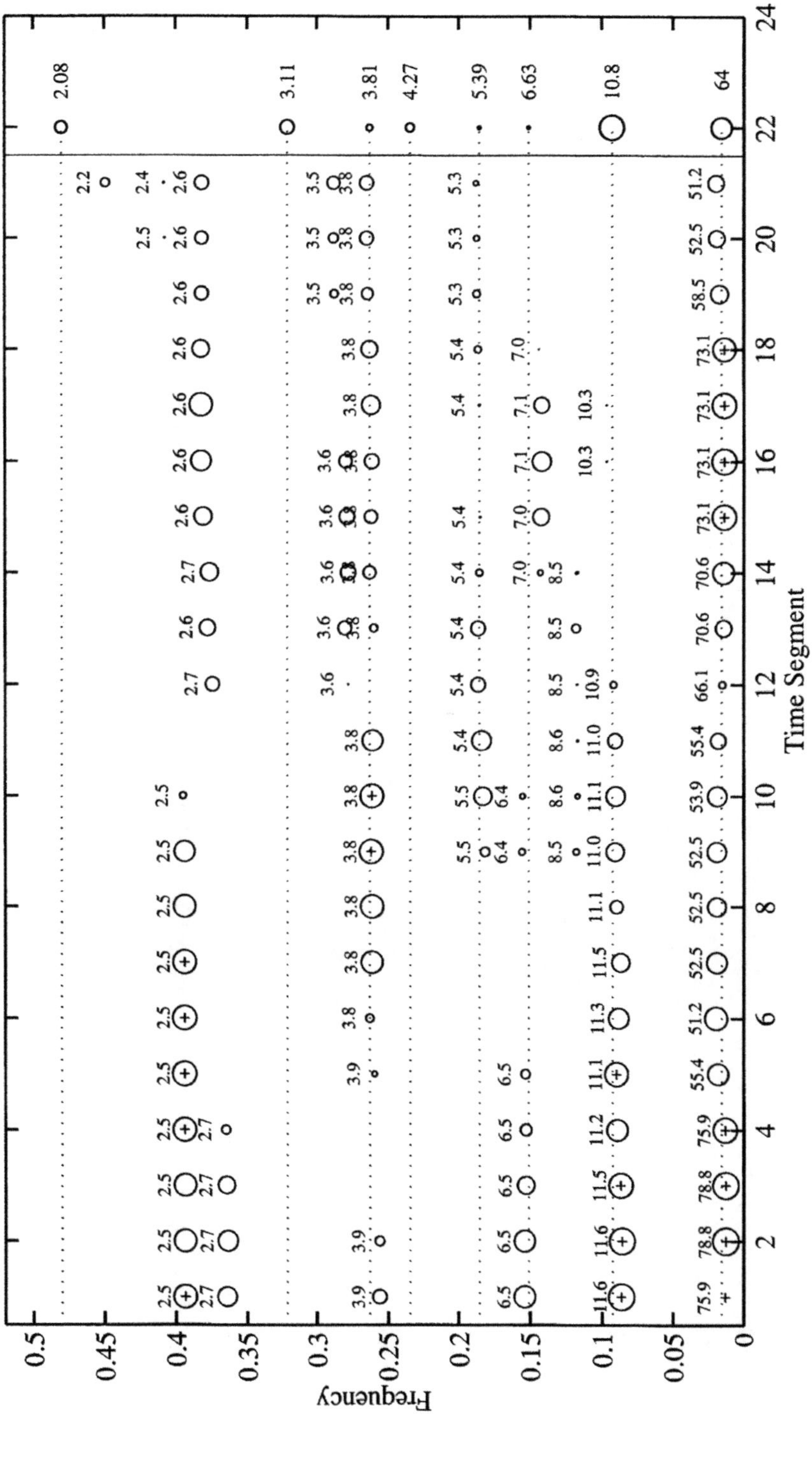

Figure 5.13. The evolution of line components in the NCDC temperature time series from station 20. A sliding window of length 60 is applied at 2-year increments.

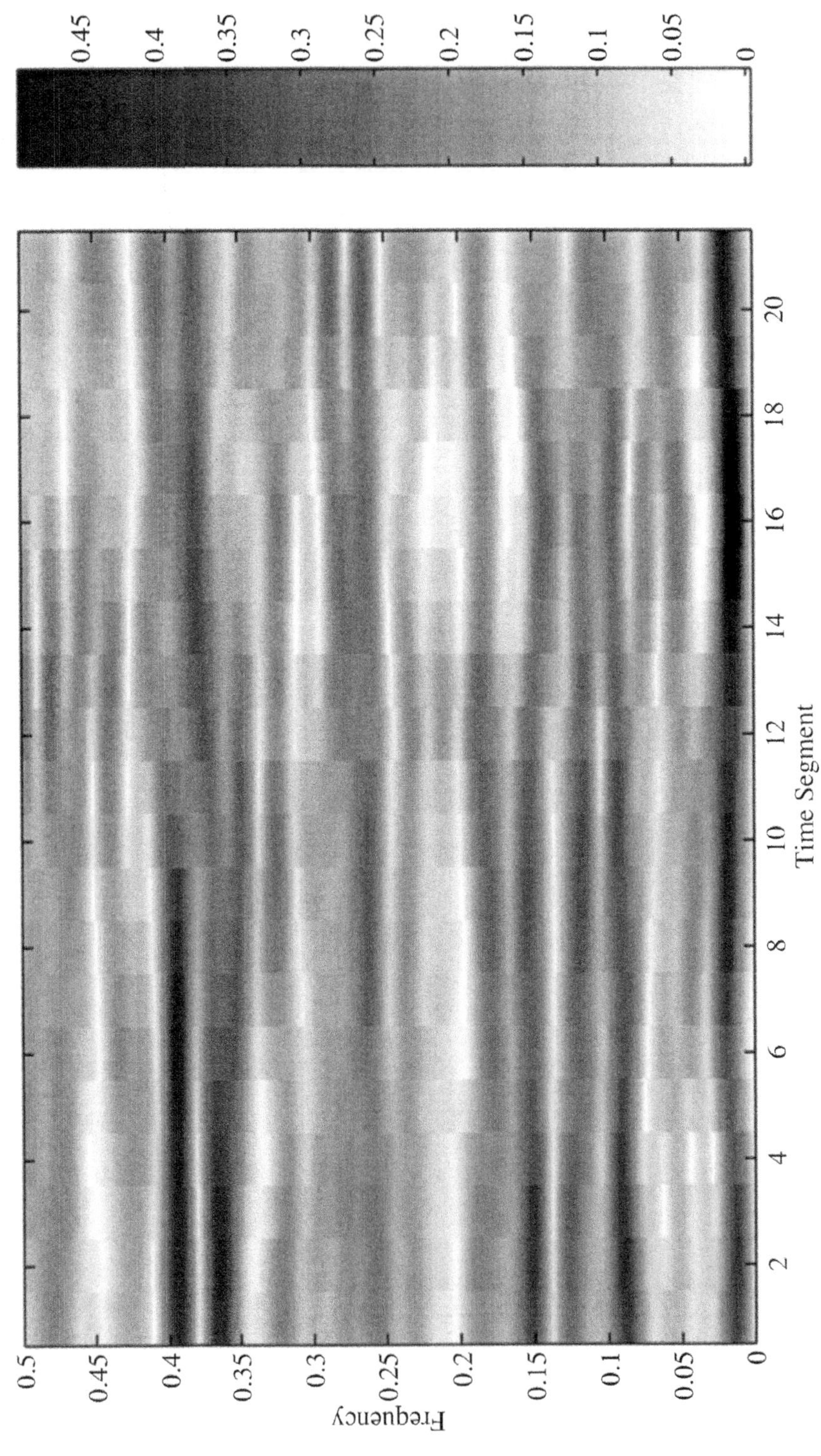

Figure 5.14. The evolution of the amplitude of line components in the NCDC temperature time series from station 20. A sliding window of length 60 is applied at 2-year increments.

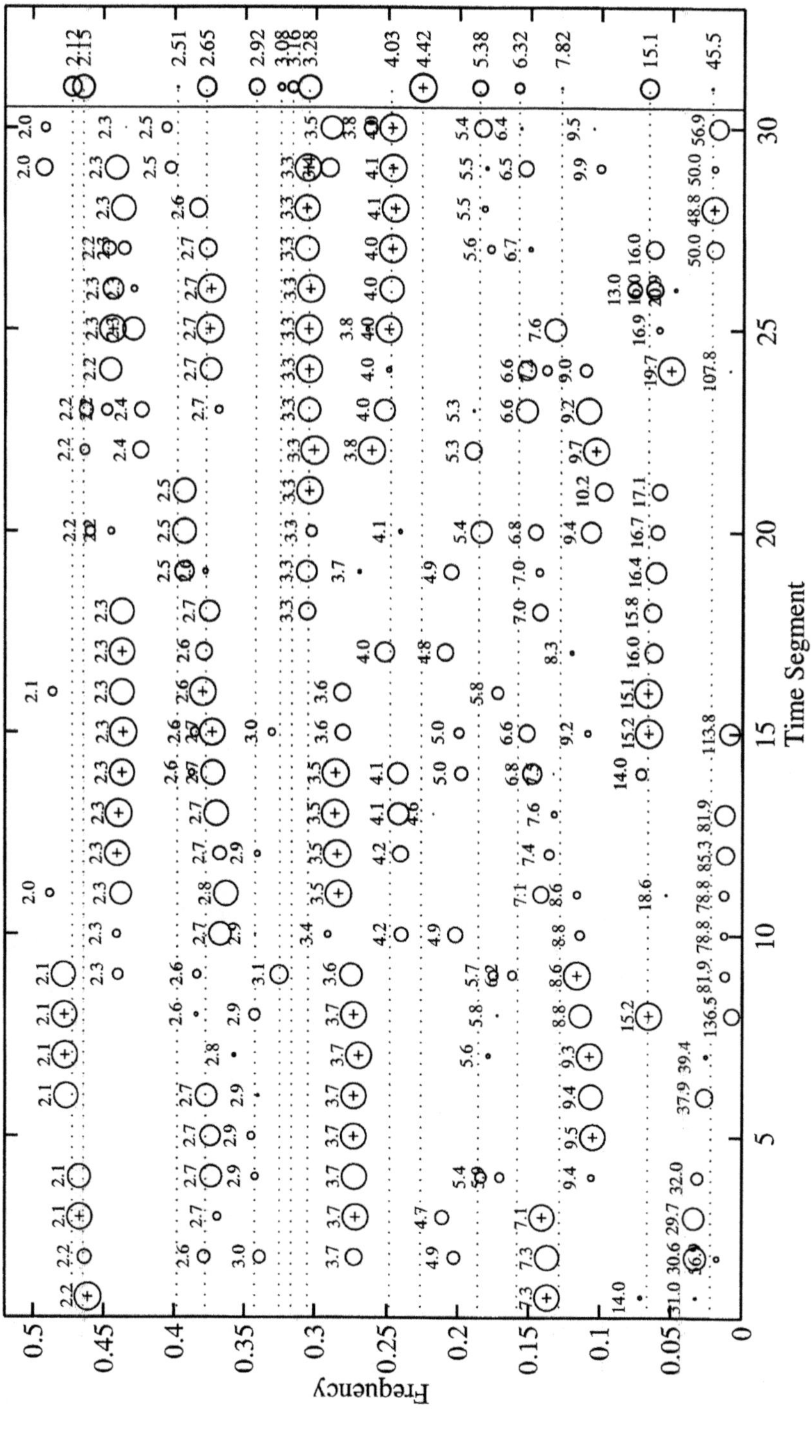

Figure 5.15: The Evolution of Line Components in the NOAA Tree–ring Time Series from Station 18. A Sliding Window of Length 100 is Applied at 14–year Increments.

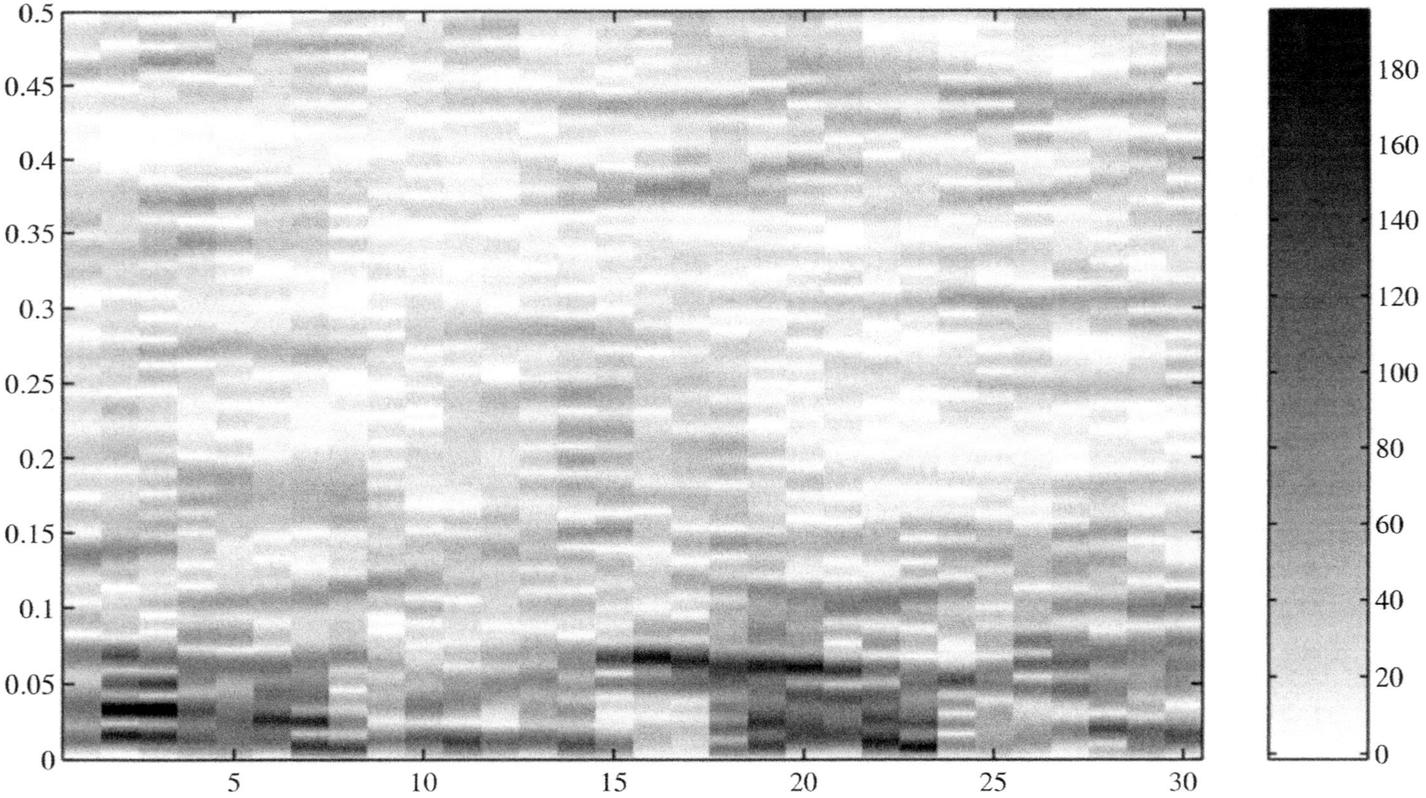

Figure 5.16. The evolution of the amplitude of line components in the NOAA tree-ring time series from station 18. A sliding window of length 100 is applied at 14-year increments.

Figure 5.17 shows the evolutionary spectra of the sunspot series at eight different time instants using a sliding window of length 72 years at 12-year increments. The spectra in Fig. 5.17 are not similar. The results of the Priestley and Subba Rao stationarity test by using the MTM technique are: $F(S_{I+R})$=0.9989, $F(S_T)$= 0.9995, and $F(S_F)$=1.0000. It is clear from these values that the sunspot series is nonuniformly modulated and highly nonstationary. Furthermore, $F(S_F)$ is highly significant, indicating that the time series is highly autocorrelated. Further testing at different frequencies and time steps of the statistics S_{FT} and S_{TF} indicates that nonstationarity is present at almost all frequencies and that the time series is autocorrelated at all time segments.

Figure 5.18 shows the evolutionary spectra of the Southern Oscillation Index (SOI) time series at four different time instants using a sliding window of length 40 years at 26-year increments. The spectra in Fig. 5.18 are not similar. The results of the Priestley and Subba Rao stationarity test by using the MTM technique are: $F(S_{I+R})$=1.0000, $F(S_T)$=1.0000, and $F(S_F)$=1.0000. It is clear from these values that the SOI time series is nonuniformly modulated and highly nonstationary. Also, the value of $F(S_F)$ is highly significant, indicating that the time series is autocorrelated. Further testing at different frequencies and time steps of the statistics S_{FT} and S_{TF} indicates that nonstationarity is present at almost all frequencies and that the time series is autocorrelated at all time segments.

Results of testing the USGS flow data series for stationarity are given in Table V.2. For this data set, four time segments each of length $N/3$ at 30 % overlapping are used. A band width of W=2 is used for the MTM to ensure three to five independent frequencies, depending on the length of the time series N. From the results in Table V.2, we conclude that 13 of the time series are nonstationary at the 0.05 level, and 11 are nonstationary at the 0.01 level. For this group of 26 stations, the 95% confidence interval for the number of stations giving false indication of nonstationarity would be 0 to 3 at the 0.05 level, and 0 to 2 at the 0.01 level. Therefore, the number of nonstationary time series in this group is larger than that due to chance. Also, from the results of the test on S_F we conclude that 21 time series at the 0.05 level and 17 time series at the 0 .01 level are autocorrelated and these spectra is significantly different than those of random processes.

Table V.2. Results of the Priestley and Subba Rao stationarity test for USGS flow data.

	Number of Significant Stations		
	Based on S_{I+R}	Based on S_T	Based on S_F
At the 0.05 Level	13	9	21
At the 0.01 Level	11	6	17

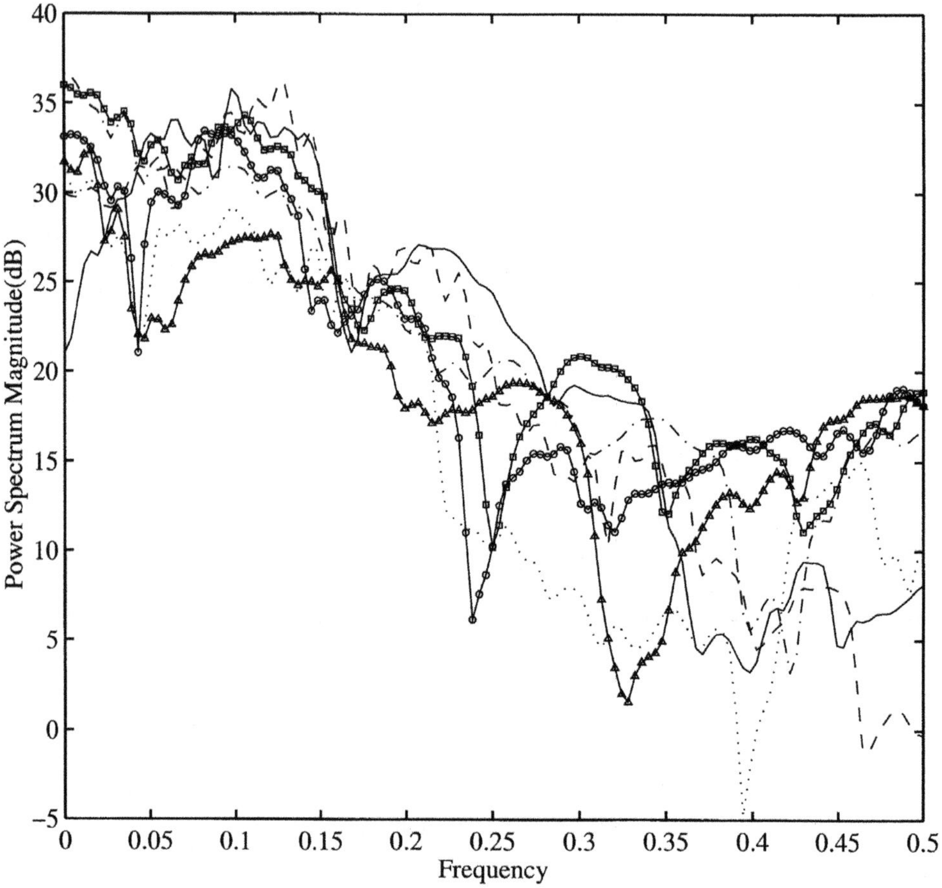

Figure 5.17. The evolutionary spectra for the sunspot time series.

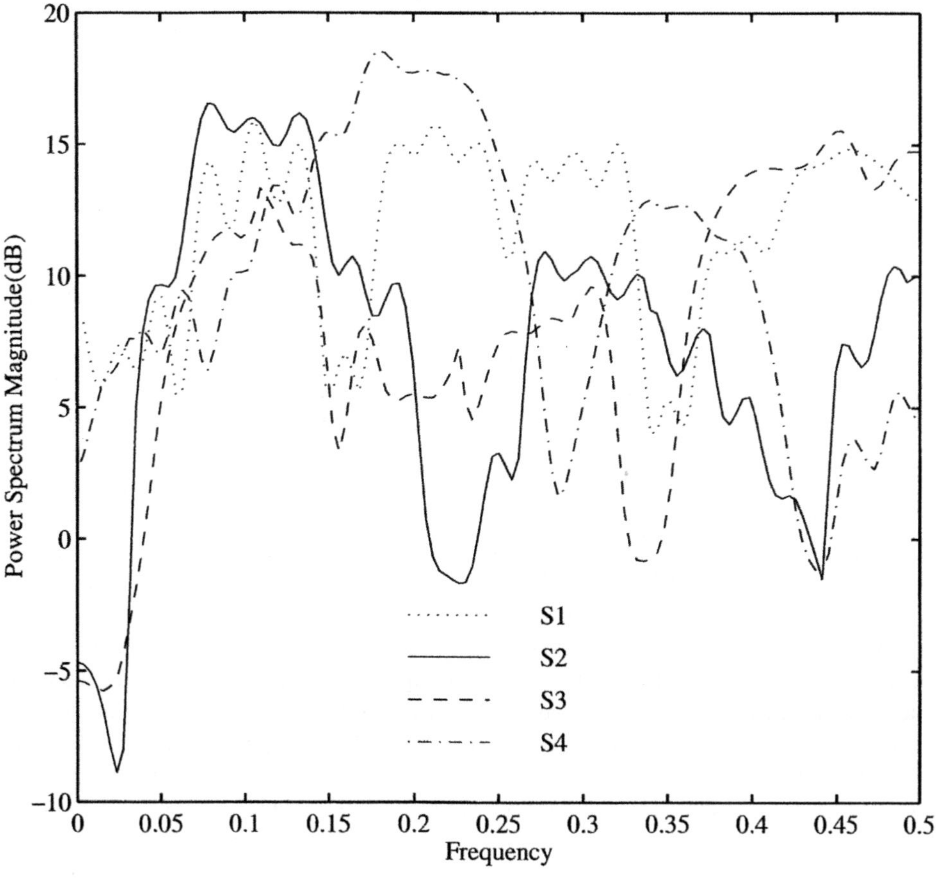

Figure 5.18. The evolutionary spectra for the Southern Oscillation Index (SOI) time series.

The results of testing the USGS flow data for stationarity at different frequencies are shown in Figure 5.19. In Figure 5.19, points which are significant at the 0.05 level are shown as solid circles. Points that are significant at the 0.01 level are further indicated by a hollow circle surrounding the solid one. This convention will be followed for the remainder of the figures in this section. From Figure 5.19, it is observed that most of the nonstationary behavior appears at high (0.4 to 0.5 cycles per year, cpy) or low frequencies (0 to 0.05 cpy), but some stations (stations 5, 6, and 22) do exhibit nonstationarity at intermediate frequencies as well.

Figure 5.20 shows the evolutionary spectra of the USGS flow data from station 22 as an example of a nonstationary time series in this group. In Figure 5.20, there are discrepancies between different spectra especially at high and low frequencies. The stationarity test results for this particular station gave $F(S_{I+R})=1.0000$, $F(S_T)=1.0000$, and $F(S_F)=1.0000$. According to these values the time series from station 22 is highly nonstationary and its spectrum is significantly different from that of a random process. Further testing at individual frequencies supports nonstationarity at high and low frequencies seen in Fig. 5.19.

Results of testing the HCN rainfall data for stationarity are given in Table V.3. For this data set, four time segments each of length $N/3$ at 30 % overlapping are used. A band width of $W=2$ is used for the MTM to ensure three to five independent frequencies, depending on the length of the time series N. The results in Table V.3 indicate that 11 time series are nonstationary at the 0.05 level, and six are nonstationary at the 0.01 level. For this data set of 36 stations, the 95% confidence interval for the number of stations giving false indication of nonstationarity is 0 to 4 at the 0.05 level, and 0 to 2 at the 0.01 level. Based on these numbers, there is some evidence of nonstationarity in the HCN rainfall data set. The number of stations with significant S_F (12 and 7 at the 0.05 and 0.01 levels, respectively) indicates that the time series are autocorrelated, but the number is not as large as that for the USGS flow data. This is typical of rainfall time series which are usually less autocorrelated than flow time series.

Results of testing the HCN rainfall data for stationarity at different frequencies are shown in Fig. 5.21. It can be concluded from Fig. 5.21 that nonstationarity is mainly at low frequencies (0 to 0.10 cpy) and to a lesser extent at high frequencies (0.45 to 0.50 cpy). Figure 5.22 shows the evolutionary spectra for the HCN rainfall data from station 12. It is clear from Fig. 5.22 that there are large discrepancies in spectra especially in the range from 0 to 0.3 cpy. The stationarity test results for this station gave $F(S_{I+R})=0.9657$, $F(S_T)=0.9968$, and $F(S_F)=0.8230$. These values indicate that this time series is nonstationary at both the 0.01 and 0.05 levels. Also, the spectrum of this time series is not significantly different from that of a random process. This is to be expected in the case of rainfall data which usually exhibit weak autocorrelations.

The results of testing the NCDC divisional rainfall data for stationarity are given in Table V.4. For this data set, four time segments each of length $N/3$ at 30 % overlapping are used. A band width of $W=2$ is used for the MTM to ensure three to five independent frequencies, depending on the length of the time series N. For this data set of 37 stations, the 95% confidence interval for the number of stations giving false indication of nonstationarity is 0 to 4 at the 0.05 level, and 0 to 2 at the 0.01 level. The results in Table

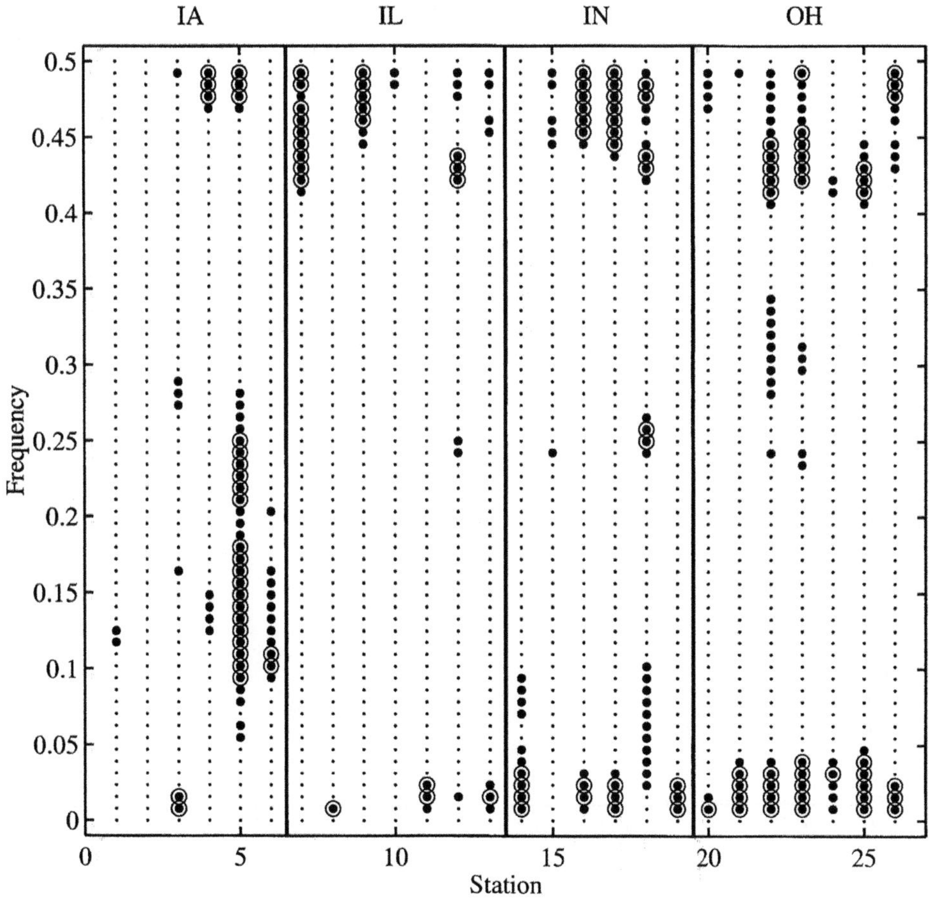

Figure 5.19. Results of the stationarity test at different frequencies for the USGS flow data.

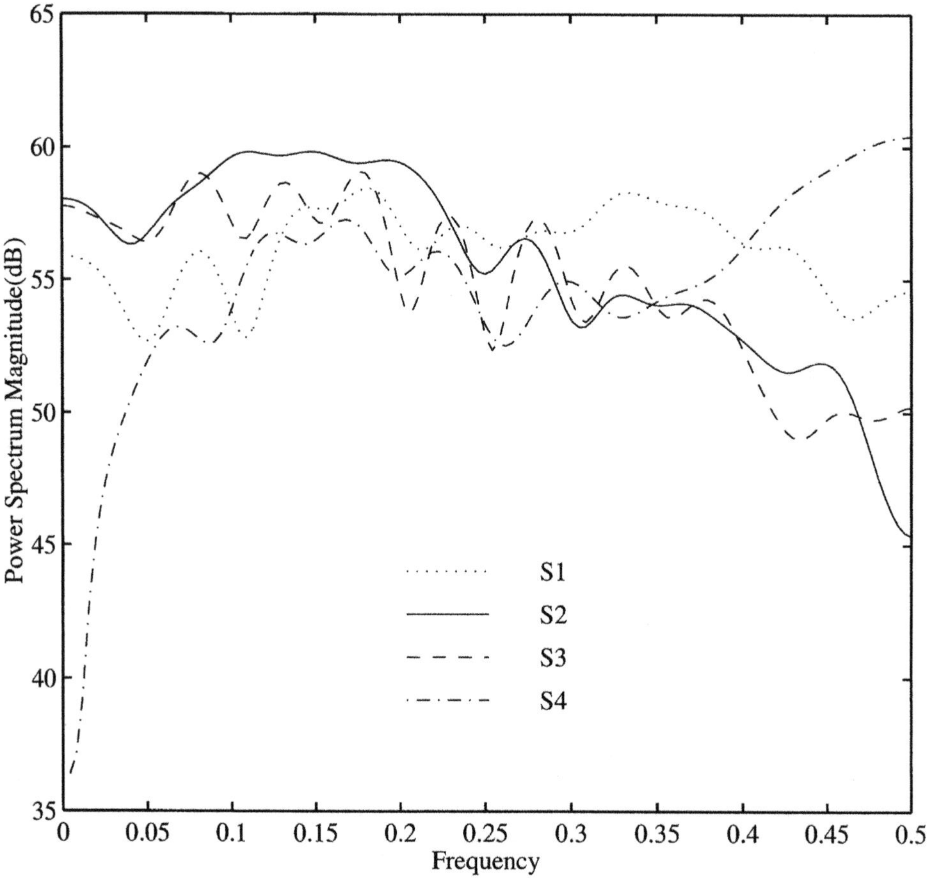

Figure 5.20. The evolutionary spectra for the USGS flow data from station 22.

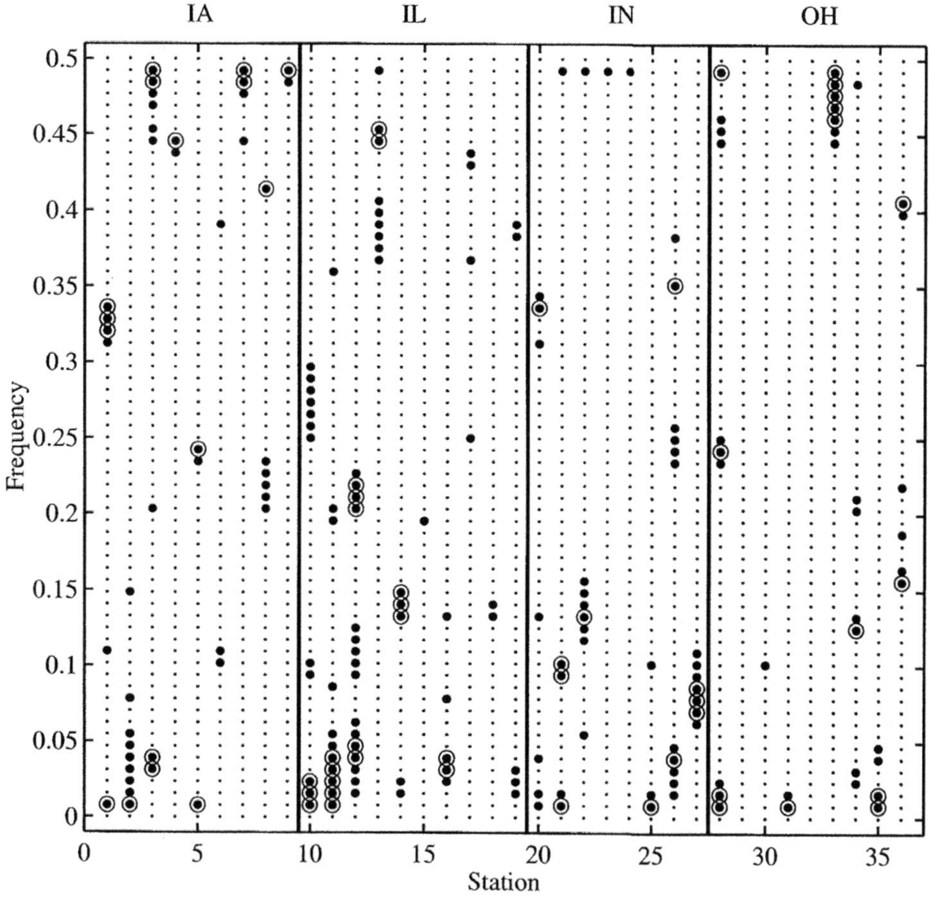

Figure 5.21. Results of the stationarity test at different frequencies for the HCN rainfall data.

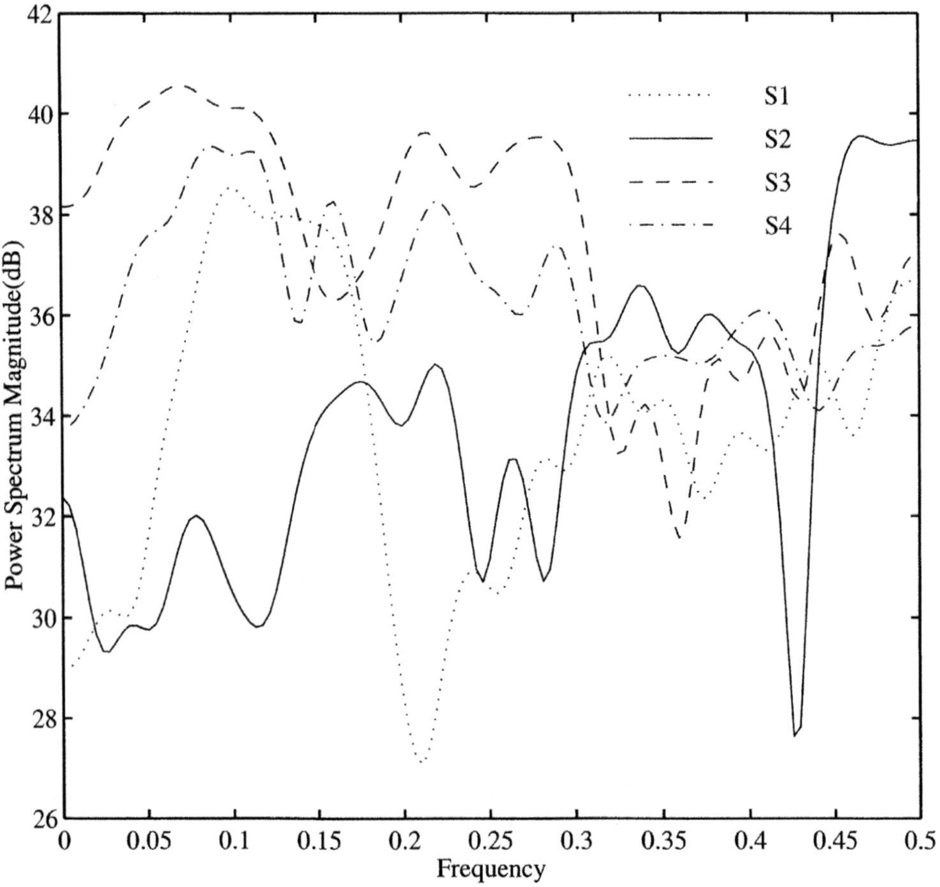

Figure 5.22.The evolutionary spectra for the HCN rainfall data from station 12.

V.4 indicate that 13 time series are nonstationary at the 0.05 level, and seven time series are nonstationary at the 0.01 level. Accordingly, there is evidence of nonstationarity in the NCDC divisional rainfall data.

Table V.3. Results of the Priestley and Subba Rao stationarity test for HCN rainfall data.

	Number of Significant Stations		
	Based on S_{I+R}	Based on S_T	Based on S_F
At the 0.05 Level	8	11	12
At the 0.01 Level	1	6	7

Table V.4. Results of the Priestley and Subba Rao stationarity test for NCDC divisional rainfall data.

	Number of Significant Stations		
	Based on S_{I+R}	Based on S_T	Based on S_F
At the 0.05 Level	12	13	16
At the 0.01 Level	5	7	9

Results of testing the NCDC divisional rainfall data for stationarity at different frequencies are shown in Fig. 5.23. It can be observed from Fig. 5.23 that nonstationarity is observed mainly at frequencies 0.25 to 0.40 cpy in Ohio stations, while for other states it is more or less at lower frequencies (0 to.25 cpy), in addition to 0.4 to 0.5 cpy in Iowa stations. The fact that Ohio stations have nonstationarity at a different band than other stations is intriguing and needs further investigation. It should also be noted that the results for the HCN rainfall data in Fig. 5.21 do not indicate nonstationarity in the range 0.25 to 0.40 cpy in Ohio. Figure 5.24 shows the evolutionary spectra for the NCDC divisional rainfall data from station 3. It is clear from Fig. 5.24 that there are large discrepancies between different evolutionary spectra, especially in the range from 0 to 0.3 cpy. The stationarity test results for this station gave $F(S_{I+R})$=1.0000, $F(S_T)$=0.9952, and $F(S_F)$=1.0000. This would indicate that this time series is highly nonstationary at both the 0.01 and 0.05 levels. Also the value of $F(S_F)$ indicates that the time series is autocorrelated and its spectrum is significantly different from that of a random process.

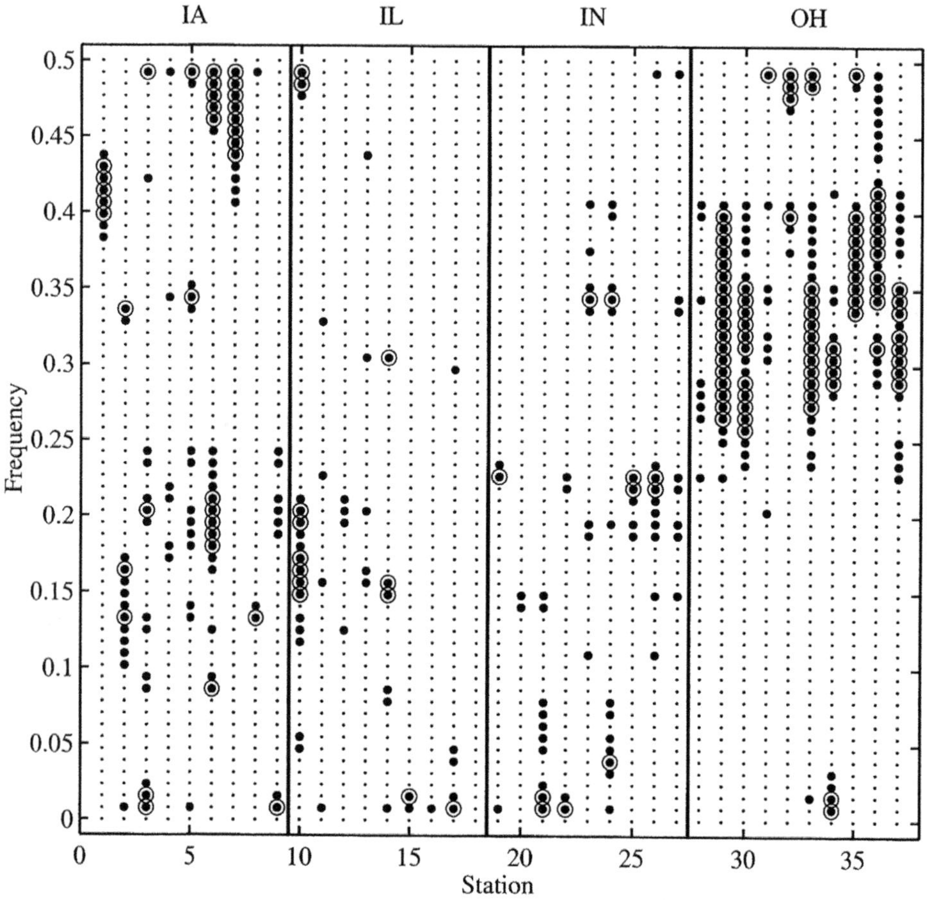

Figure 5.23. Results of the stationarity test at different frequencies for the NCDC divisional rainfall data.

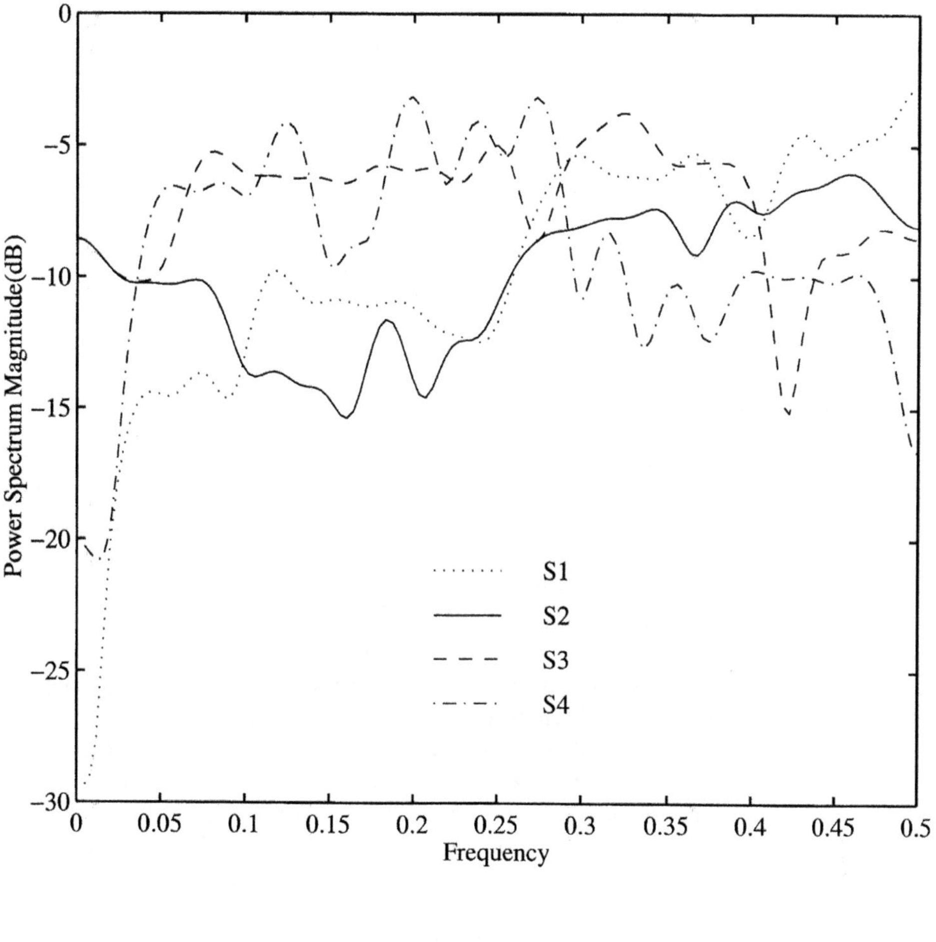

Figure 5.24. The evolutionary spectra for the NCDC divisional rainfall data from station 3.

Results of testing the HCN temperature data for stationarity are given in Table V.5. For this data set, three time segments each of length *N*/3 are used. A band width of W=2 is used for the MTM to ensure three to five independent frequencies, depending on the length of the time series *N*. For this data set of 36 stations, the 95% confidence interval for the number of stations giving false indication of nonstationarity is 0 to 4 at the 0.05 level, and 0 to 2 at the 0.01 level. The results in Table V.5 indicate that 19 time series are nonstationary at the 0.05 level, and eight time series are nonstationary at the 0.01 level. This would indicate nonstationarity in the HCN temperature data. The number of stations with significant S_F (22 and 15 at the 0.05 and 0.01 levels, respectively) indicates highly significant autocorrelation in the data, which is typical of temperature time series.

Table V.5. Results of the Priestley and Subba Rao stationarity test for HCN temperature data.

	Number of Significant Stations		
	Based on S_{I+R}	Based on S_T	Based on S_F
At the 0.05 Level	5	19	22
At the 0.01 Level	3	8	15

Results of testing the HCN temperature data for stationarity at different frequencies are shown in Fig. 5.25. It can be observed from Fig. 5.25 that nonstationarity is somewhat concentrated at 0 to 0.05 cpy, 0.20 to 0.25 cpy, and 0.37 to 0.50 cpy. Figure 5.26 shows the evolutionary spectra for the HCN temperature data from station 21. Large discrepancies in spectra can be observed between 0.2 to 0.45 cpy. The stationarity test results for this station gave $F(S_{I+R})$=0.8874, $F(S_T)$=0.9867, and $F(S_F)$=0.9253. This would indicate that this time series is nonstationary at the 0.05 level but not at the 0.01 level. Also the value of $F(S_F)$ indicates that the time series is uncorrelated and its spectrum is not significantly different from that of a random process.

Results of testing the NCDC divisional temperature data for stationarity are given in Table V.6. For this data set, three time segments each of length *N*/3 are used. A band width of W=2 is used for the MTM to ensure three to five independent frequencies, depending on the length of the time series *N*. For this data set of 37 stations, the 95% confidence interval for the number of stations giving false indication of nonstationarity is 0 to 4 at the 0.05 level, and 0 to 2 at the 0.01 level. A large number of stations (20) have significant values of S_{I+R} at the 0.05 level. This number drops to four stations at the 0.01 level, which is twice the upper limit of the 95% confidence interval of 2 stations. Therefore there is an indication of nonstationarity for this group of time series. The number of stations with significant S_F (9 and 7 at the 0.05 and 0.01 levels, respectively) indicates that autocorrelation in the NCDC divisional temperature data is less than that in the HCN temperature data. Again, this may be related to the spatial averaging process used to obtain the NCDC divisional temperature data.

Table V.6. Results of the Priestley and Subba Rao stationarity test for NCDC divisional temperature data.

	Number of Significant Stations		
	Based on S_{I+R}	Based on S_T	Based on S_F
At the 0.05 Level	20	2	9
At the 0.01 Level	4	1	7

Results of testing the NCDC temperature data for stationarity at different frequencies are shown in Fig. 5.27. It can be observed from Fig. 5.27 that nonstationarity is somewhat concentrated at frequencies 0.35 to 0.4 cpy, which is different than the case for the HCN temperature data in Fig. 5.25. Spatial averaging may have introduced or emphasized some type of nonstationarity at frequencies 0.35 to 0.4 cpy. Figure 5.28 shows the evolutionary spectra for the NCDC divisional average temperature data from station 18. Large discrepancies in spectra can be observed in the range from 0.3 to 0.45 cpy. The stationarity test results for this station gave $F(S_{I+R})$=0.9567, $F(S_T)$=0.7847, and $F(S_F)$=0.4322. This would indicate that this time series is nonstationary at the 0.05 level but not at the 0.01 level. Also the value of $F(S_F)$ indicates that the time series is uncorrelated and its spectrum is not different from that of a random process.

The results of testing the NOAA tree-ring data for stationarity are given in Table V.7. For this data set of 22 stations, the 95% confidence interval for the number of stations giving false indication of nonstationarity is 0 to 3 at the 0.05 level, and 0 to 2 at the 0.01 level. In Table V.7, 16 stations are nonstationary at the 0.05 level, while 13 stations are nonstationary at the 0.01 level. The large number of nonstationary time series indicates that these tree-ring time series are highly nonstationary. The number of stations with significant S_F (19 and 14 at the 0.05 and 0.05 levels, respectively) indicates that the NOAA tree-ring data are autocorrelated and their spectra are significantly different from those of random processes.

Table V.7. Results of the Priestley and Subba Rao stationarity test for NOAA tree-ring data.

	Number of Significant Stations		
	Based on S_{I+R}	Based on S_T	Based on S_F
At the 0.05 Level	11	16	19
At the 0.01 Level	6	13	14

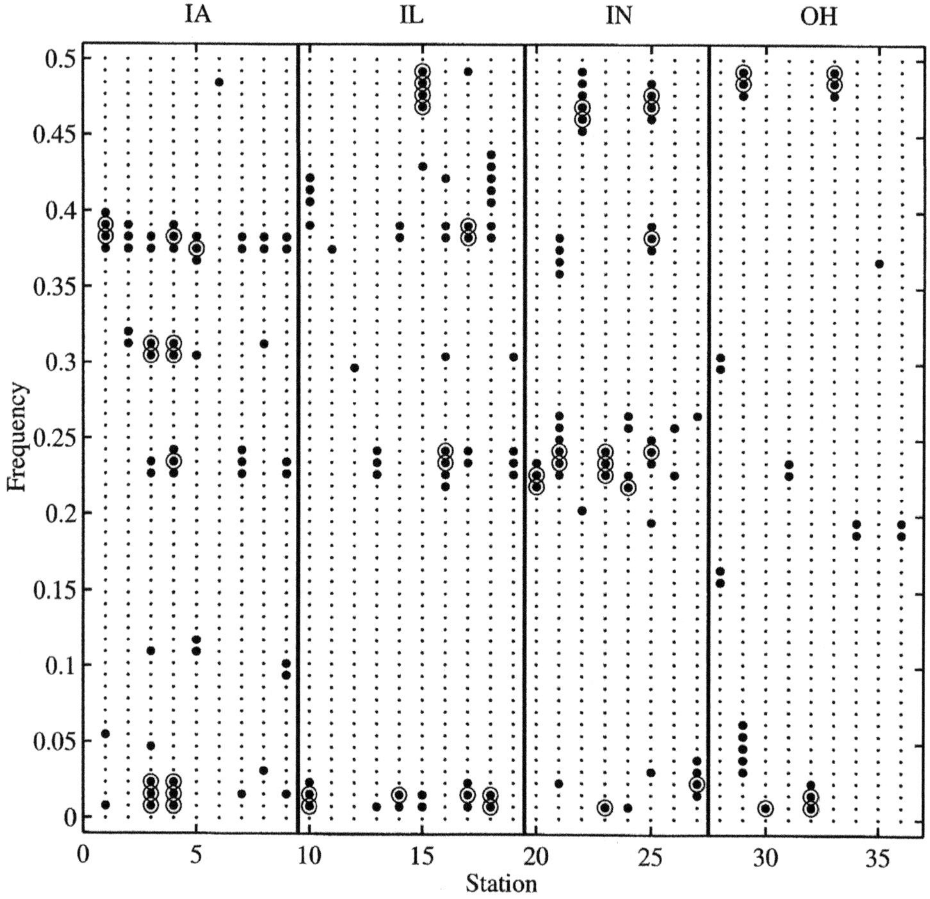

Figure 5.25. Results of the stationarity test at different frequencies for the HCN temperature data.

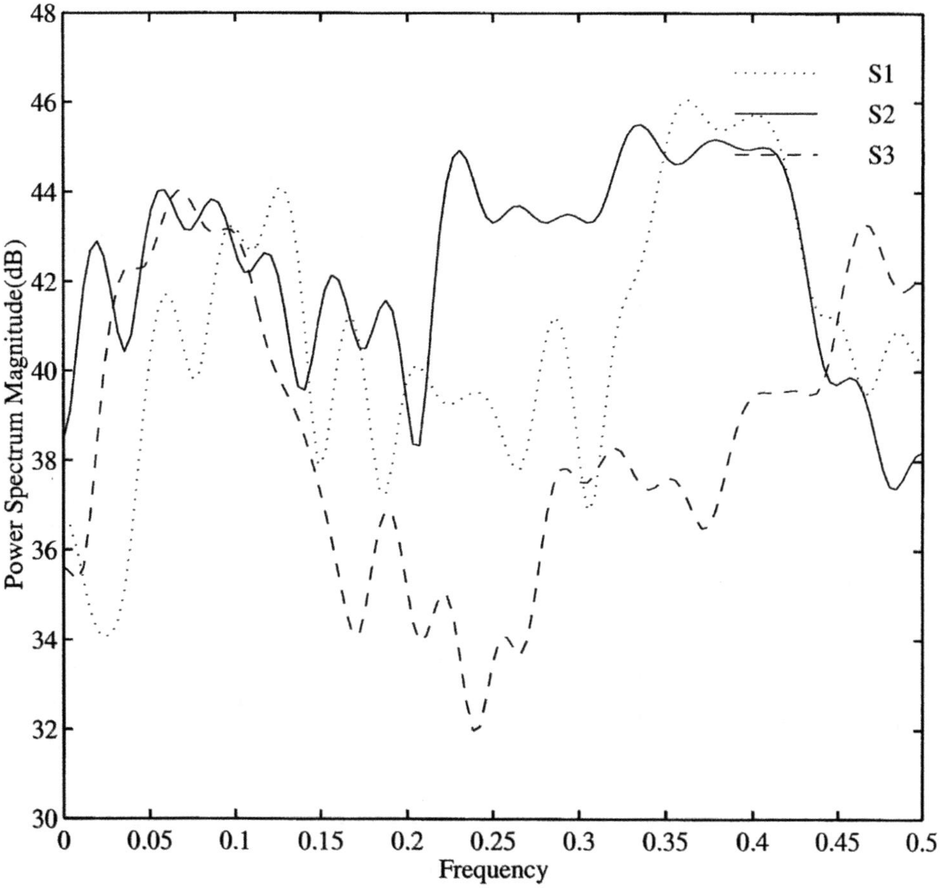

Figure 5.26. The evolutionary spectra for the HCN temperature data. from station 21.

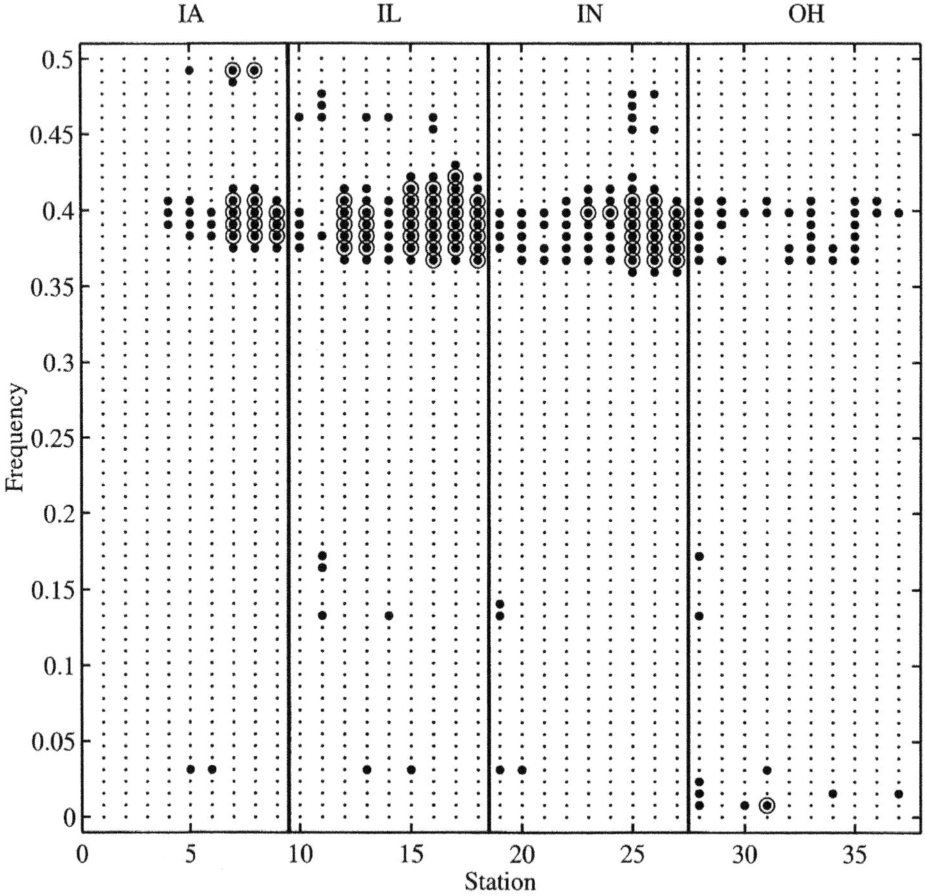

Figure 5.27. Results of the stationarity test at different frequencies for the NCDC divisional temperature data.

Results of testing the NOAA tree-ring series for stationarity at different frequencies are shown in Fig. 5.29. It can be observed from Fig. 5.29 that nonstationarity is somewhat concentrated at low frequencies (0 to 0.2 cpy), while some stations are nonstationary at intermediate frequencies (station 2, 4,10, and 13) and high frequencies (stations 6,9,10,11, and 20). In particular, station 10 seems to be nonstationary at almost all frequencies. Figure 5.30 shows the evolutionary spectra for the NOAA tree-ring data from station 10. Large discrepancies between different evolutionary spectra can be observed overall frequencies. The stationarity test results for this station gave $F(S_{I+R})$=0.9693, $F(S_T)$=1.0000, and $F(S_F)$=0.9999. This would indicate that this time series is highly nonstationary at both the 0.05 and 0.01 levels. Also the value of $F(S_F)$ indicates that the time series is autocorrelated and its spectrum is significantly different from that of a random process.

5.5. CONCLUSIONS

In this chapter the evolution of the spectrum of hydrologic and climatic data was investigated. The analysis was divided into two parts. The first part of the analysis dealt with the evolution of line components in the spectra regarding their frequencies, amplitude, and significance. Results discussed in this chapter lead to the following conclusions:

- The analyzed hydrologic as well as climatic time series exhibit a behavior which is characterized by frequency drifts as well as splitting and aggregation of frequencies.
- This behavior is not typical of stationary linear systems. However, due to the shortness of the available records and the lack of a statistical test to assess the significance of the observed changes, conclusions about the nature of these changes is necessarily tentative.

The second part of the analysis dealt with the continuous part of the spectrum. Evolutionary spectra were calculated for each of the stations and were used to test for stationarity of the time series. The conclusions from this part of the analysis are as follows:

- Based on the Priestley and Subba Rao (1969) test there is evidence of nonstationarity in the analyzed time series. For all data sets, the number of stations with significant nonstationarity exceeds the upper limit of the 95% confidence interval for the expected number of stations falsely identified as nonstationary under the null hypothesis of stationarity, both at the 0.05 and 0.01 significance levels.

The most evident nonstationarity is that of the NOAA tree-ring series. The results of the Priestley and Subba Rao stationarity test indicate that 16 stations out of 22 are nonstationary at the 0.05 level. Furthermore, 13 out the 22 stations are nonstationary at the 0.01 level. More than half of the USGS flow time series are nonstationary at the 0.05 level, while about 40% are nonstationary at the 0.01 level. About 60% of the temperature data at the 0.01 level, and 10 to 20% at the 0.01 level were found nonstationary. Also, about 30% of the rainfall time series at the 0.05 level, and about 20% at the 0.01 level were found nonstationary.

- The NOAA tree-ring data are the longest records in this study, with record lengths of 250 to 500 years. The fact that most of the available 22 time series are nonstationary may be related to the length of the time series. The reason that nonstationarity in other

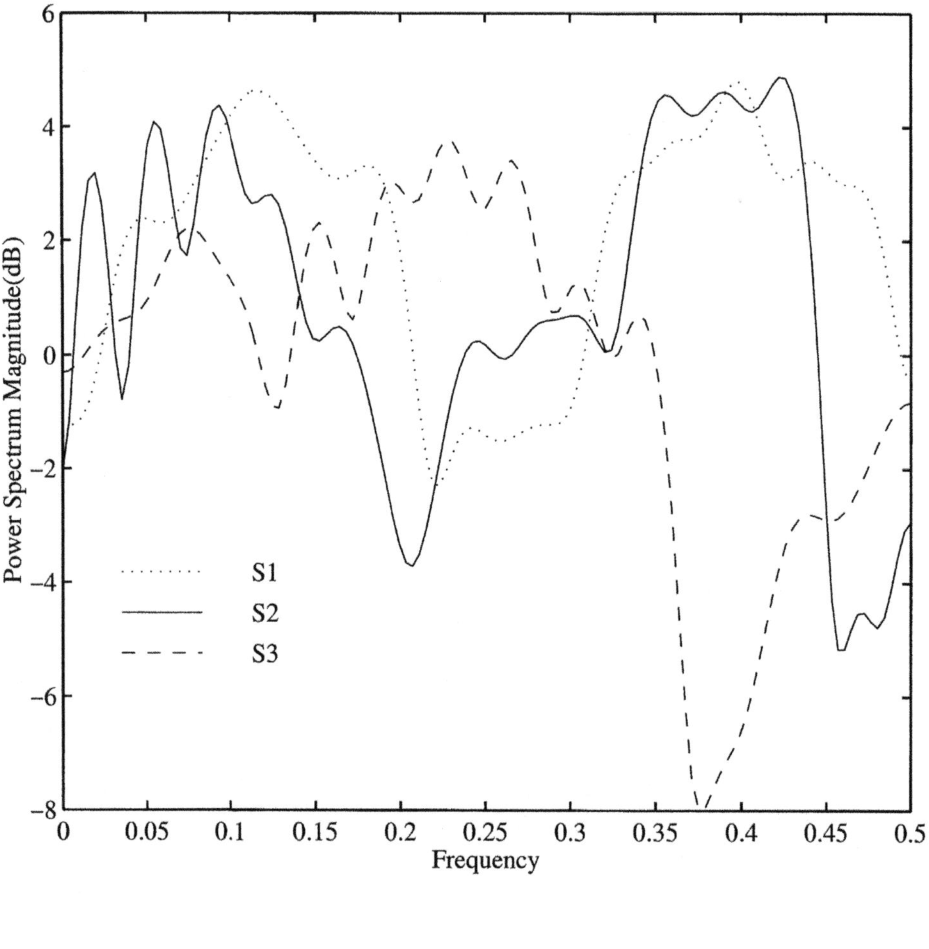

Figure 5.28. The evolutionary spectra for the NCDC divisional temperature data from station 18.

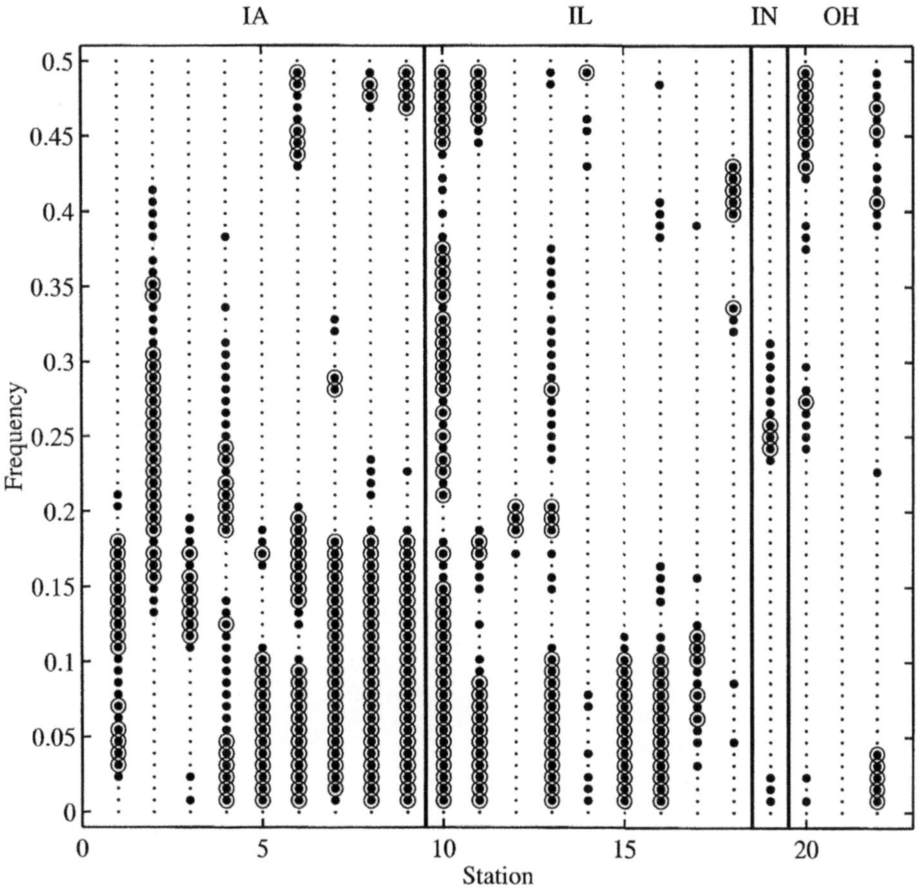

Figure 5.29. Results of the stationarity test at different frequencies for the NOAA tree-ring data.

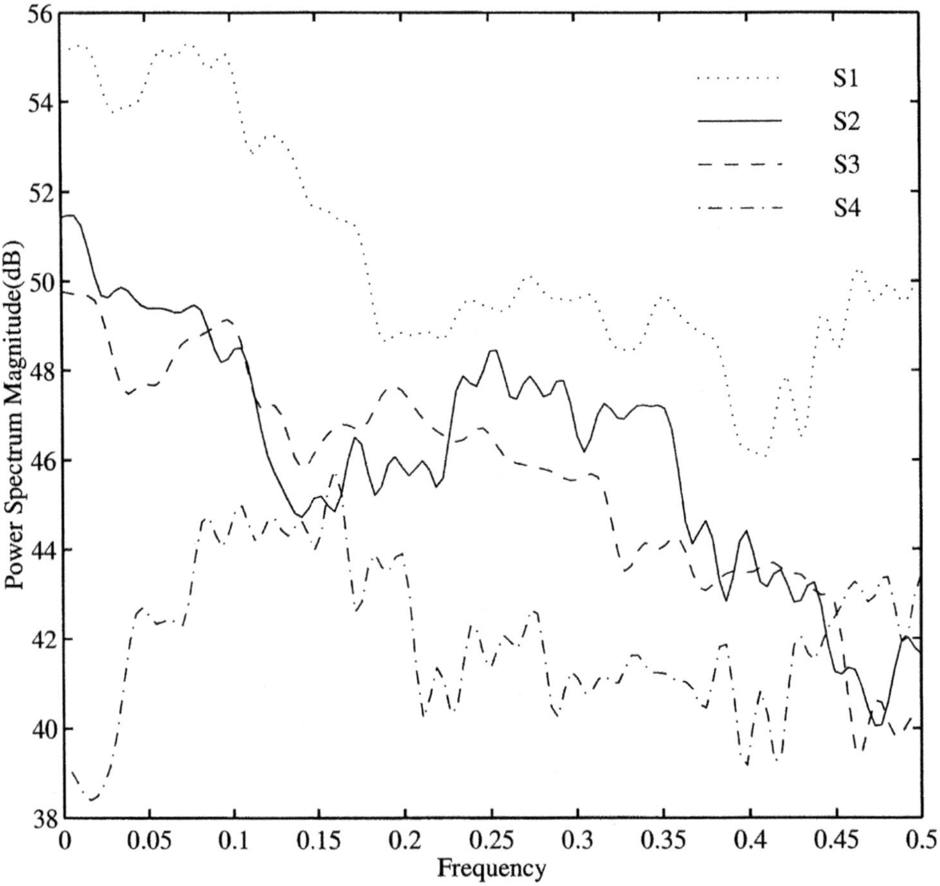

Figure 5.30. The evolutionary spectra for the NOAA tree-ring data from station 10.

time series is not as highly identified may simply be due to the lack of records longer than 100 years for the other data.

- A difference in the stationarity test results was observed between the HCN and NCDC divisional average temperature as well as rainfall data. This suggests that the divisional averaging process may have introduced or emphasized some type of nonstationarity in the averaged data.

The results of the analysis of both components of the mixed spectra agree in that the analyzed hydrologic and climatic time series have spectra that are changing with time, similar to other natural phenomena such as the sunspots and the Southern Oscillation. This would indicate that the processes generating such time series are nonlinear systems. Nonlinear dynamical systems have the ability to undergo bifurcations (i.e. changes in spectrum) as a result of the change of one of their parameters (Guckenheimer and Holmes, 1983; Ghil and Childress, 1987). Therefore, the use of nonlinear models may be more efficient in the modeling of hydrologic and climatic time series. This will, hopefully, translate into more accurate forecasting of such time series.

6. Time-Scale Analysis

6.1. INTRODUCTION

In conventional spectral analysis (Fourier analysis), the signal is compared with a number of basis functions which are composed of sines and cosines of different frequencies. A sinusoid with a given frequency ω can be represented in the frequency domain as a delta function at $\pm\omega$, and is therefore well localized in frequency. However, a sinusoid in the time domain has an infinite extent, and is therefore poorly localized in time. In order to improve the performance of Fourier analysis when time localization is desired, the Short Time Fourier Transform (STFT) is used. STFT effectively reduces the extent of the sinusoids to the size of the window used in the analysis. However, the choice of a window size that is suitable for a given signal involves a tradeoff. On the one hand a short time window will capture high frequency components and allow for more time localization. On the other hand, a longer window is required to precisely capture low frequency information. Accordingly, a more effective method of analysis would require a variable-size window to be applied to the data.

To overcome these difficulties, a different approach known as Wavelet analysis has been introduced during the past 10 years (Daubechies, 1992; Combes et al., 1990; Mallat, 1996; Meyer, 1993; Strang and Nguyen, 1996; Chui, 1992). In Wavelet analysis, a different type of basis functions, called Wavelets, are used. Wavelets, as opposed to sine and cosine functions, have a limited extent in time (a limited support), and are therefore localized in time as well as in frequency. In the theory of Wavelet transforms, the notion of 'frequency' is changed to an equivalent one of 'scale', where large scales correspond to low frequencies and small scales correspond to high frequencies. Wavelet analysis provides a technique for analyzing the structure of a time series by using variable size windows. By using Wavelet analysis, longer time windows can be used to detect low frequency (large scale) information, while shorter time windows can be used to detect high frequency (small scale) information. Wavelet analysis offers a different approach to looking at signals. In effect, by using the Wavelet transform the signal is decomposed into components of different scales. This offers a method for a local examination of the signal, a multi-scale outlook, and a time-scale analysis (Misiti et al., 1996). Wavelet analysis can be used as a tool for performing a wide range of tasks which include: detecting discontinuities in signals, detecting long-term evolution (trends), detecting self similarity (fractal signals), identifying pure frequencies (time-frequency analysis), and suppression, deionizing, and compression of signals.

Wavelet analysis is used in this chapter to investigate the stationarity of hydrologic and climatic time series as well as tree-ring series as proxy to hydrologic time series. The time-scale structures in these time series are studied and the relationships between them are investigated. Also, Wavelet analysis is used for assessing similarities between different time series at different scales in the region under study.

6.2. WAVELET ANALYSIS

Similar to the Fourier transform, the wavelet transform of a signal $S(t)$ can be calculated as the inner product of the signal with a sequence of functions $\psi_{m,n}$ with m indicating frequency localization and n time localization, as in Eq. 6.1 (Daubechies, 1993).

$$W_{m,n} = \int S(t)\ \overline{\psi_{m,n}(t)}\ dt \tag{6.1}$$

The functions $\psi_{m,n}$ in Eq. 6.1 replace the sinusoids in the conventional Fourier transform and are known as Wavelets. The $\psi_{m,n}$ are generated based on a basic (or 'mother') wavelet ψ which is well concentrated in both time and frequency. Once a mother wavelet has been chosen, the wavelet family $\psi_{m,n}$ can be generated by successive dilations and translations of the mother wavelet as in Eq. 6.2,

$$\psi_{m,n}(t) = a_o^{-m/2}\ \psi(a_o^{-m} t - nb_o) \tag{6.2}$$

where $a_o > 1$ and $b_o > 0$ are fixed parameters (Daubechies, 1993). By changing the value of m the width of the wavelet can be changed to correspond to high or low frequency. Similarly, by changing the value of n the wavelet is translated by amounts proportional to its width. Applying Eq. 6.1 to a signal results in the continuous wavelet transform. While this type of transform is redundant, it is often easier to interpret since its redundancy tends to make information more visible (Misiti et al., 1996) and can reveal subtle structures in the data such as self-similarity.

Alternatively, if the values $a_o = 2$ and $b_o = 1$ are considered and m and n are integers the resulting transformation is of the discrete type. In this case the coefficients are given by Eq. 6.3,

$$C(j,k) = \int S(t)\ \psi_{j,k}(t)\ dt \tag{6.3}$$

where $\psi_{j,k}$ in this case is given by Eq. 6.4.

$$\psi_{j,k}(t) = 2^{-j/2}\ \psi(2^{-j} t - k) \tag{6.4}$$

The discrete wavelet transform obtained by using Eq. 6.3 allows a multi-resolution decomposition of a signal at different levels j, where the scale is given by 2^j (Daubechies, 1992; Misiti et al., 1996). At each level j, a detail $D_j(t)$ of the signal is given by the inverse transform in Eq. 6.5.

$$D_j(t) = \sum_k C(j,k)\ \psi_{j,k}(t) \tag{6.5}$$

Also, at a given level J, the sum of details for $j > J$ (i.e. the sum of coarser details) defines what is called an approximation to the signal at level J.

$$A_J(t) = \sum_{j>J} D_j(t) \tag{6.6}$$

The result of this multi-resolution analysis is decomposing the signal into a hierarchial tree of details and approximations in the form of Eq. 6.76,

$$S = A_1 + D_1 = A_2 + D_2 + D_1 = ... \tag{6.7}$$

This process of multi-resolution decomposition is equivalent to using a series of high-pass and low-pass filters to extract details and approximations, respectively, where the coefficients for these filters are related to the family of wavelets used for analyzing the signal. The wavelet function ψ is related to the series of high-pass filters and is therefore directly associated with the details of the signal. For some wavelets, an additional function called the scaling function φ can be defined, which is related to the series of low-pass filters and is therefore directly associated with the approximations of a signal, as well as its details and approximations. Figure 6.1 shows the wavelet function and the scaling function as well as the associated decomposition and reconstruction filters for the type 4 Daubechies wavelet (db4).

There exists a number of wavelets with different properties. These include the Mexican hat wavelet which is the second derivative of the Gaussian curve, Morlet, symlets (approximately symmetric wavelets), Coiflets (Coifman wavelets), Meyer, and Daubechies wavelets. All of these wavelets except the first two allow orthogonal analysis which is required to insure proper reconstruction of the signal as well as its details. Another useful property of some wavelets is the property of vanishing moments, which gives some wavelets the ability to suppress polynomial signals. For example, the Daubechies wavelets of order N have N vanishing moments and can therefore suppress polynomial signals up to degree N-1.

6.3. WAVELET TREND ANALYSIS

By definition a trend in a signal is the slowest part of that signal. Based on wavelet analysis, a signal can be decomposed into components at different scales. Trends can therefore be detected by inspecting details of the signals at large scales (Misiti et al., 1996). In this section, wavelet decomposition is used as a tool for identifying trends in Hydrologic and climatic time series. The result of this analysis are compared to those obtained through trend analysis techniques applied in Chapter 3. For the purpose of decomposition a Daubechies wavelet of order 4 (db4) is used. Other wavelets were also investigated and gave similar results. For the data in this study, details of the signals at scales which are close to the length of the data are investigated for the existence of slow components representing trends in the time series. Figure 6.2 shows an example of the decomposition of a signal into its details and approximations at different levels by using the db4 wavelet. The decomposed signal in Figure 6.2 is the flow time series from USGS station 2. For trend analysis we focus on the large scales such as levels 8 and 9 which represent the slowest changing part of the signal.

Figure 6.3 shows the details at a scale of about 42 years for flow time series from 26 USGS stations. Figure 6.4 is a representation of trends in Figure 6.3, where increasing trends are depicted as black lines, while decreasing trends are depicted as white lines. It can be observed from these figures that there is a decreasing trend in the first part of the record up to about 1950 then an increasing trend afterwards. These trends are consistent among different stations and are similar to those obtained in Chapter 3, Figure 3.7.

Figure 6.5 shows the details at a scale of about 84 years for rainfall time series from 36 HCN stations. Figure 6.6 shows the corresponding trends in the time series. The trends in this case are not very consistent. However, Figure 6.6 is similar to Figure 3.8 in Chapter 3, giving essentially the same conclusions. Similarly, Figure 6.7 shows the details at a scale of about 84 years for rainfall time series from 37 NCDC divisions. Figure 6.8 shows the corresponding trends in the time series. Trends in this case are more consistent among stations than in the case of HCN rainfall time series, and indicate, in general, a decreasing trend followed by an increasing one. These results also agree with those obtained in Chapter 3, where Figure 3.9 compares very well to Figure 6.8.

Figure 6.9 shows the details at a scale of about 84 years for temperature time series from 36 HCN stations. Figure 6.10 shows the corresponding trends in the time series. Increasing trends are observed in Figure 6.10 from the beginning of the record till 1940 with a decreasing trend afterwards. This behavior is the opposite of that of flow and rainfall data as can be seen by comparing Figure 6.10 to Figures 6.4 and 6.8. The results are consistent among all stations and are comparable to those obtained in Chapter 3. Trends in Figure 6.10 are comparable to those in Figure 3.10 in Chapter 3. Similarly, Figure 6.11 shows the details at a scale of about 84 years for temperature time series from 37 NCDC divisions. Figure 6.12 shows the corresponding trends in the time series. The same observation for the HCN holds here also: an increasing trend up to 1945, followed by a decreasing trend till the end of the record. Figure 6.12 is comparable to Figure 3.11 in Chapter 3, confirming the existence of these tends. Trends are also consistent among different stations indicating regional trends.

Figure 6.13 shows the details at a scale of about 256 years for tree-ring time series from 22 NOAA locations. Figure 6.14 shows the corresponding trends in the time series. No trend of such scale could be identified in Figure 3.12. However the results in Figure 6.14 indicate a decreasing trend for the first part of the record followed by an increasing trend for the second part, but these trends are not consistent among all sites.

In general, the results from investigating details of the time series in this study at the largest scales indicate the existence of trends of opposite directions in the data. These results agree to a large extent with those from trend analysis in Chapter 3. However, the limited length of the available time series makes it difficult to infer whether these are pure trends linked to some physical effects, or if they are parts of periodicity in the data with periods much larger than that of the available data.

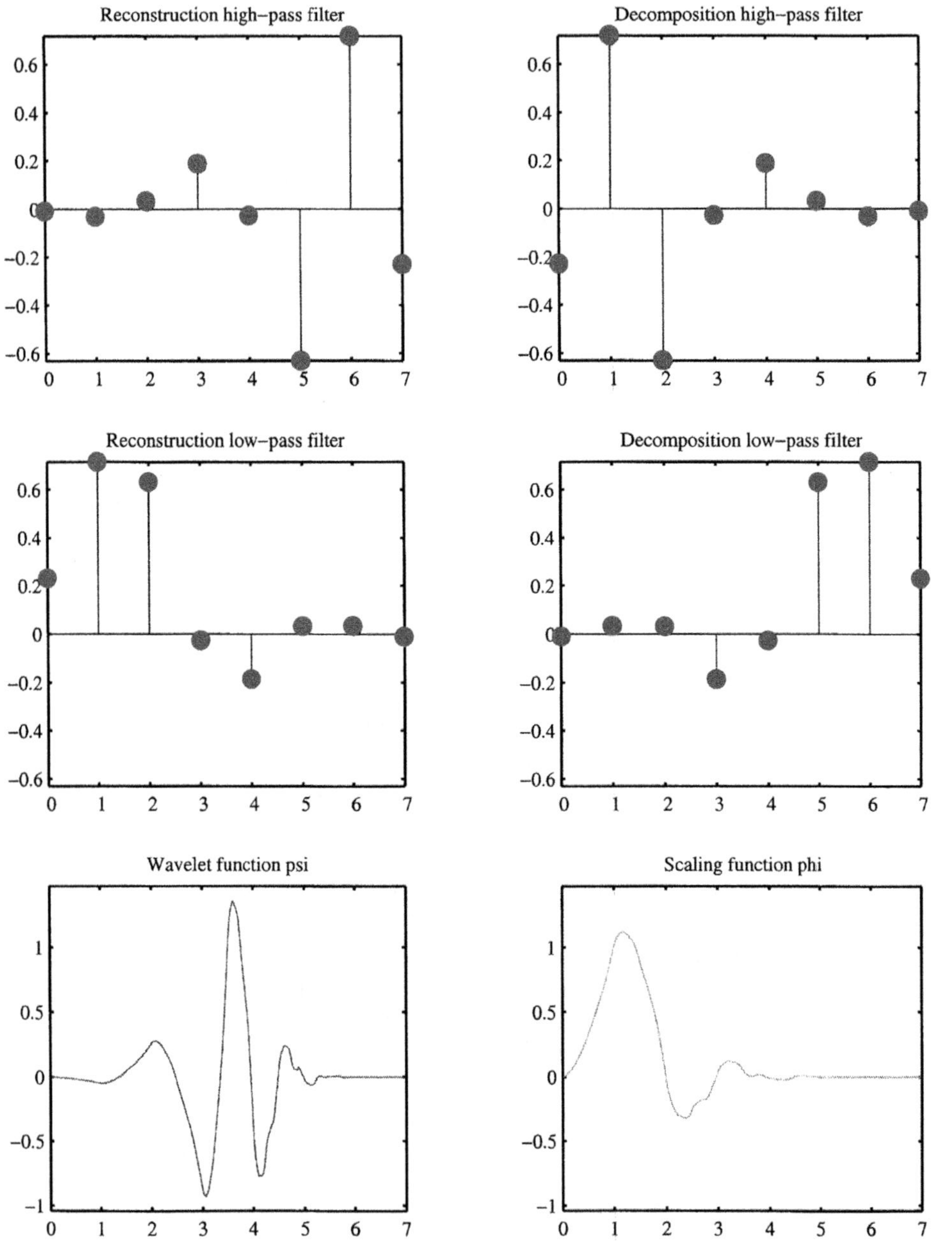

Figure 6.1. Daubechies Wavelet of Order 4 (db4), the corresponding scaling function, and the associated decomposition and reconstruction filters.

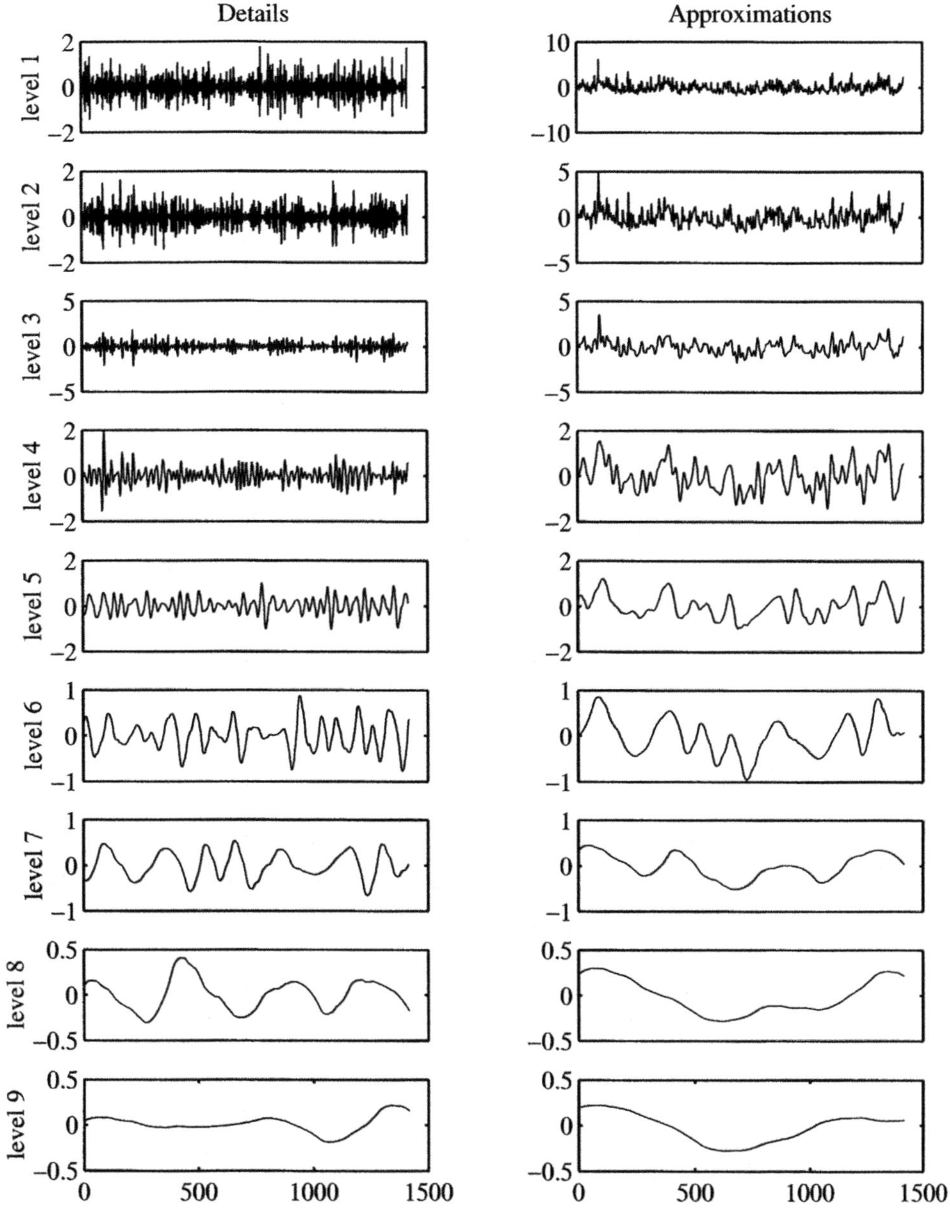

Figure 6.2. Details and approximations of the flow time series from USGS station 2.

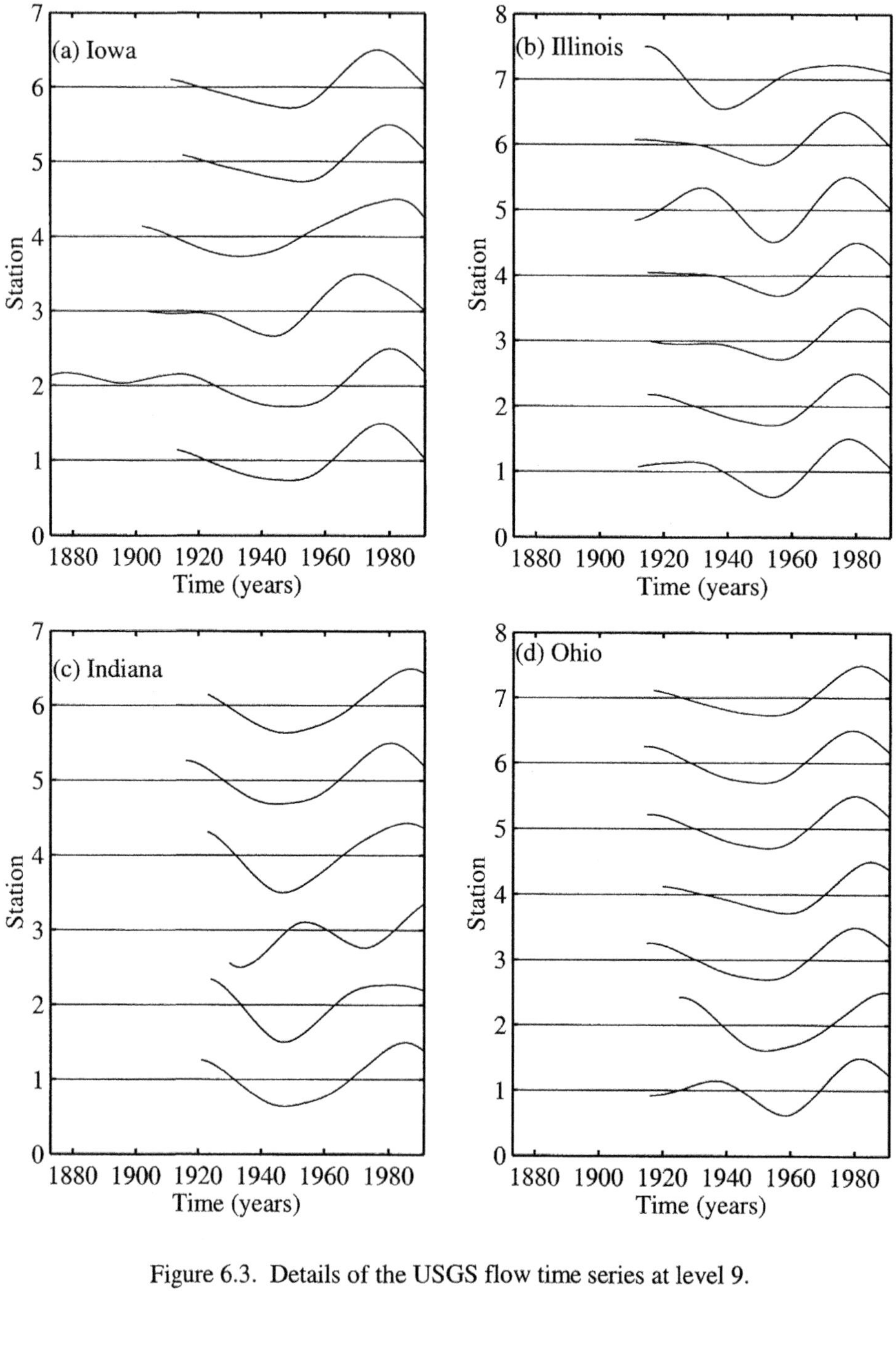

Figure 6.3. Details of the USGS flow time series at level 9.

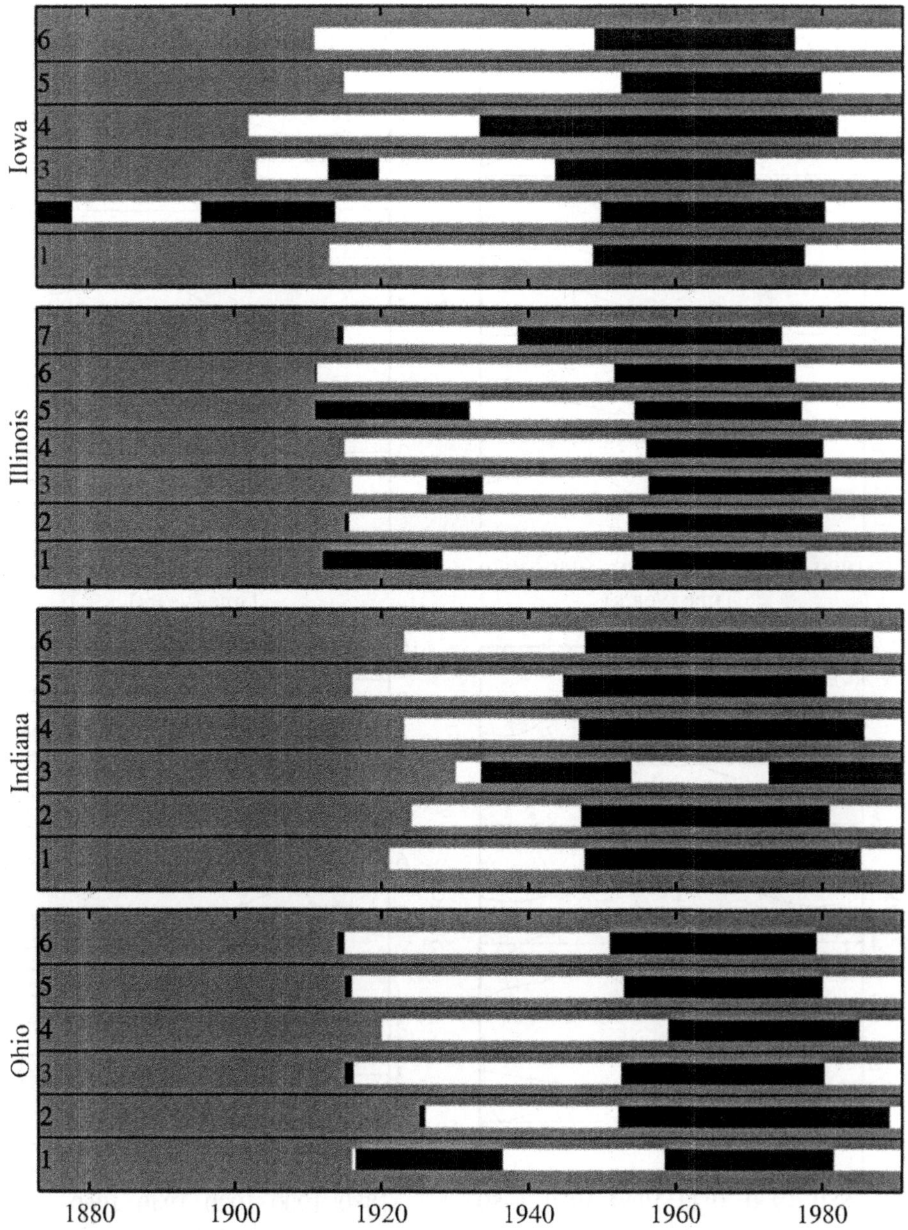

Figure 6.4. Directions of trends in USGS flow time series based on details at level 9.

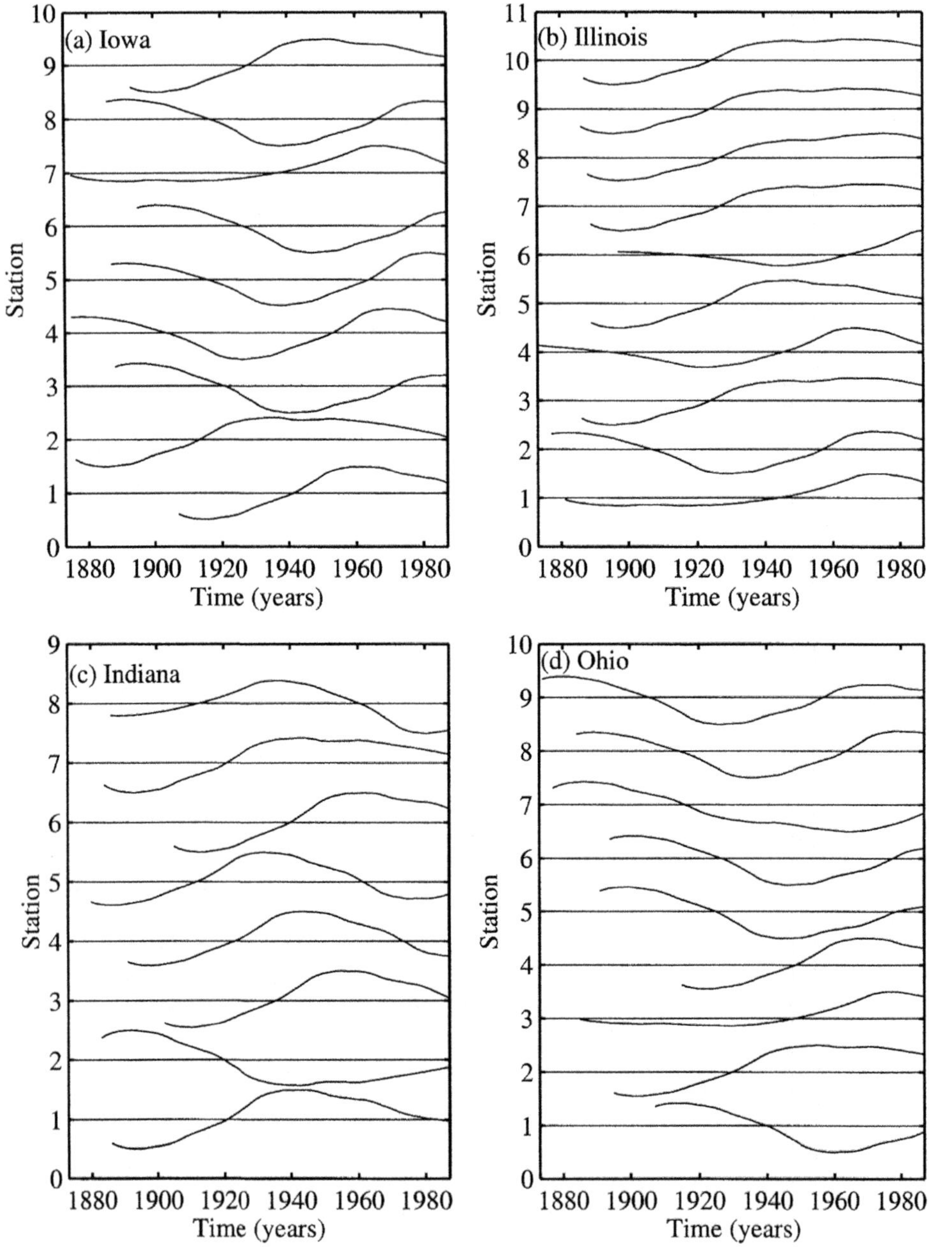

Figure 6.5. Details of the HCN rainfall time series at level 10.

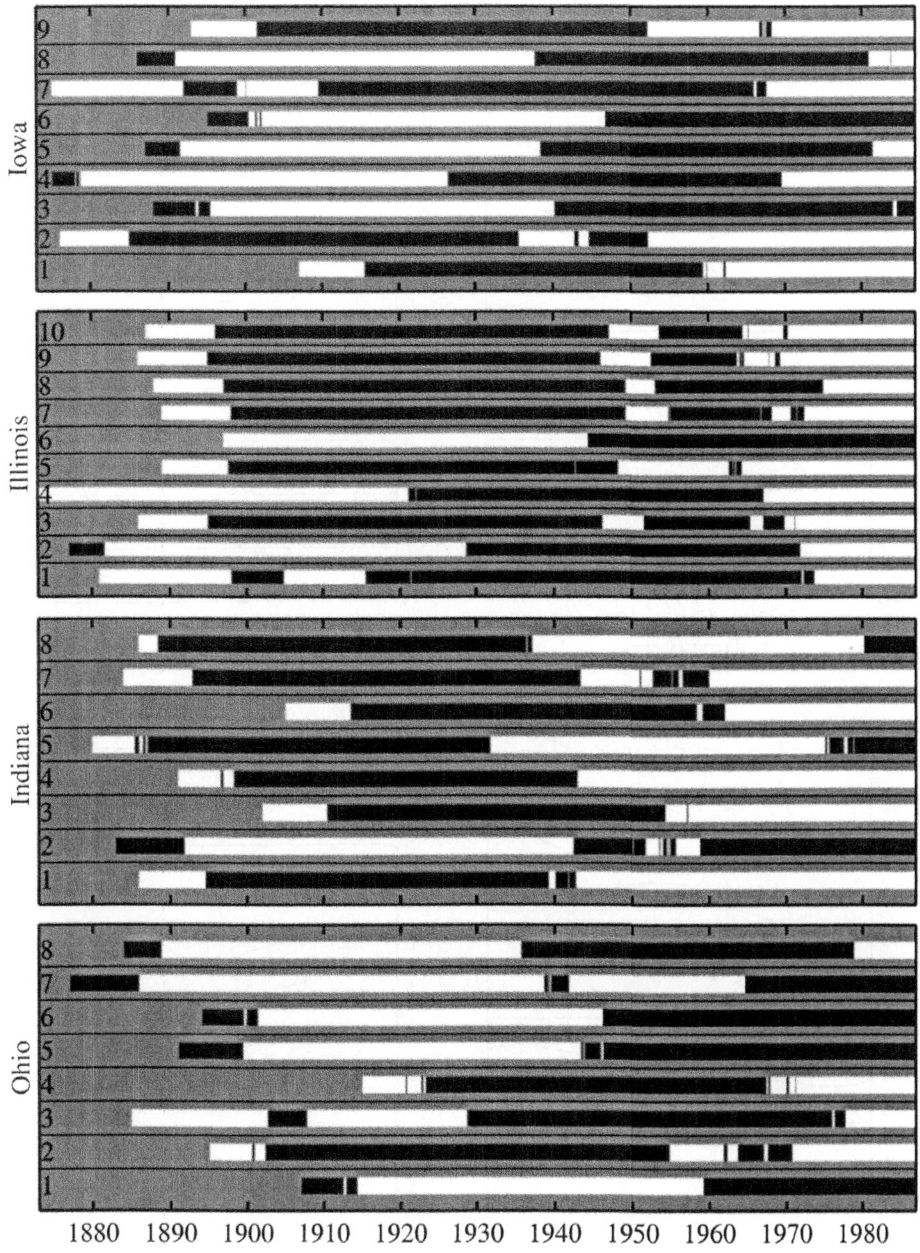

Figure 6.6. Directions of trends in HCN rainfall time series based on details at level 10.

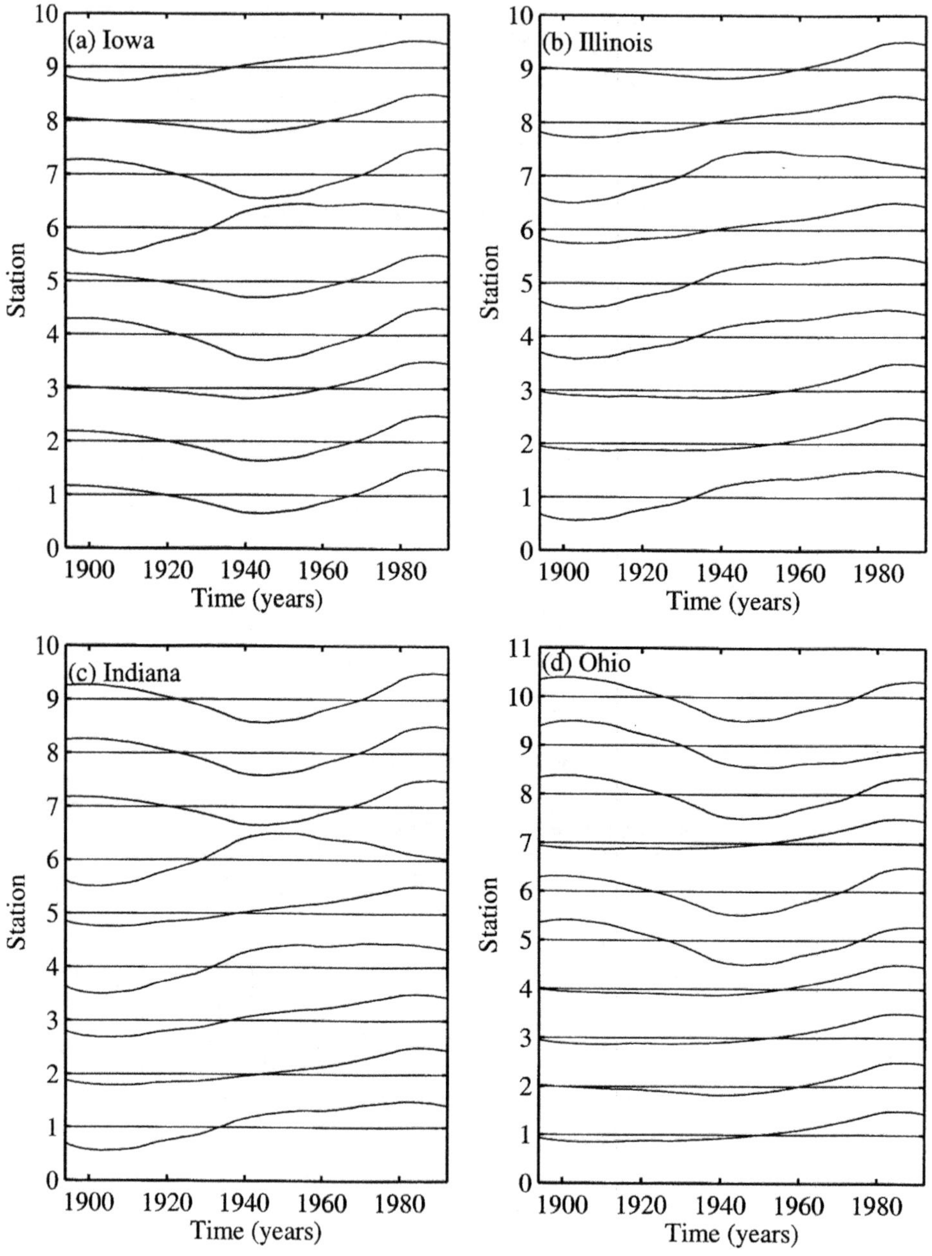

Figure 6.7. Details of the NCDC divisional rainfall time series at level 10.

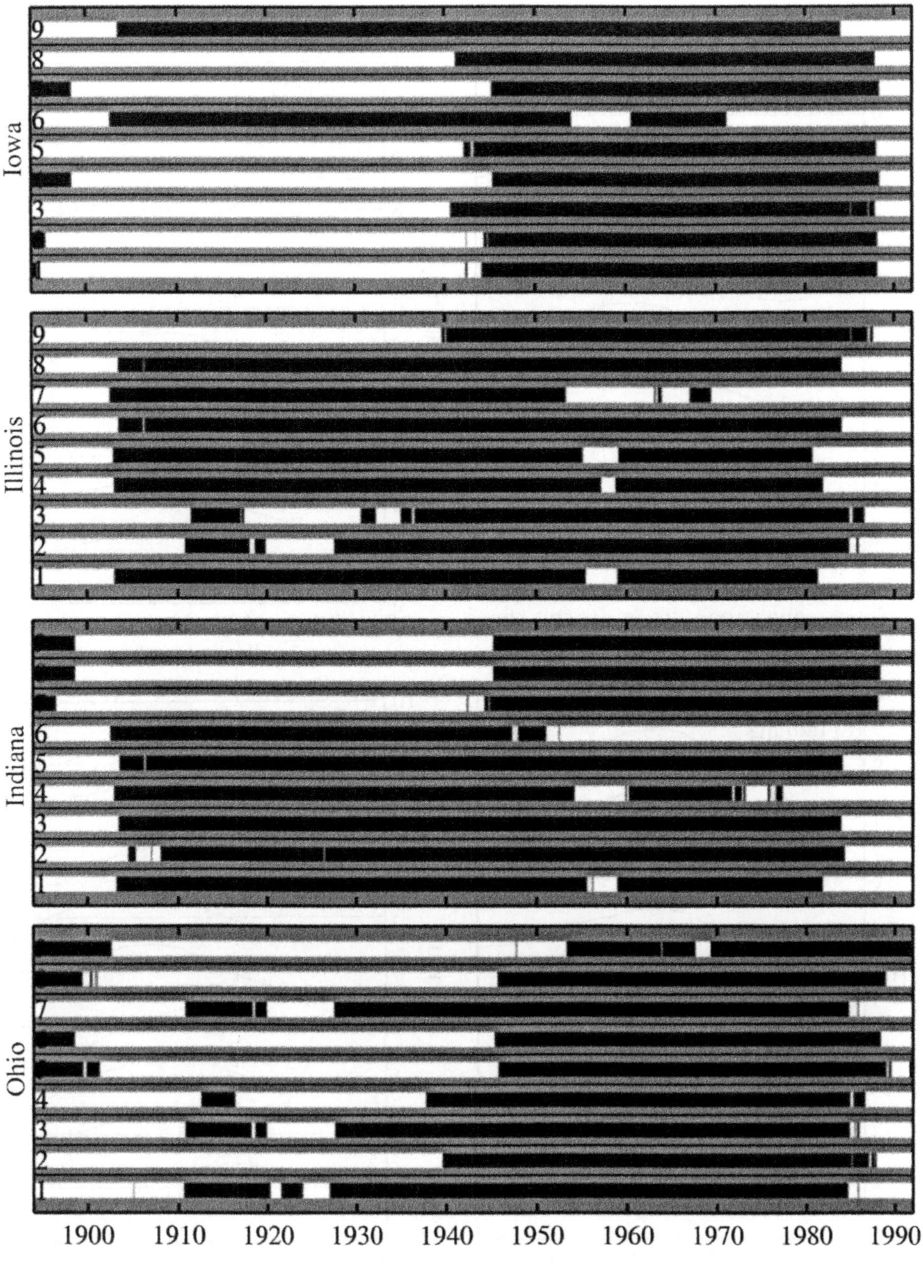

Figure 6.8. Directions of trends in NCDC divisional rainfall time series based on details at level 10.

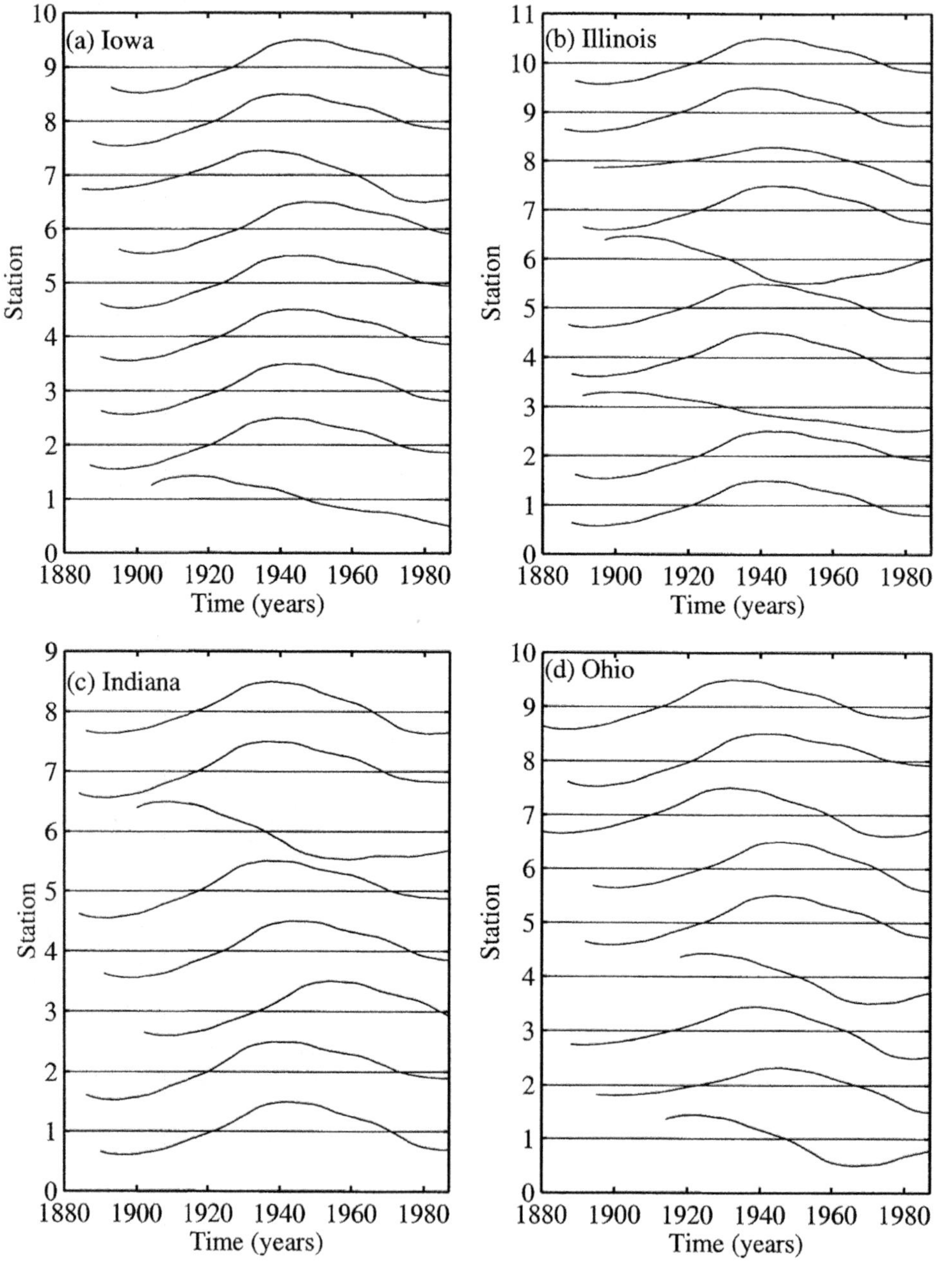

Figure 6.9. Details of the HCN temperature time series at level 10.

Figure 6.10. Directions of trends in HCN temperature time series based on details at level 10.

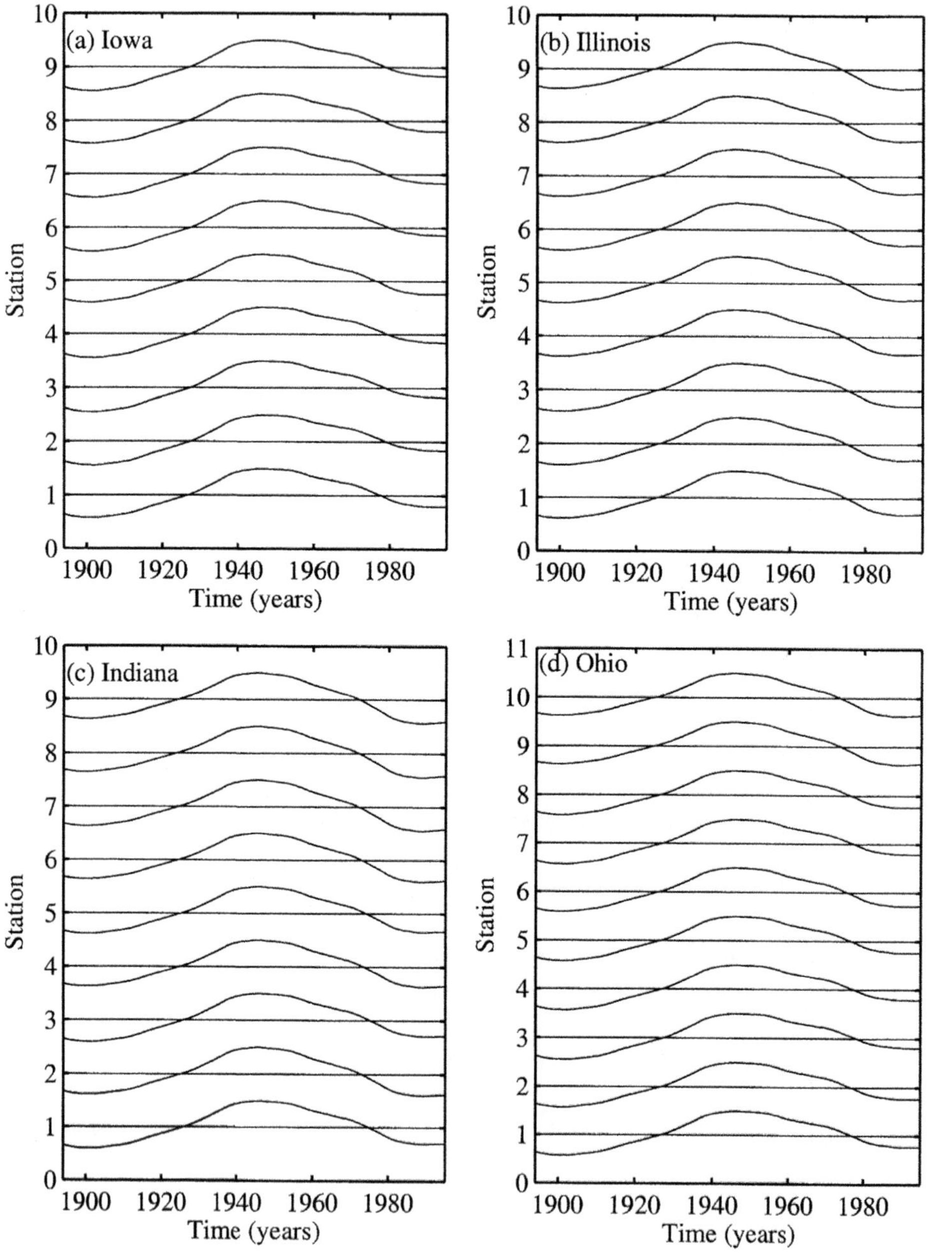

Figure 6.11. Details of the NCDC divisional temperature time series at level 10.

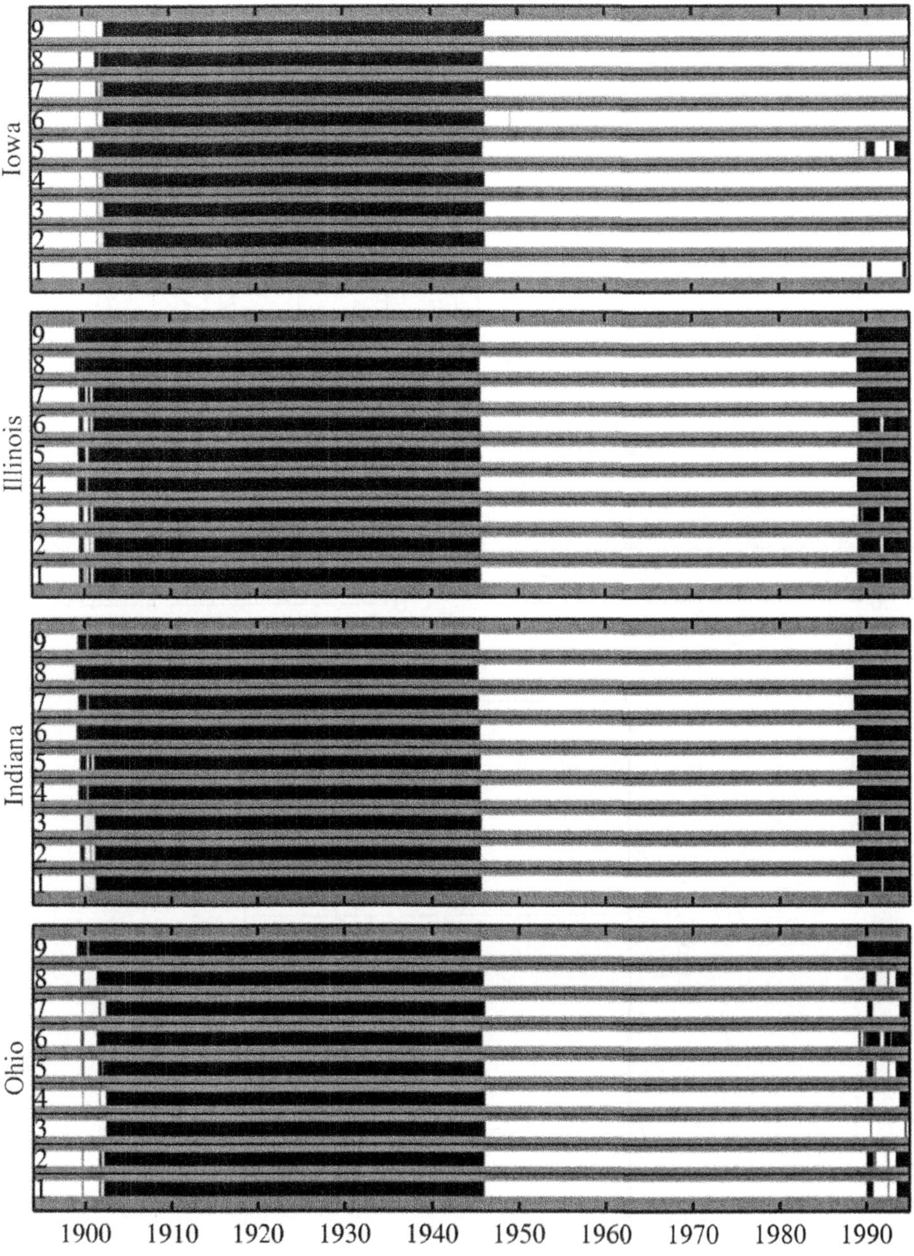

Figure 6.12. Directions of trends in NCDC divisional temperature time series based on details at level 10.

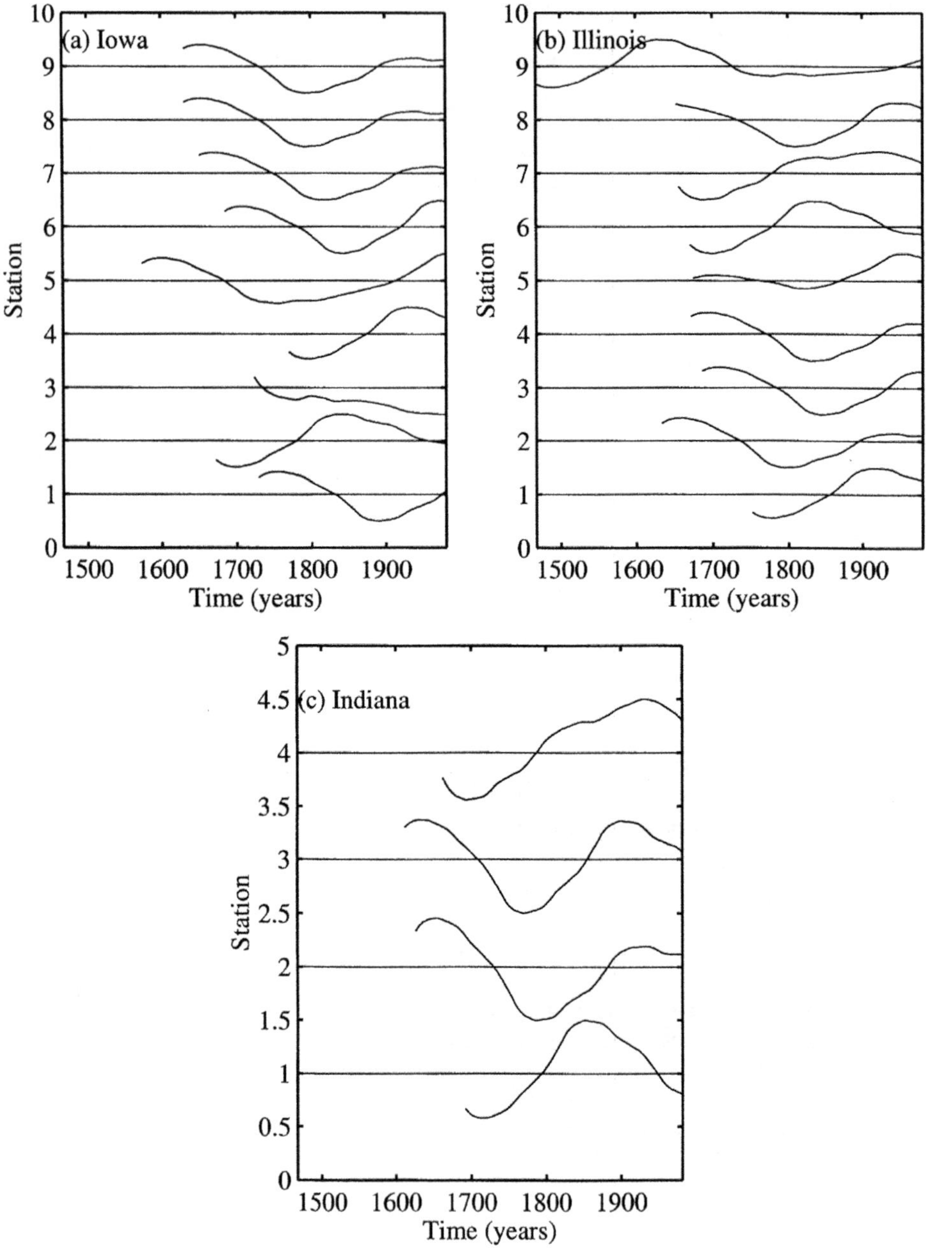

Figure 6.13. Details of the NOAA tree-ring time series at level 8.

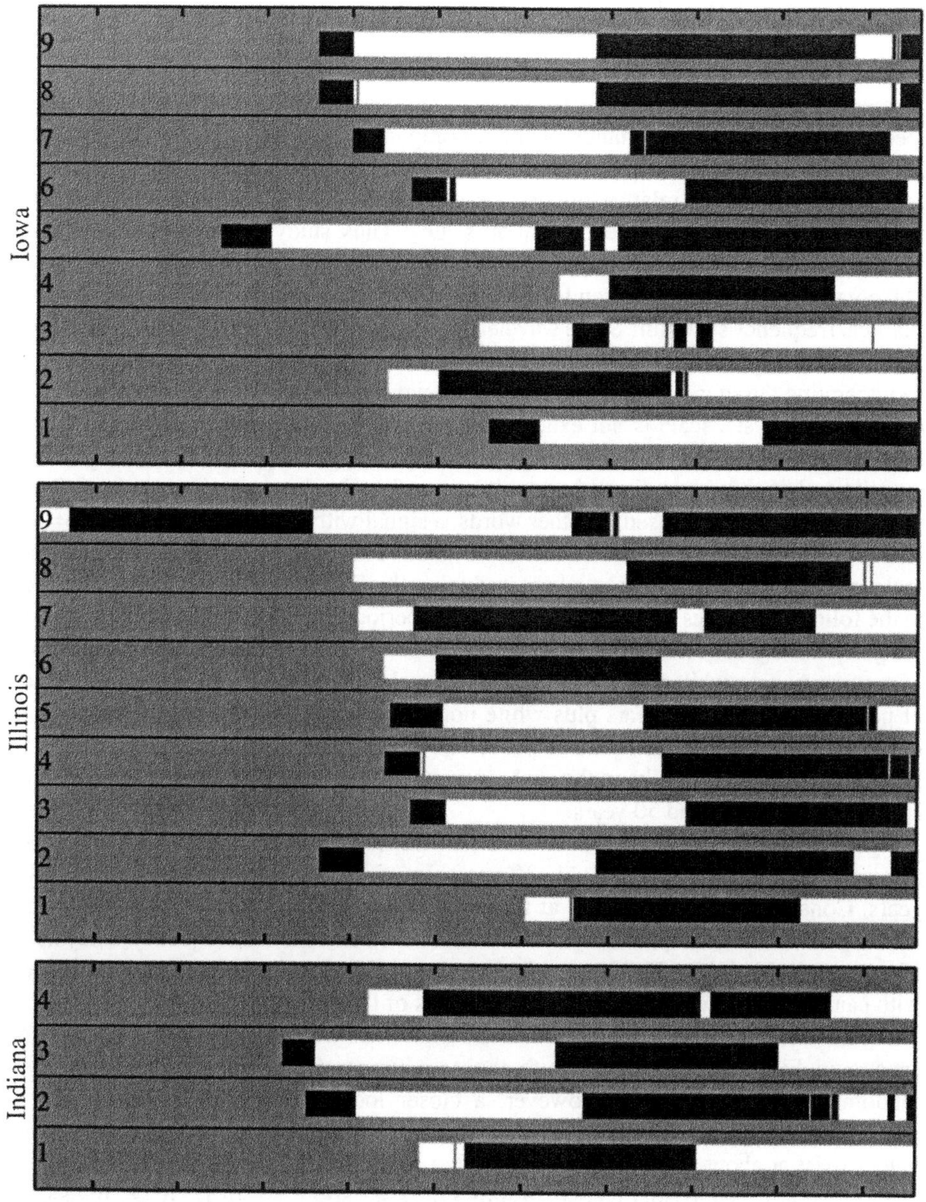

Figure 6.14. Directions of trends in NOAA tree-ring time series based on details at level 8.

6.4. IDENTIFICATION OF DOMINANT SCALES

In classical spectral analysis, a signal is decomposed into a number of components at different frequencies. By observing the values of the decomposition coefficients (the power spectrum) the relative importance of these components can be assessed and different periodicities in the signal can be identified. In wavelet analysis a similar approach can be used to investigate the relative importance of components. In this case, however, the notion of 'frequency' is replaced by that of 'scale'. Thus, studying the relative amplitude of the coefficients at different scale gives an idea about the dominant scales in the signal. The obtained plot in this case can be interpreted as a scale 'spectrum' as opposed to the classical frequency spectrum. Since wavelet decomposition gives a time-scale distribution of the coefficients, the magnitude of such 'spectrum' at a given scale can be constructed by integrating the squared values of these coefficients over time. It should be noticed that in wavelet analysis, scale is not exactly the inverse of frequency (Misiti et al., 1996). However, based on studying a number of sinusoidal components at different frequencies the relation between scale S and frequency ω was found to be approximately $S = 0.7\ \pi/\omega$ when the db4 wavelet is used. In other words, a signal with a period of ten years ($\omega = 0.1\ \pi$) would appear at a wavelet scale of 7, and so on. This relation between frequency and scale is taken into consideration when producing the figures in this section and periods in the following figures correspond to the actual periodicities. Examples of scale spectra are given in Figures 6.15 and 6.16. Figure 6.15 shows the scale spectra for a collection of random signals, while Figure 6.16 shows the scale spectra for a sum of three sinusoids of periods 3, 10, and 50 years plus white noise. It is clear from Figure 6.15 that for a random signal none of the scales is particularly prominent. For the three sinusoids in Figure 6.16, however, peaks in the scale spectrum can be easily identified at periods of approximately 3, 10, and 50 years.

Figure 6.17 shows the scale spectra for 26 USGS flow time series. Peaks in Figure 6.17 appear at approximately 6 to 8, 11, 30 years, in addition to large values at about 100 years. Consequently, components at these scales are the most prominent of all scales. These periodicities compare very well with the results from the frequency analysis in Chapter 4 (Table IV.7). The large values at about 100 years, however, must be treated with caution since the length of the time series is of that order.

Figure 6.18 shows the scale spectra for 36 HCN rainfall time series. The shapes of the spectra in Figure 6.18 are similar to those in Figure 6.15, reflecting the increased randomness of rainfall data. However, a closer look at Figure 6.18 reveals that a considerable number of peaks exist at a periodicity of about 11 years. Also some large values exist at about 100 years, but again, due to the limited extent of data this period cannot be confirmed. Also, Figure 6.19 shows the scale spectra from 37 NCDC divisional rainfall time series. In Figure 6.19 there is an indication of periods of 6 to 8, 11, and 30 to 40 years. These periods are consistent with the frequency analysis findings in Chapter 3.

Figure 6.20 shows the scale spectra for 36 HCN temperature time series. The main periods identified in Figure 6.20 are 11 and 60 to 90 years. Similarly, Figure 6.21 shows the scale spectra for 37 NCDC divisional temperature time series. The main periods identified in Figure 6.21 are 11 and 60 to 90 years, similar to the case of the HCN temperature time series. These periodicities compare well with those identified through frequency analysis in Chapter 4 in Tables IV.10 and IV.11.

Figure 6.22 shows the scale spectra for 22 NOAA tree-ring time series. The spectra in Figure 6.22 indicate a prominent scale of 20 to 30 years, in addition to a few peaks at about 90 to 120 years. The 20 to 30 years period is consistent with the MEM frequency analysis results (Table IV.6) which indicates the existence of that periodicity in 19 stations out of 22, and is also partially supported by the MTM results in Table IV.12. The 90 to 120 period was not detected by the frequency analysis.

6.5. TIME-SCALE DISTRIBUTION

Similar to the time-frequency plots in Chapter 5, coefficients obtained from the continuous wavelet transform can be represented as a time-scale plot. This time-scale plot is known as a 'scalogram' (Strang and Nguyen, 1996). Similar to the spectrogram in time-frequency analysis, the scalogram reveals changes in the relative contribution of different scales, as reflected by the absolute value of the wavelet coefficients at a given time and scale. Fig. 6.23 shows the scalogram of sinusoid with a period of 10 sampling intervals. High coefficient values at 10 years indicate the existence of such component. Figure 6.24 shows the same sinusoid with added white noise with eight times the variance of the sinusoid. The coefficients in Fig. 6.24 are disturbed but the high values are in general near 10 years.

Figure 6.25 shows the scalogram of the flow time series from USGS station 2 which is the longest of all flow time series. It can be observed that there are changes in the scale of prominent features as time progresses. This is similar to what was observed in the time-frequency plots. Similar behavior was detected for other flow time series as well. Compared to the results for a single component in Figures 6.23 and 6.24 the interaction between different scales is evident for these time series. Another interesting feature for this time series is that the scalogram for the first part from 1872 till 1930 is similar to the part from 1930 to 1980. However, the limited extent of the available records does not offer any support to the existence of such similarity in the data.

Figure 6.26 shows the scalogram of the rainfall time series from HCN station 13. In Figure 6.26 a period of around 20 years is prominent for the first part of the time series and then becomes weaker for the second part. Similar behavior was also found in other HCN rainfall time series. Figure 6.27 shows the scalogram of the time series from NCDC division 35. A period of around 40 years changes steadily to about 20 years towards the end of the record. Smaller scale components do not seem to be consistent over the length of the record.

Figures 6.28 and 6.29 show the scalograms for temperature time series from HCN station 34 and NCDC division 20, respectively. In Figure 6.28, components at 20 years drift to about 10 years at the end of the record, while the component at 10 years shows strongly only at the middle of the record. In Figure 6.29, the components at 60 to 70 years are not very stable. Also small scale components seem to be more dominant at the beginning of the record than at the end.

Figure 6.30 shows the scalogram for the tree-ring time series from station 18 in the NOAA data set. It can be observed in Figure 6.30 that the role of different scales changes as time changes. Periods of around 60 years at the beginning and end of the record are not as strong in the middle part of the record, where periods of 80 to 100 years are more prominent. Also periods of around 30 years appear to be mostly strong in the middle of the record but not at the beginning or end of the record.

6.6. BEHAVIOR OF HYDROLOGIC AND CLIMATIC TIME SERIES AT DIFFERENT SCALES

It was shown in section 6.3 that the behavior of the largest scale details (trend components) of the studied hydrologic and climatic time series are more or less similar among all stations in the region for a given category, i.e. flow, rainfall, or temperature. By using wavelet analysis to decompose a noisy signal into its details, smaller scale details will capture most of the noise, while larger scale details will suffer from less distortion by the noise. Ivanov et al. (1996), used wavelet decomposition to show that, when suitably rescaled, variations in heartbeat intervals in healthy subjects can be described by a single distribution which is stable over a wide range of time scales. In this section, hydrologic and climatic time series are investigated at different scales for the purpose of identifying similarities between different time series at different scales. For this purpose, deseasonalized monthly data are used which offer a sufficient number of observations to construct histograms of the details of time series at different scales. In this case the scale in years for a detail at a given level j will be equal to $2^j / 12$. For example, a detail at level 7 represents a scale of approximately 10.66 years, while a detail at level 4 represents a scale of 1.33 years. The only exception is tree-ring data which are yearly data. In this case the scale in years corresponding to level j is equal to 2^j, i.e level 2 represents a scale of 4 years, and so on. To construct a histogram of a detail at a given scale, the data points for that scale are first standardized by subtracting the mean and dividing by the standard deviation. The standardized values are then grouped into 40 equal intervals and the frequency for each group is calculated. The frequency values are finally rescaled to obtain a unit area histogram which can be compared to the frequency distribution function of any standardized distribution.

Figure 6.31 shows the histograms of time series from each of the data groups used in this study. The histograms for the time series are shown in Figure 6.31 as scatter plots, while the thick curves represent the density function of the standard Normal distribution which is used here as a reference distribution. It can be observed in Figure 6.31 that both the flow and rainfall time series are positively skewed. Also, flow, rainfall, and tree-ring time series seem to have larger peaks than the standard Normal distribution (smaller coefficient of kurtosis). Temperature data, on the other hand seem to be more or less symmetric and normally distributed. For all data series the large scatter of points around their average indicates differences in the individual distributions for each time series, which are most pronounced in tree-ring time series (Figure 6.31(b)).

Figure 6.32 shows the distribution of the details for USGS flow time series at levels 1, 3, 5, and 7, corresponding to scales of 1 month, 8 months, 2.67 years, and 10.67 years. It can be observed from Figure 6.32 that the smallest scale detail at level 1 has the least scatter of all details, indicating the similarity between distributions of noise at all stations. It is also evident that this component is responsible for the increased peakedness of the density function. The scatter of the distributions increases as the scale increases. Figure 6.33 shows the superimposed details for USGS flow time series at levels 7, 8 , 9, and 10, corresponding to scales of 10.67, 21.33, 42.67, and 85.33 years, respectively. Details at these levels are consistent between all time series in the region, but the consistency decreases as the level decreases. What can be concluded from this behavior is that both large and small components of the different time series are similar. Similarity in large scales stems from their shape, while similarity in small scales is due to similarity in their distribution. Intermediate scales such as level 7, however, are less similar in shape and distribution, which gives rise to differences in the overall time series.

Figure 6.34 shows the distribution of the details for HCN rainfall time series at levels 1, 3, 5, and 7. The component at level 1 is the most consistent among all time series as indicated by the limited scatter of points. This component is also normally distributed, as opposed to the first component for the flow time series, which is more peaked than the normal distribution. Such change in the distribution between rainfall and flow is expected due to the nonlinear mechanisms involved in the generation of runoff from rainfall. The scatter of the distribution increases as the scale increases indicating changes in distribution. Figure 6.35 shows the superimposed details for the HCN rainfall time series at levels 7, 8, 9, and 10. Similarity between details is best at level 8 and deteriorates in both directions afterwards. Figure 6.36 shows the distribution of the details for NCDC divisional rainfall time series at levels 1, 3, 5, and 7. The behavior of components at these scales is similar to that of the HCN components in Figure 6.34 and the scatter increases with the scale of details. Figure 6.37 shows the superimposed details for the NCDC divisional rainfall time series at levels 7, 8, 9, and 10. As in Figure 6.35, the best agreement of details in Figure 6.37 is found at level 8, while components at level 7 are still in good agreement. Details at levels 9 and 10 are similar for the period from 1950 to the end of the record. The conclusion for rainfall data is that larger components are not very consistent between different stations, while smaller details are similar as far as their statistical distribution is concerned. This is to be expected given the weak autocorrelations in rainfall time series, which makes structures at smaller scales more dominant in the time series.

Figure 6.38 shows the distribution of the details for the HCN temperature time series at levels 1, 3, 5, and 7. Details at level 1 are the most consistent among different stations, while consistency in distribution decreases as the scale increases. Figure 6.39, on the other hand, shows the superimposed details for the HCN temperature time series at levels 7, 8, 9, and 10. The details at these levels are more consistent than flow or rainfall time series. Figure 6.40 shows the distribution of the details for NCDC divisional temperature data.. The most consistent details are at level 1, while scatter increases as scale increases. One interesting feature in Figure 6.40 is that the detail at level 7 is positively skewed which is not found at any other level. Level 7 corresponds to a scale of 10.66 years, which may suggest a relation to the sunspot effect on the temperature time series, an effect which is evident from frequency analysis in Chapter 4 (Tables IV.4, IV.5, IV.10, and IV.11). However, no such effect was detected in the HCN temperature data in Figure 6.38. Figure 6.41 shows the superimposed details for the NCDC divisional temperature data at levels 7, 8, 9, and 10. The details are consistent at levels 7, 9, and 10, while at level 8 the second part from 1950 to the end of the record suffers from discrepancies. The conclusion for temperature time series is that smaller scales are similar in distribution while large scales are similar in form.

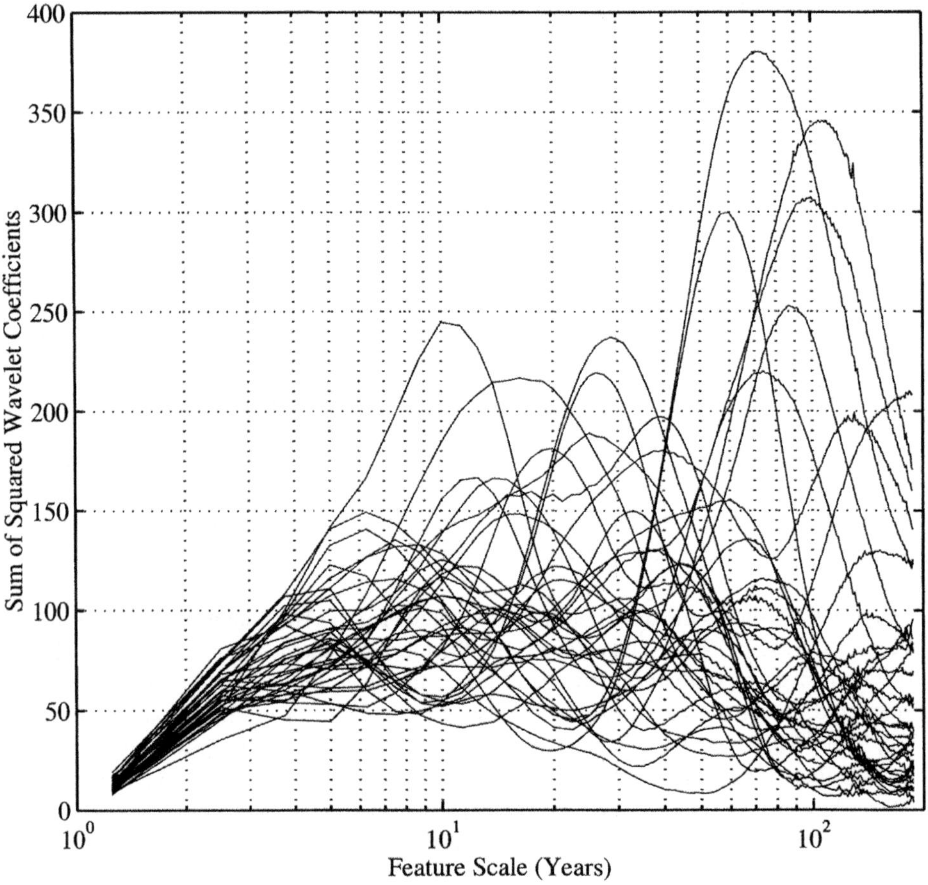

Figure 6.15. Scale spectra of 36 random time sequences.

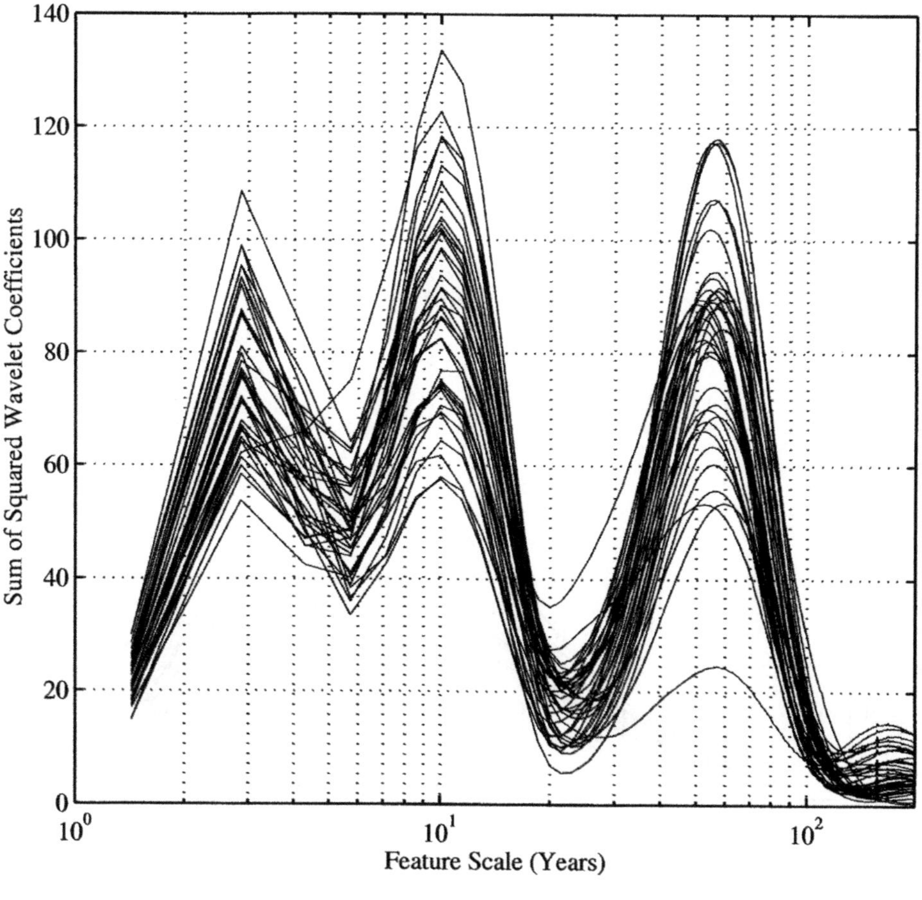

Figure 6.16. Scale spectra of a time series consisting of three sinusoids at periods 3, 10, and 50 years with added white noise.

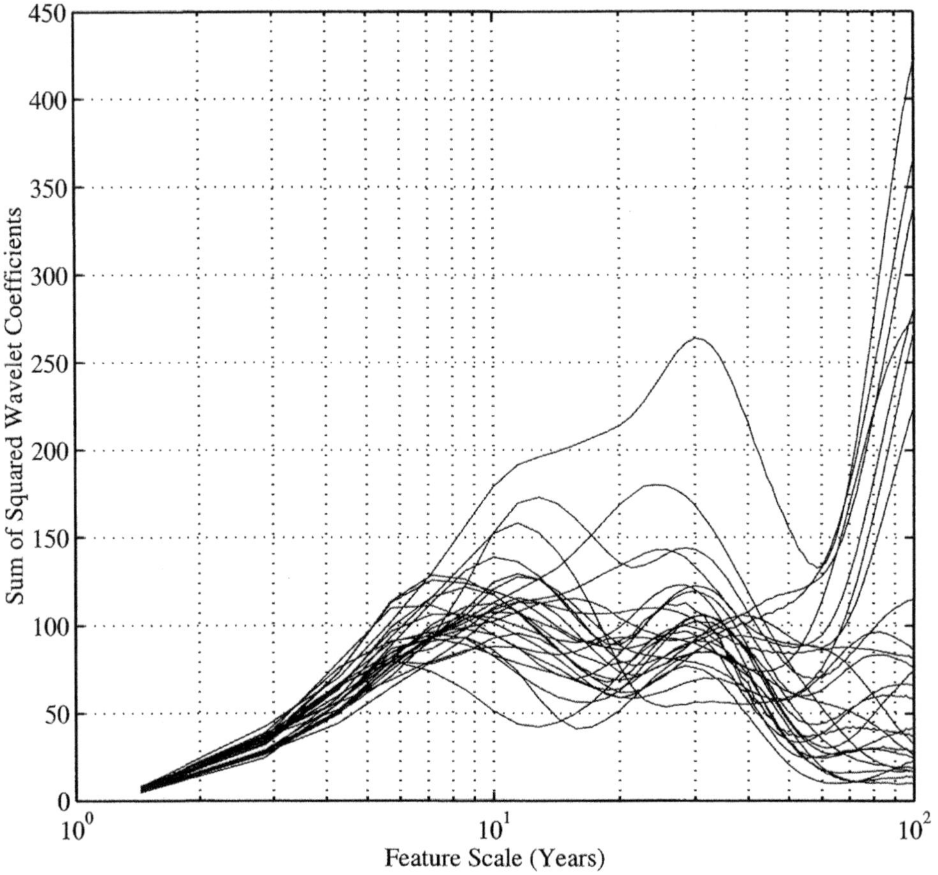

Figure 6.17. Scale spectra for 26 USGS flow time series.

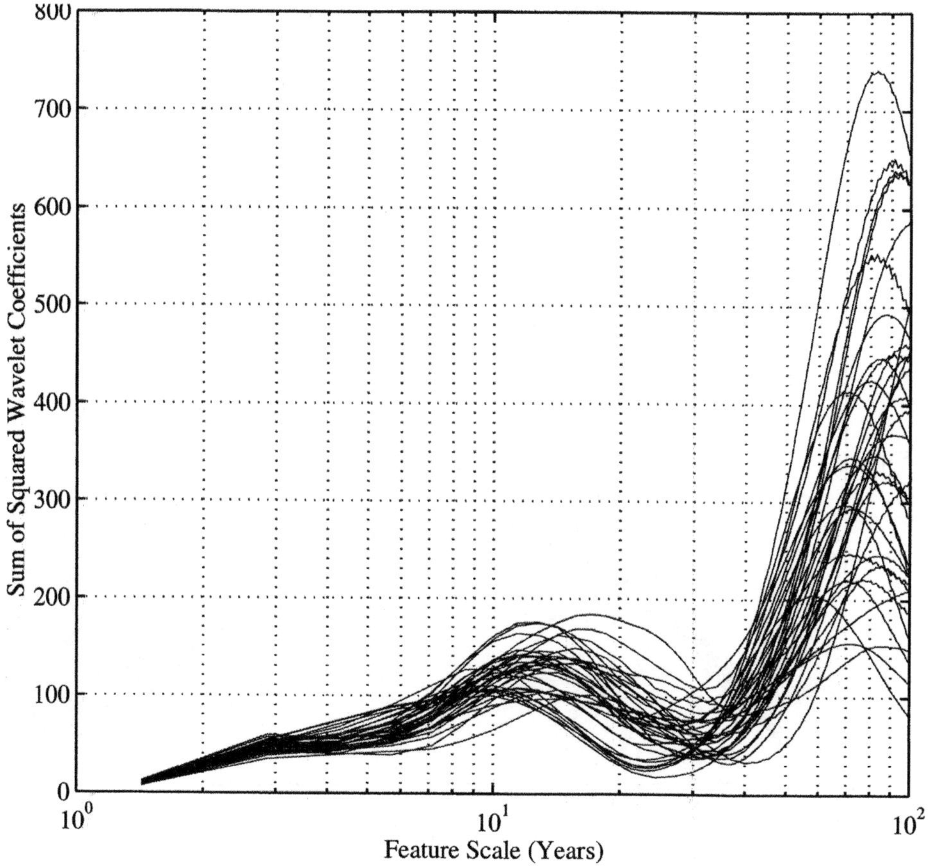

Figure 6.18. Scale spectra for 36 HCN rainfall time series.

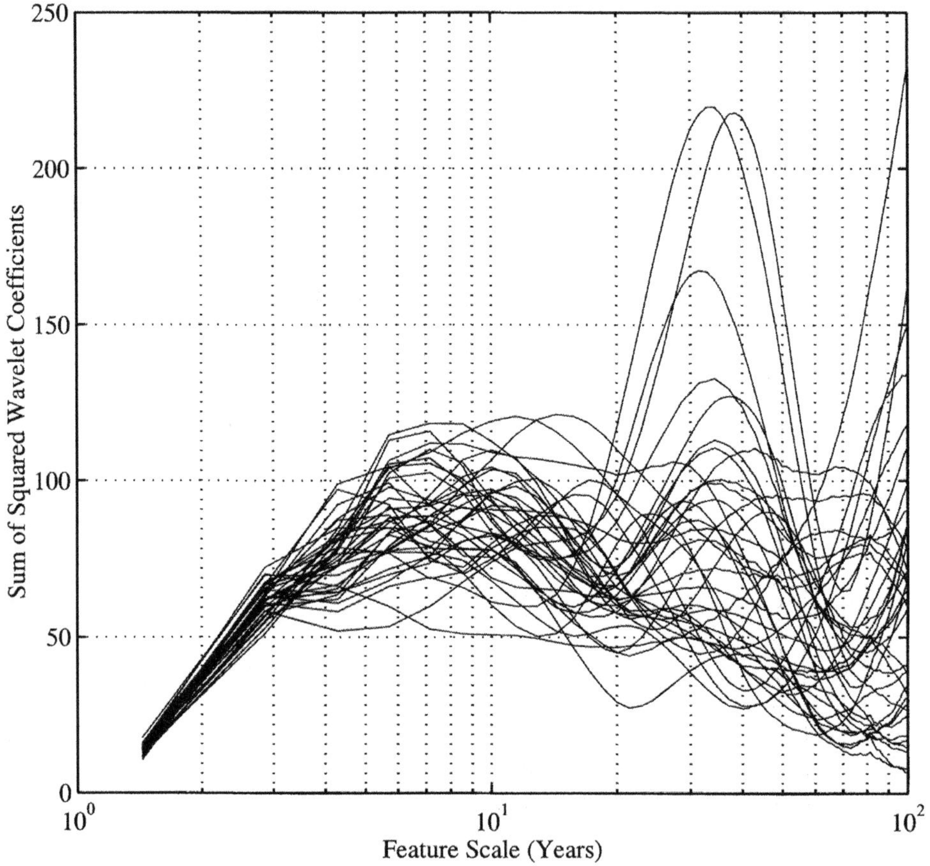

Figure 6.19. Scale spectra for 37 NCDC divisional rainfall time series.

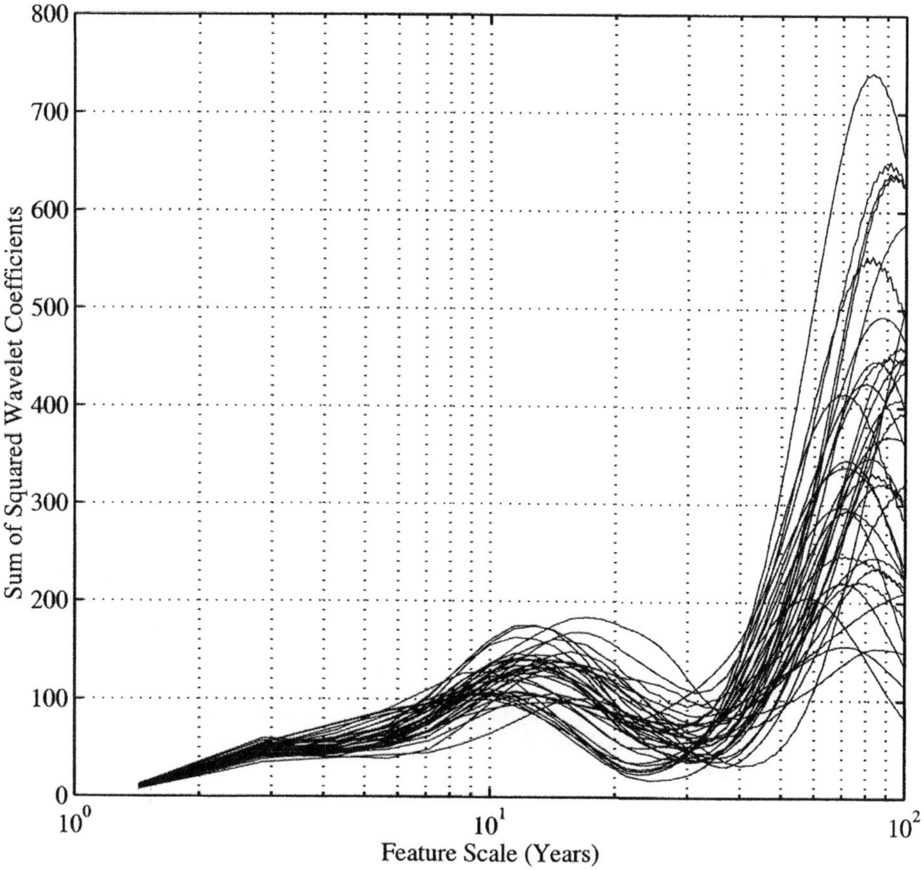

Figure 6.20. Scale spectra for 36 HCN temperature time series.

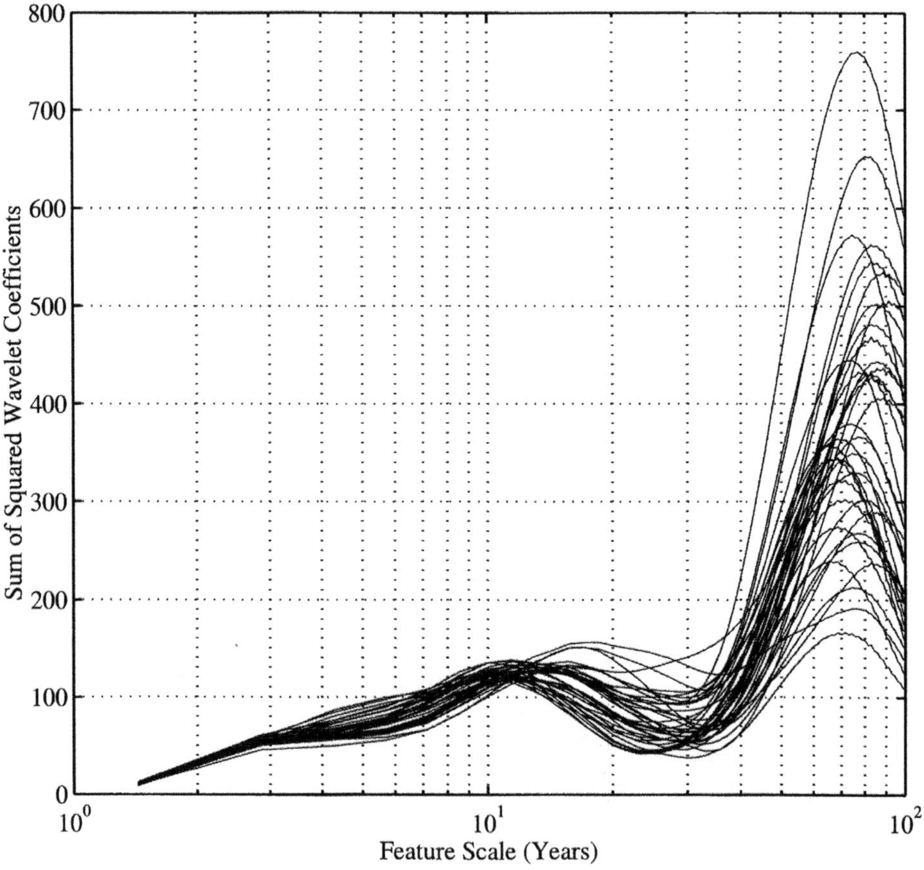

Figure 6.21. Scale spectra for 37 NCDC divisional temperature time series.

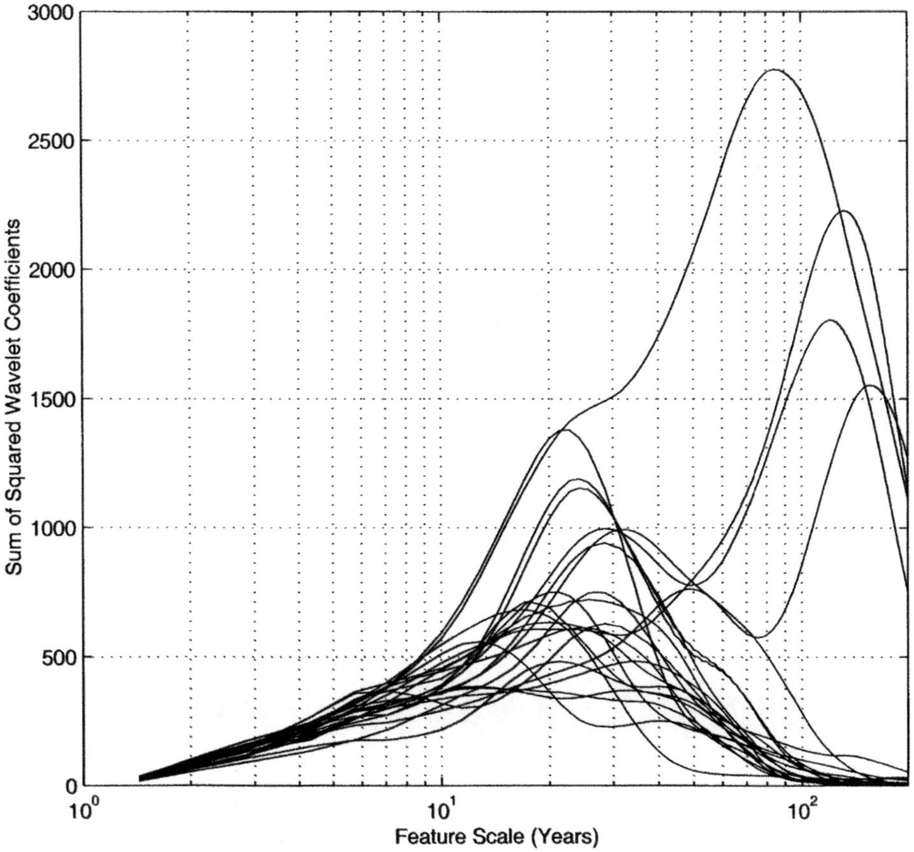

Figure 6.22. Scale spectra for 22 NOAA tree-ring time series.

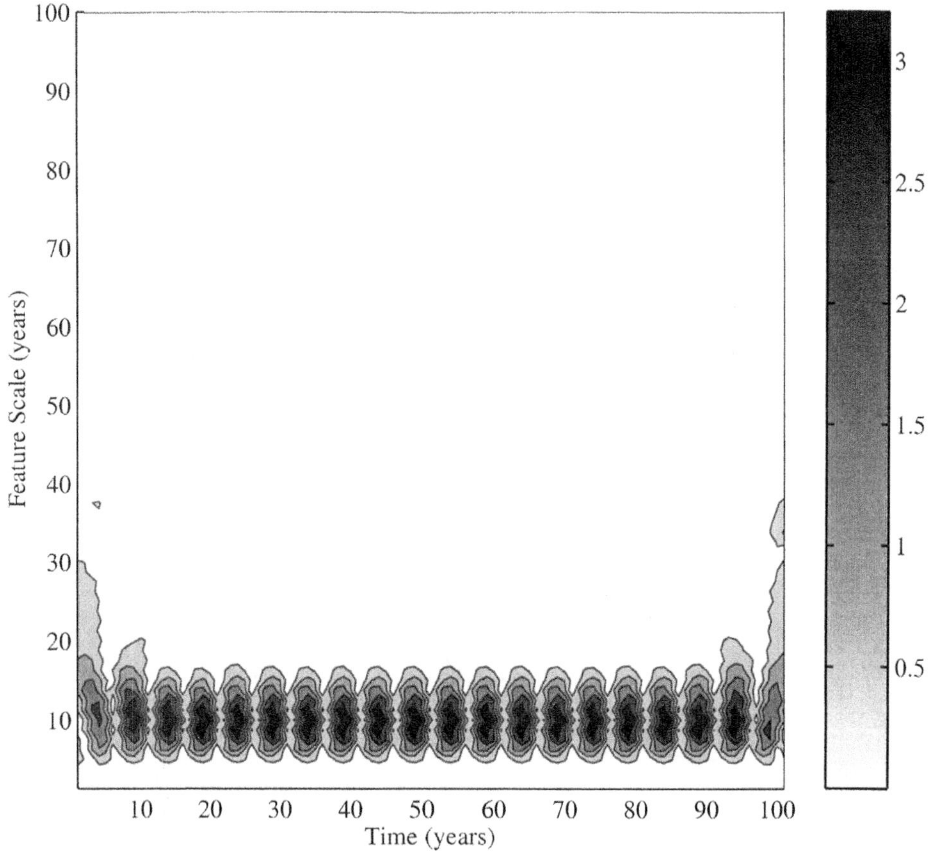

Figure 6.23. Scalogram of a pure sinusoid with period 10 years.

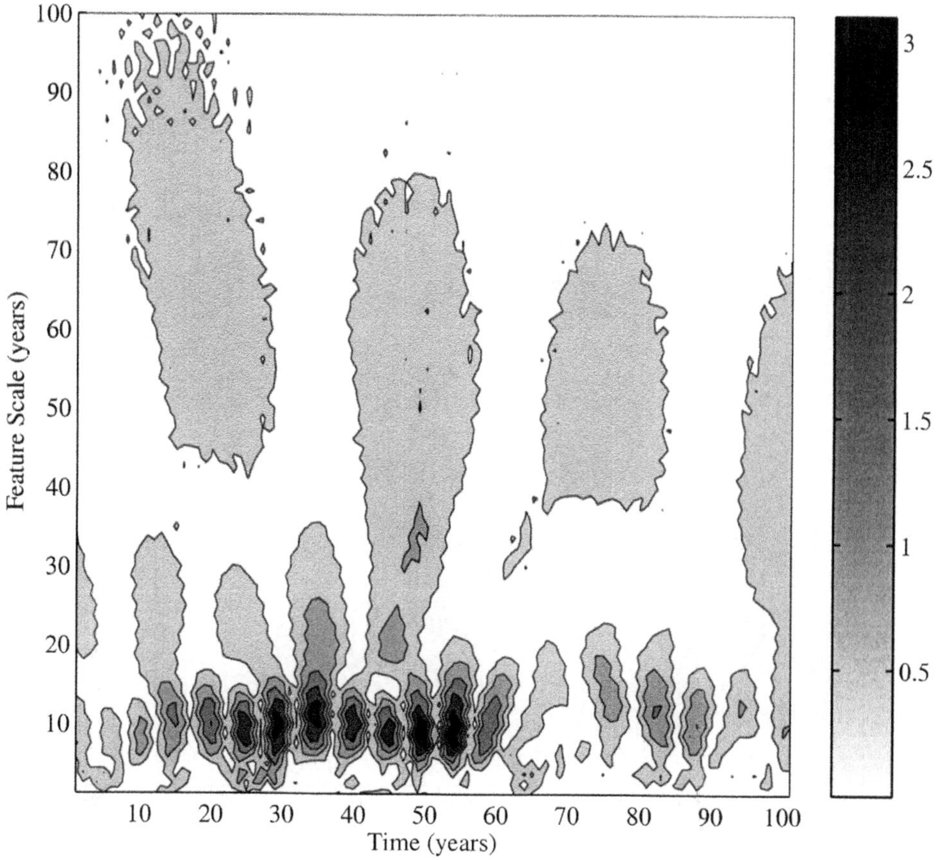

Figure 6.24. Scalogram of a pure sinusoid with period 10 years with added noise.

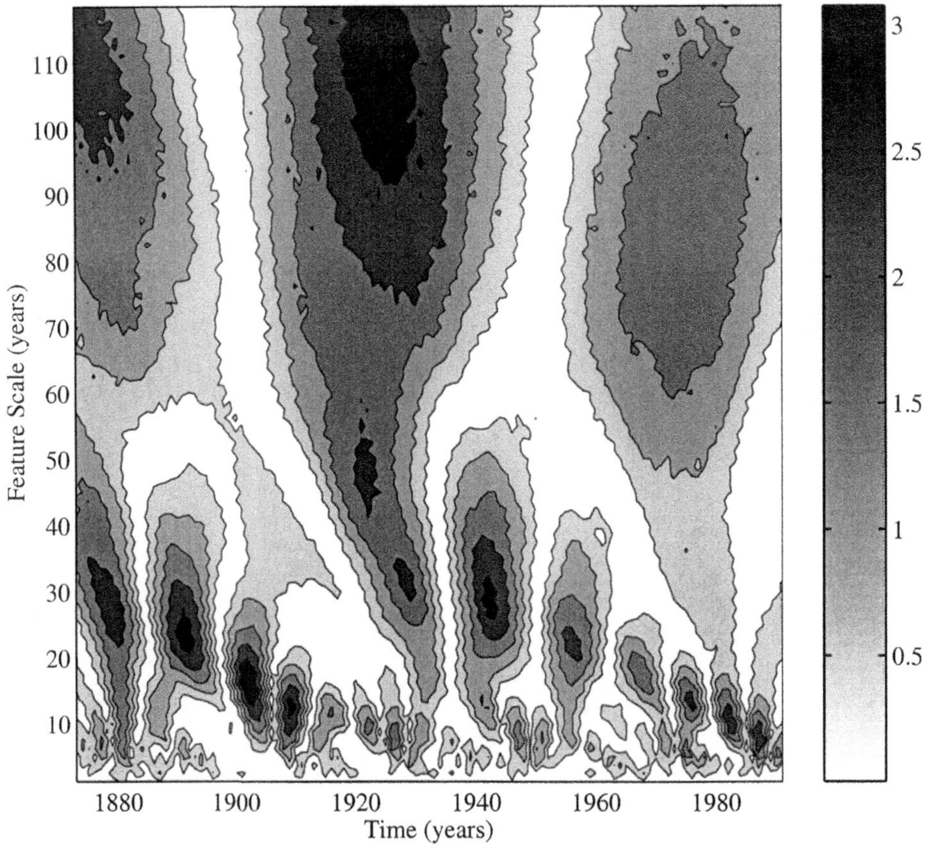

Figure 6.25. Scalogram of the flow time series from USGS station 2.

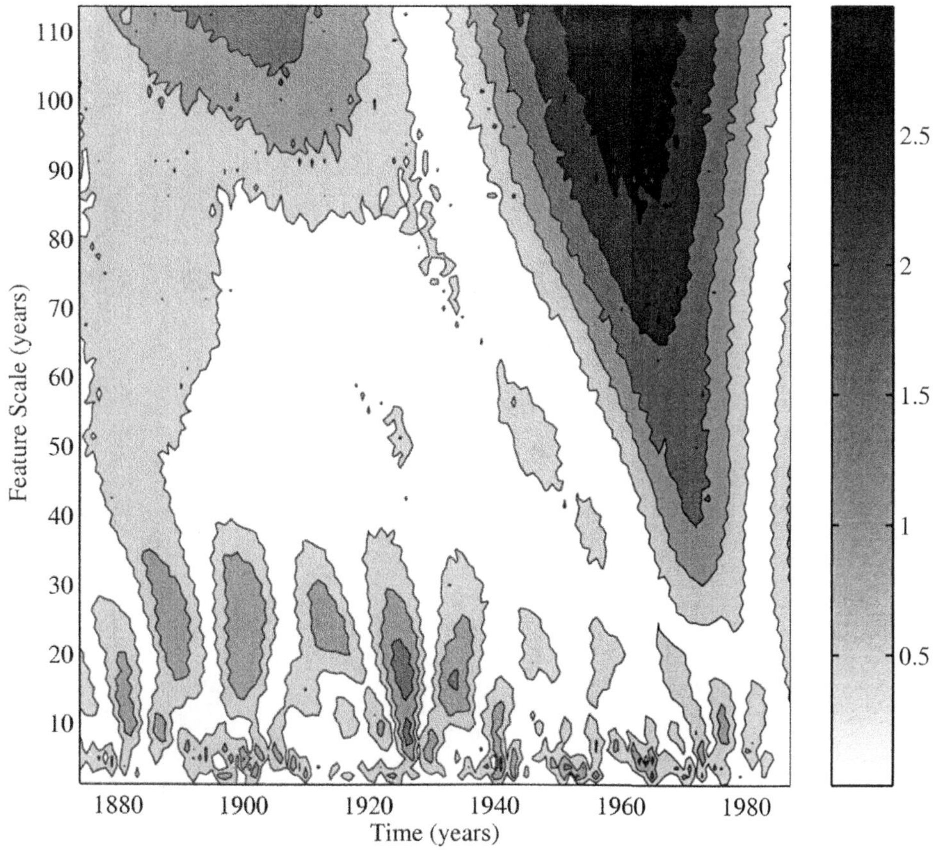

Figure 6.26. Scalogram of the rainfall time series from HCN station 13.

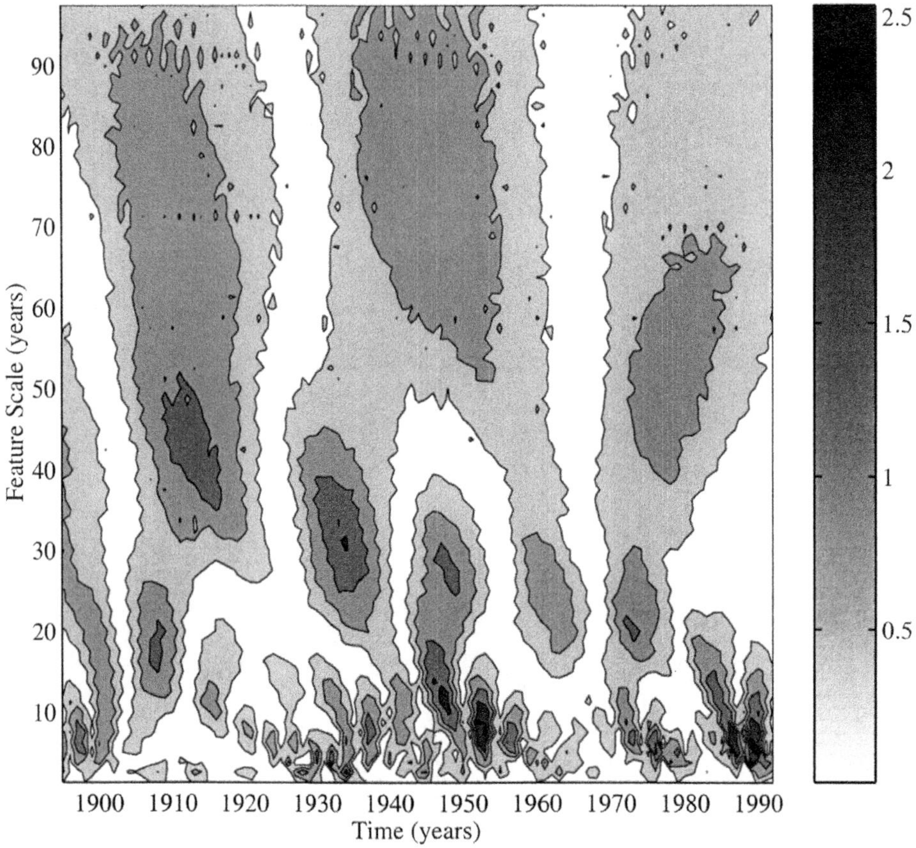

Figure 6.27. Scalogram of the rainfall time series from NCDC division 35.

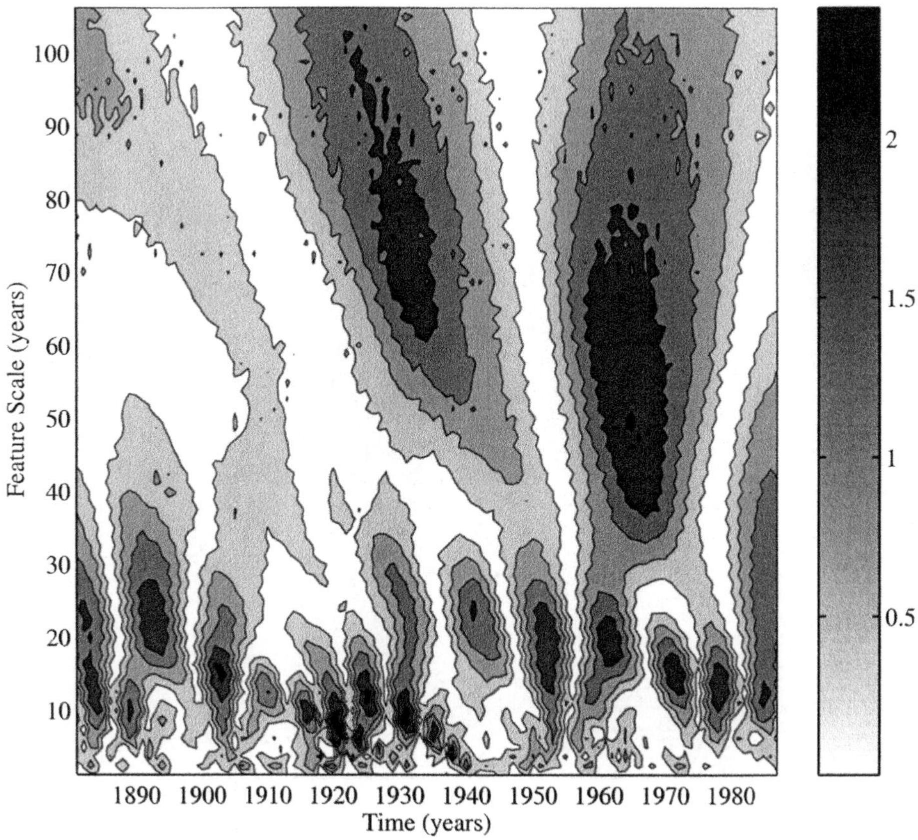

Figure 6.28. Scalogram of the temperature time series from HCN station 34.

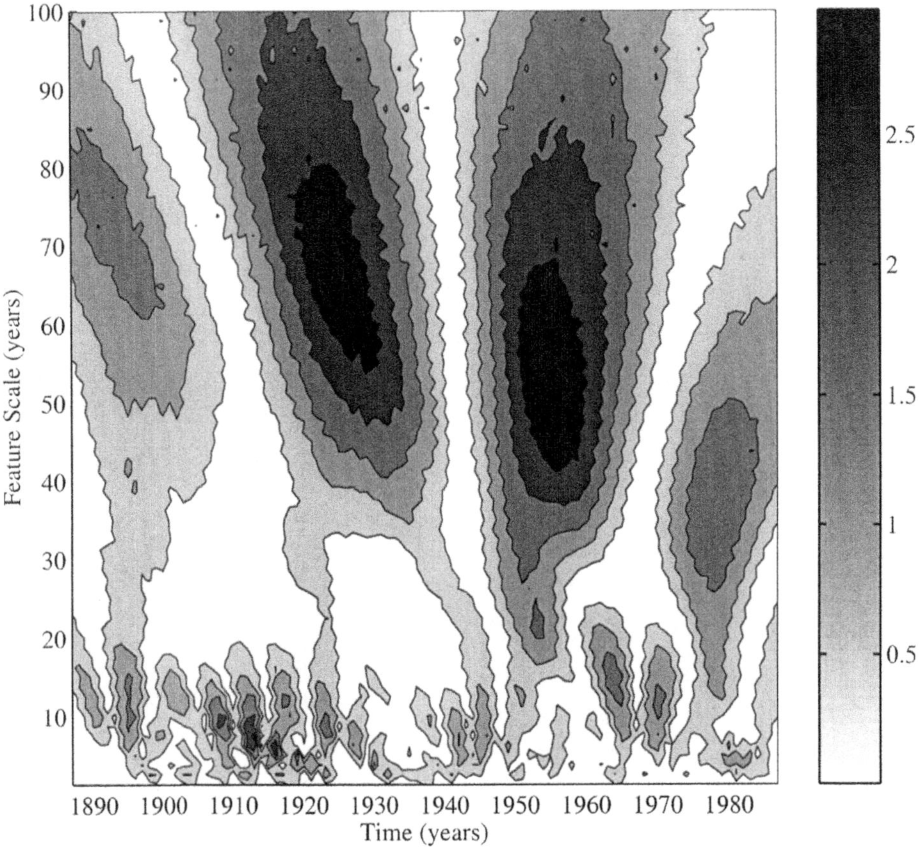

Figure 6.29. Scalogram of the temperature time series from NCDC division 20.

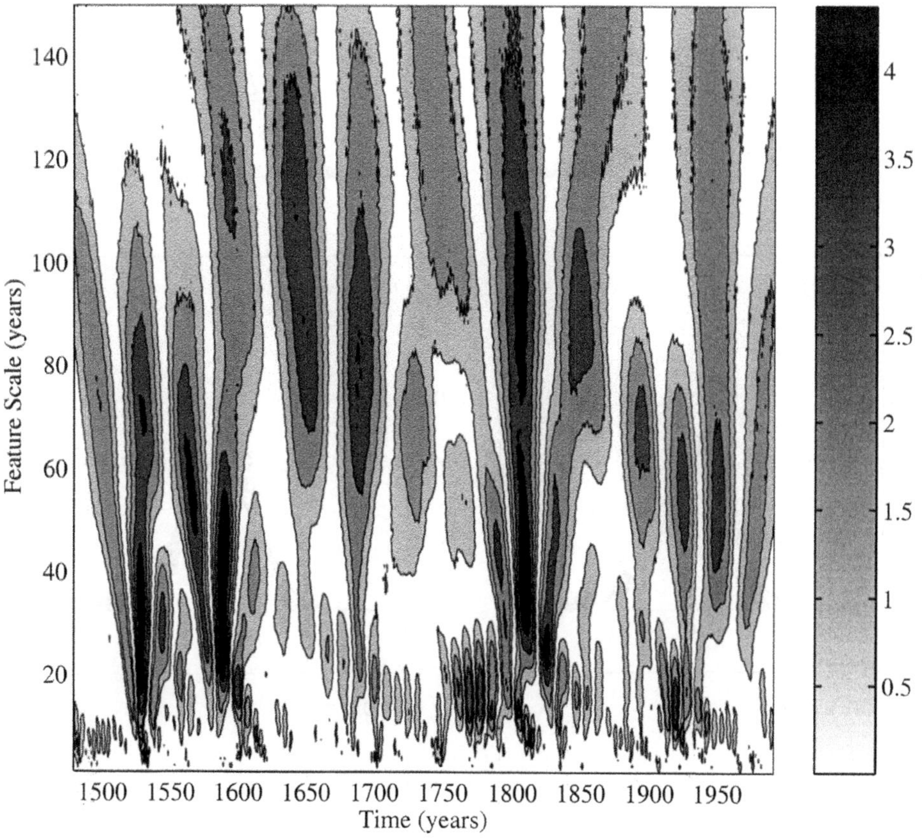

Figure 6.30. Scalogram of the tree-ring time series from NOAA location 18.

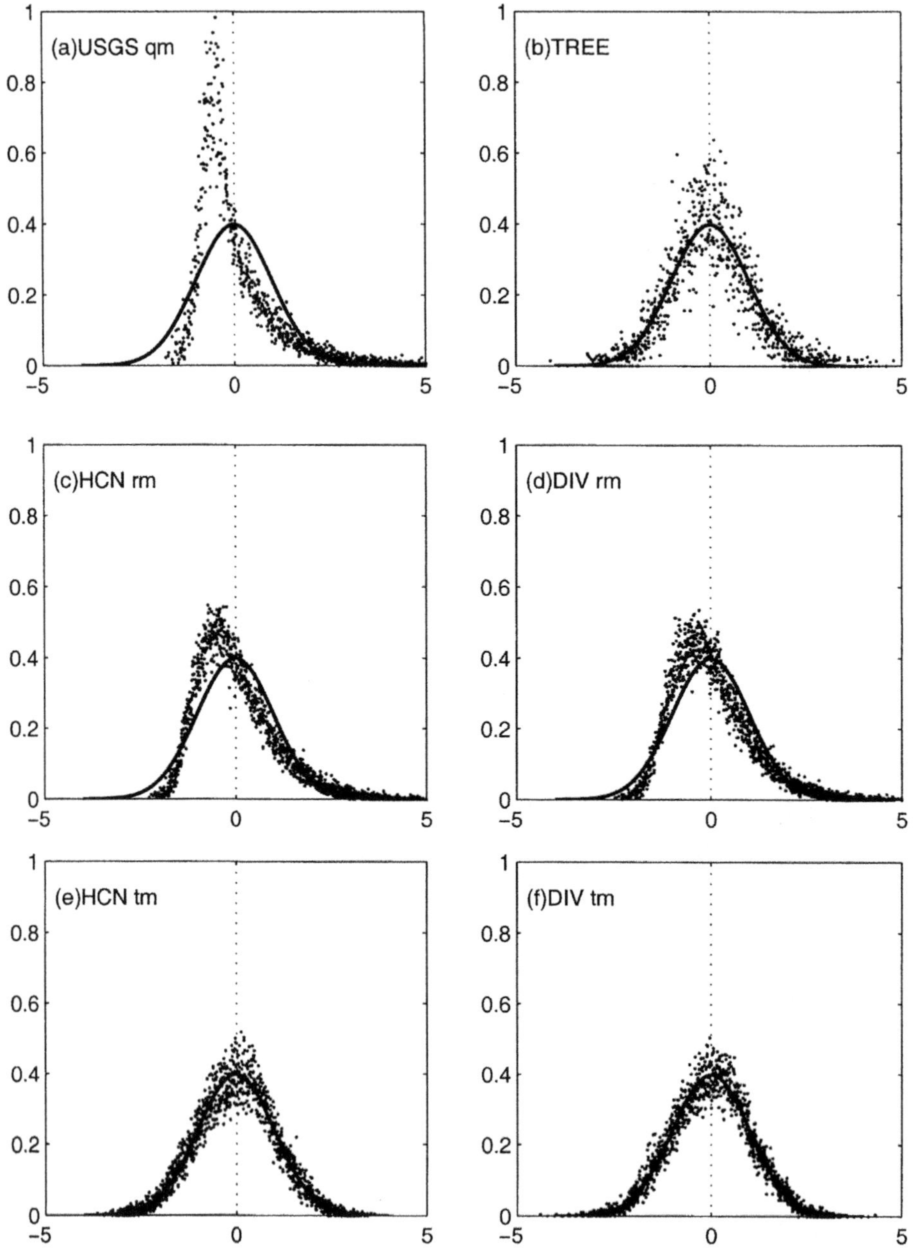

Figure 6.31. Histograms of the data used in this study.

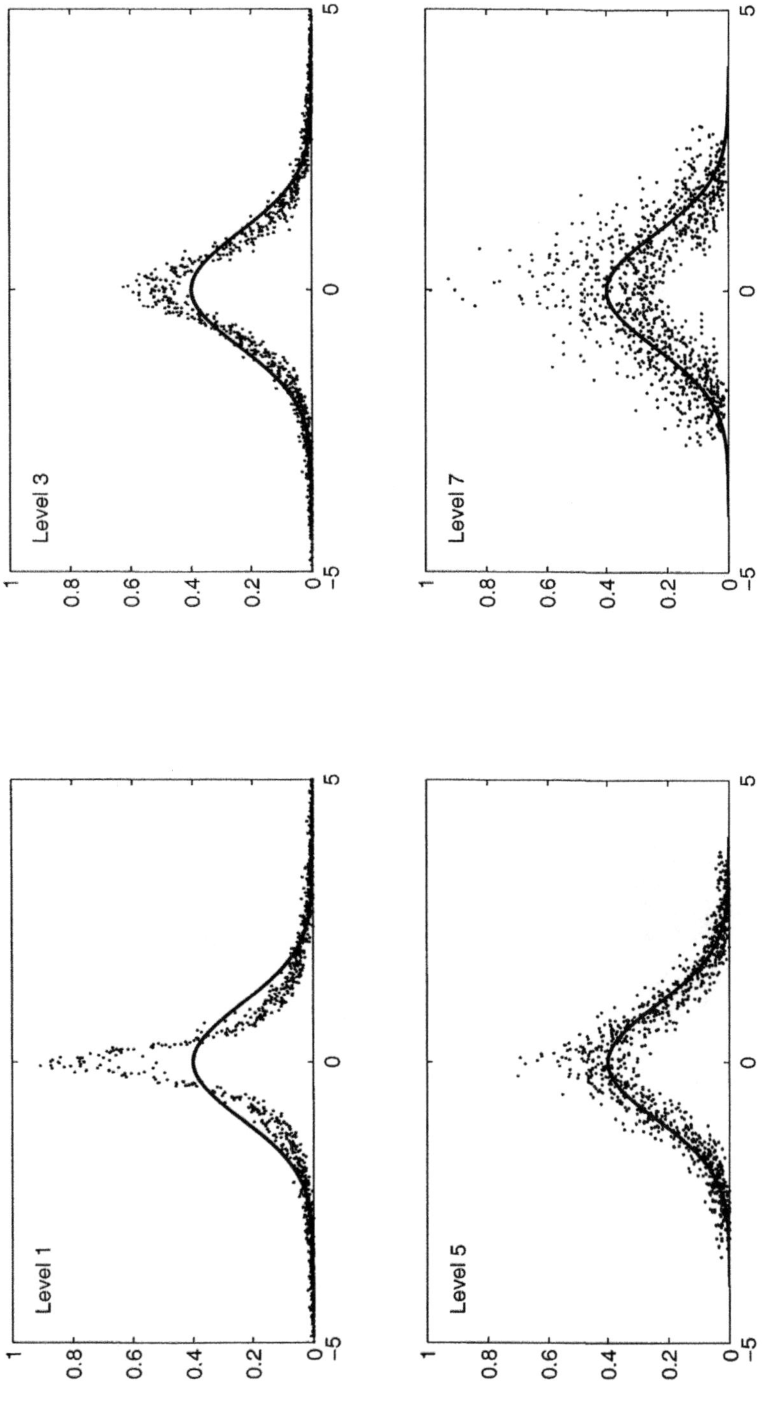

Figure 6.32. Histograms of the details of USGS flow time series at levels 1, 3, 5 and 7.

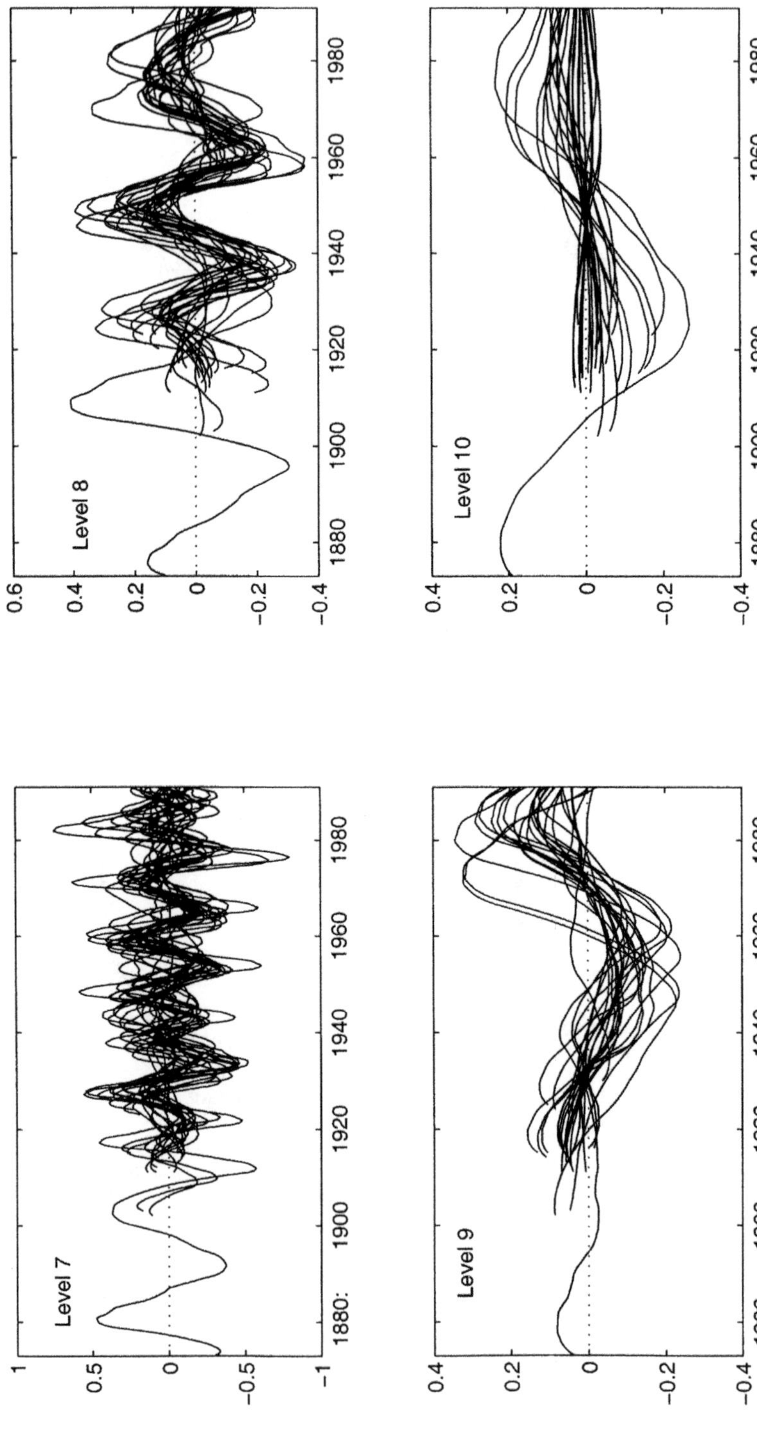

Figure 6.33. Superimposed details of USGS flow time series at levels 7 to 10.

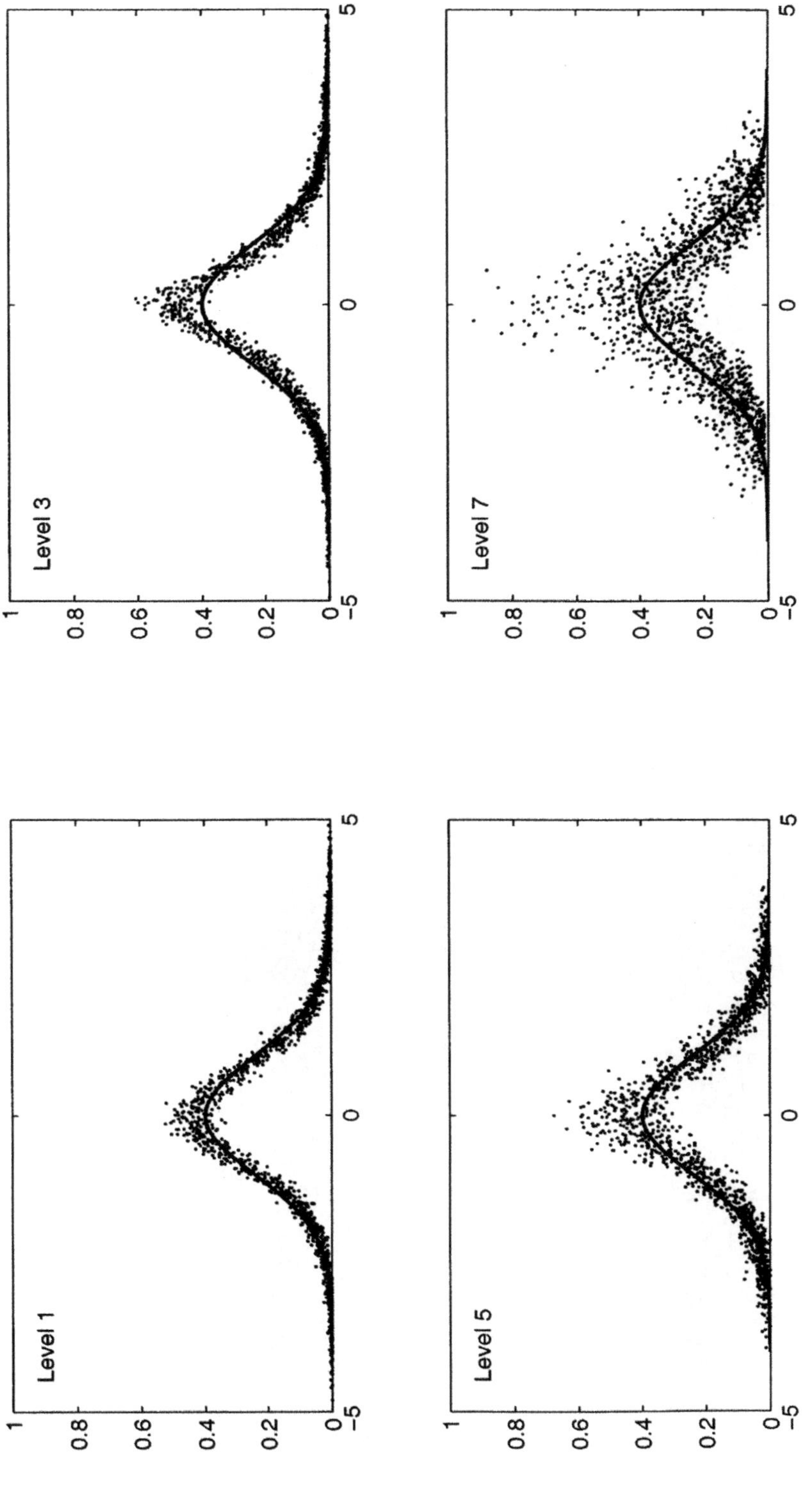

Figure 6.34. Histograms of the details of HCN rainfall time series at levels 1, 3, 5, and 7.

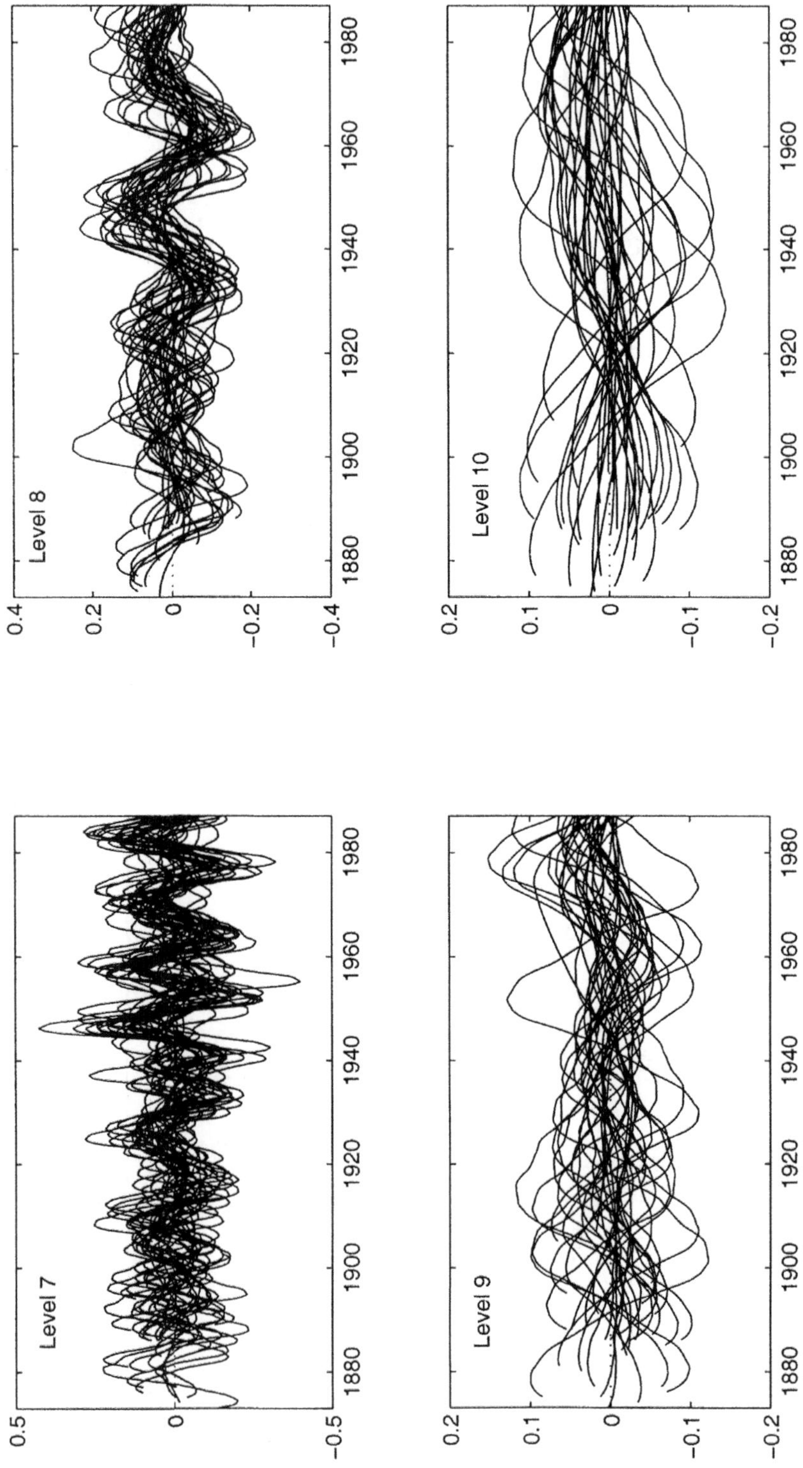

Figure 6.35. Superimposed details of HCN rainfall time series at levels 7 to 10.

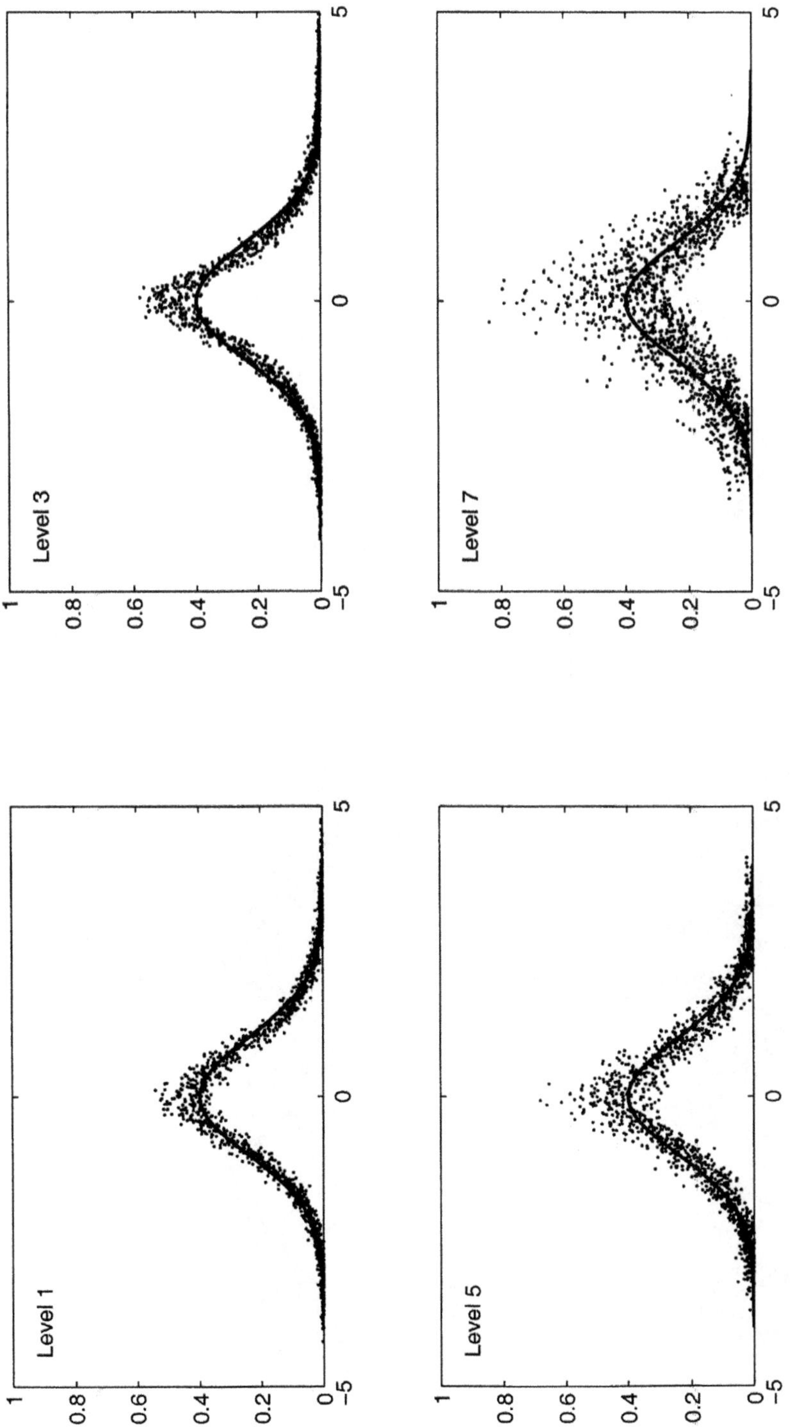

Figure 6.36. Histograms of the details of NCDC divisional rainfall time series at levels 1, 3, 5 and 7.

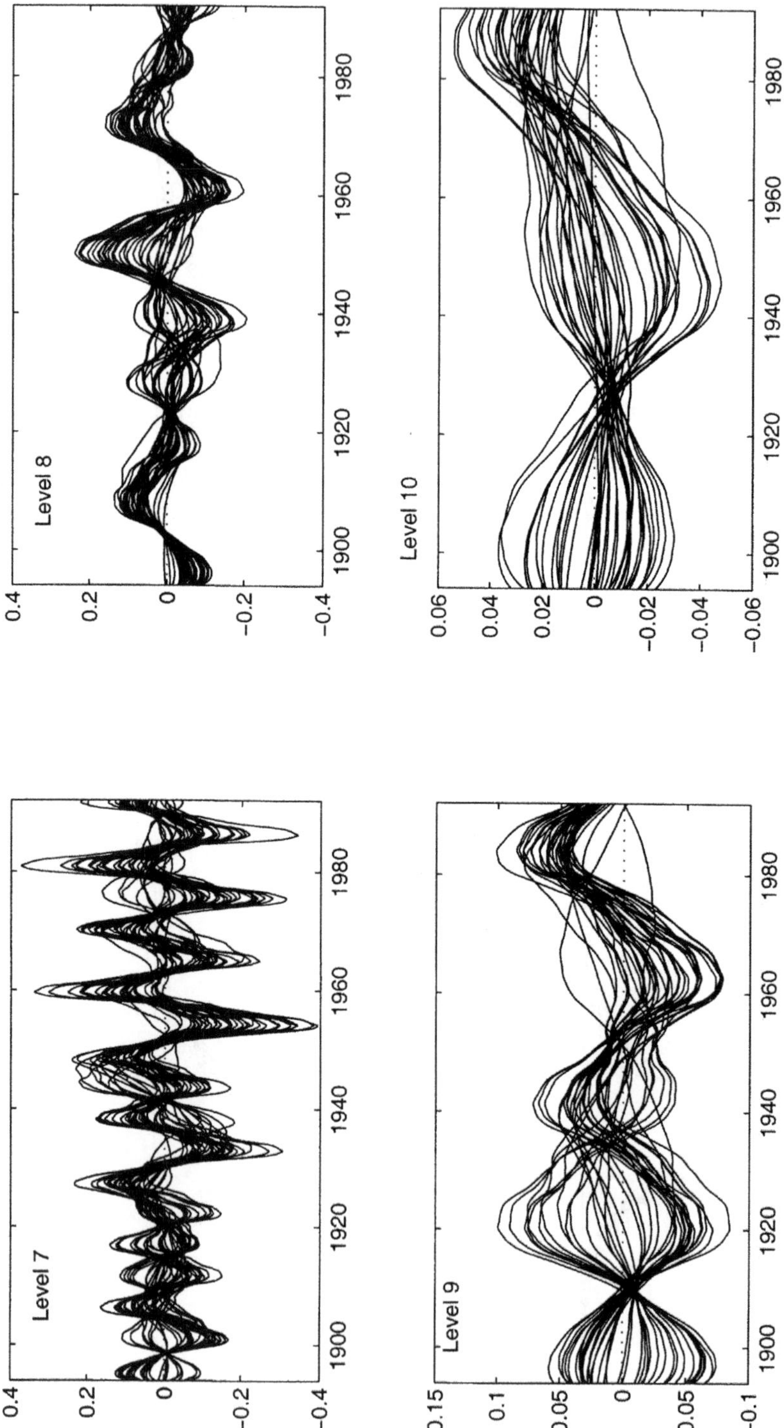

Figure 6.37. Superimposed details of NCDC divisional rainfall time series at levels 7 to 10.

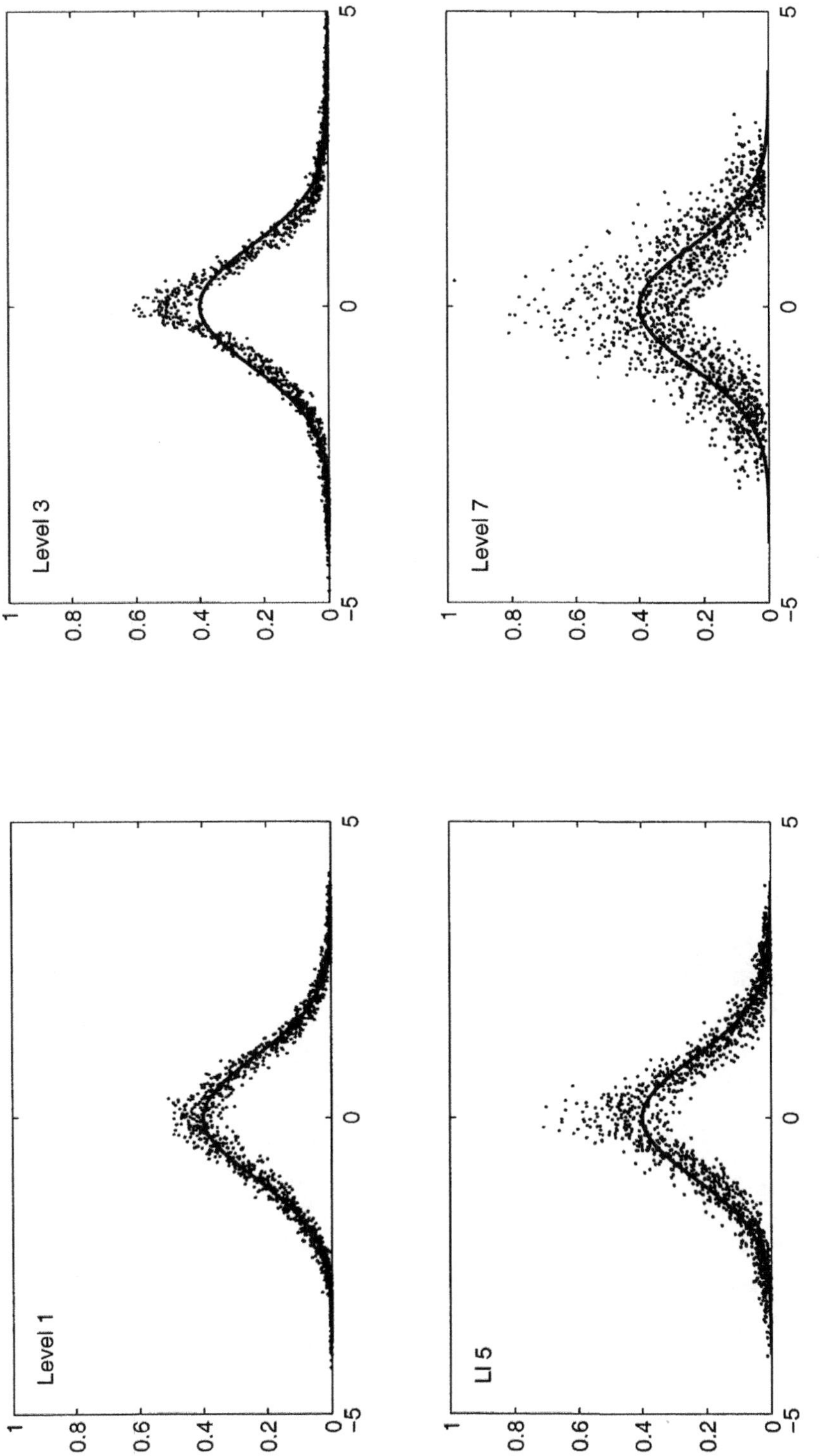

Figure 6.38. Histograms of the details of HCN temperature time series at levels 1, 3, 5 and 7.

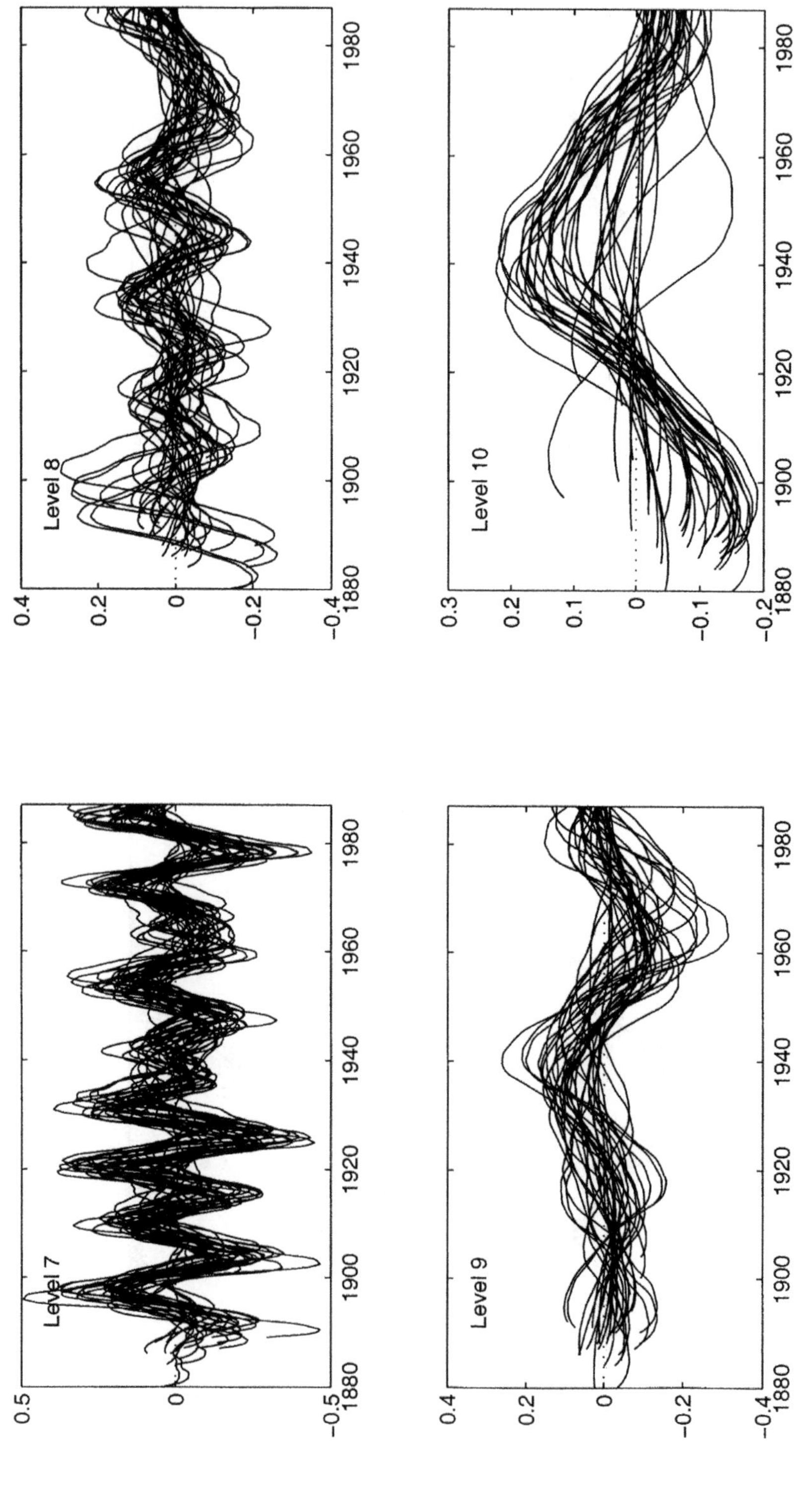

Figure 6.39. Superimposed details of HCN temperature time series at levels 7 to 10.

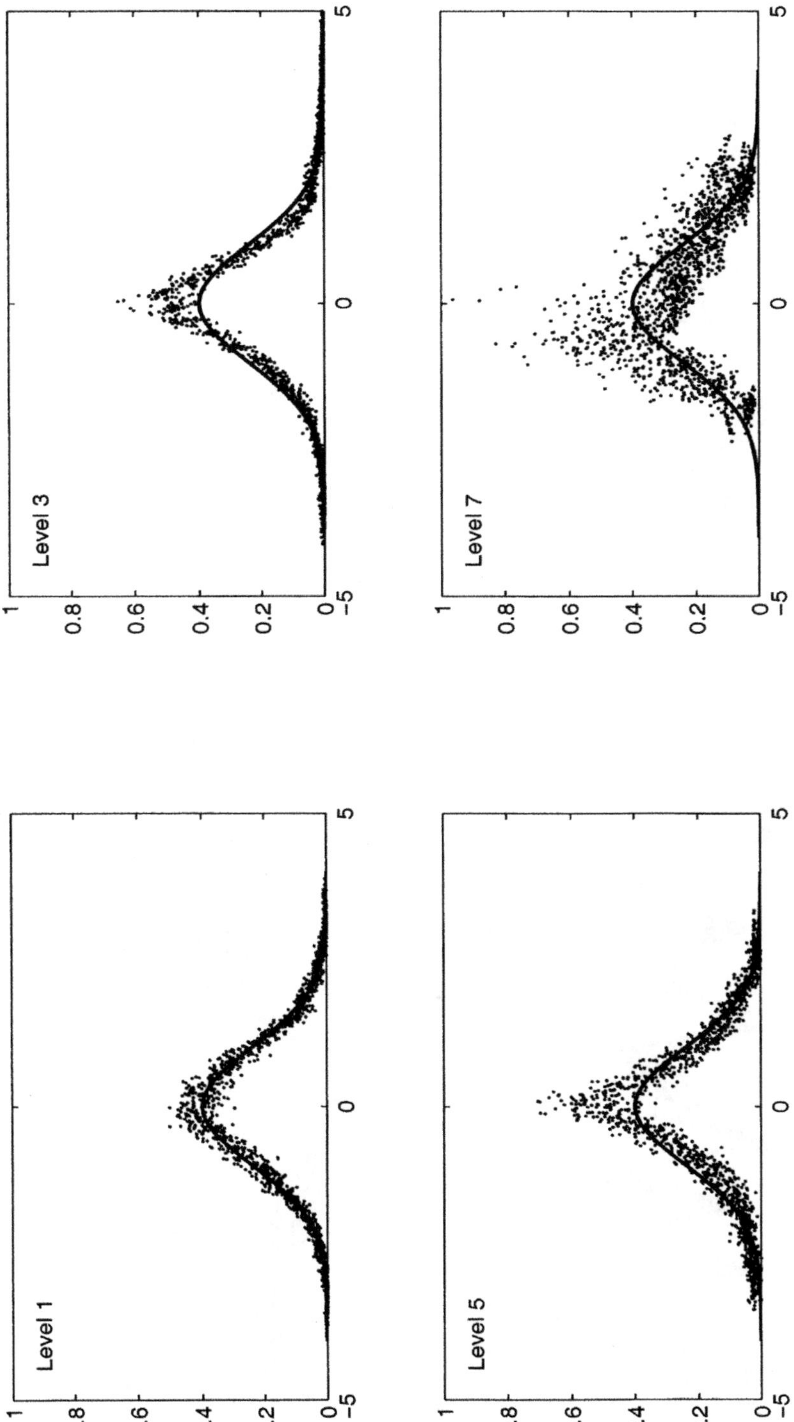

Figure 6.40. Histograms of the details of NCDC divisional temperature time series at levels 1, 3, 5 and 7.

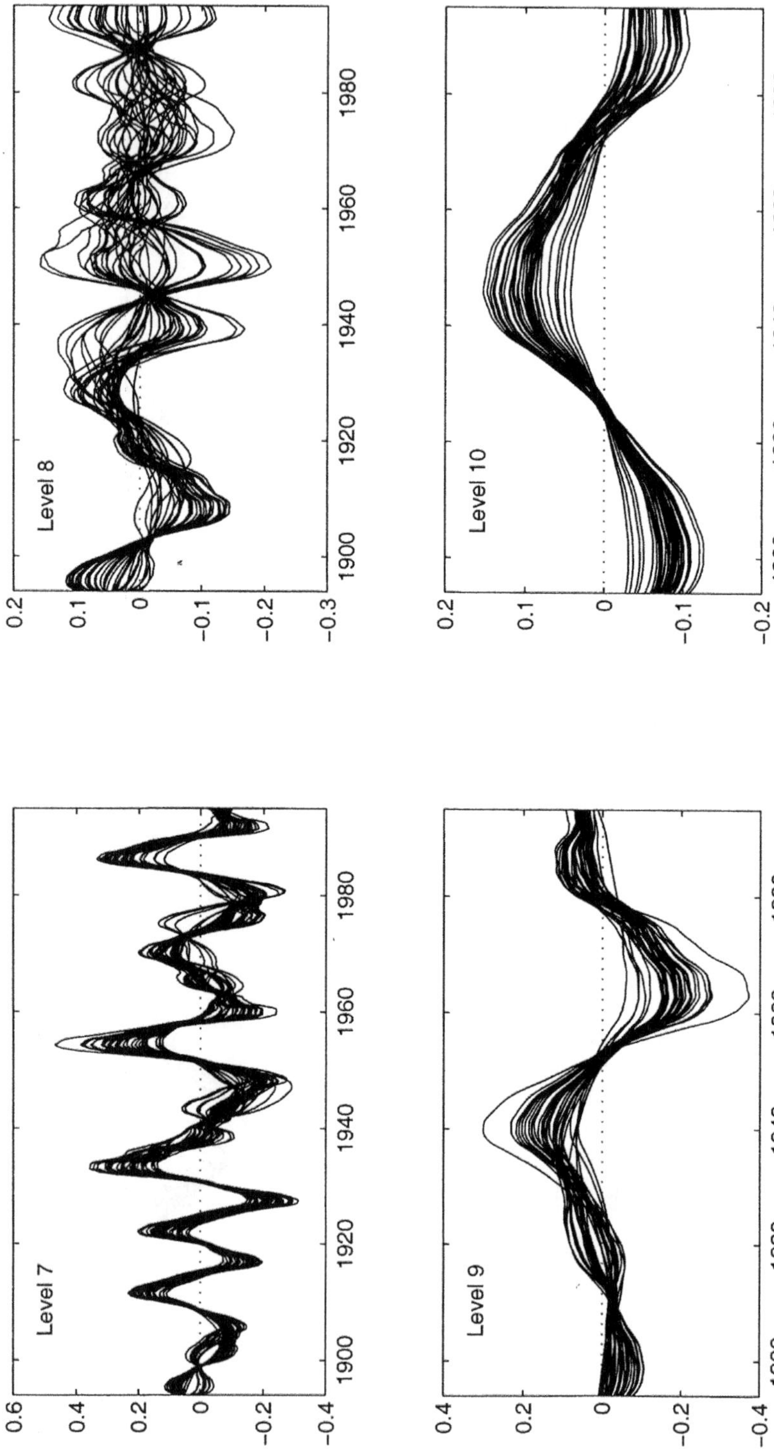

Figure 6.41. Superimposed details of NCDC divisional temperature time series at levels 7 to 10.

Figure 6.42 shows the distribution of the details for the NOAA tree-ring time series at levels 1, 3, 5, and 7. It can be noticed that the distribution of the details are not consistent at any scale, even the smallest scale which was found consistent in flow, rainfall, and temperature time series. Similarly, the superimposed details at levels 7, 8, 9, and 10 in Figure 6.43 are far from similar. This would indicate that the time series are not similar at any time scale, although these time series come from the same region. These results raise doubts about the consistency of tree-ring time series and their suitability as proxy to hydrologic time series.

6.7. CONCLUSIONS

Based on wavelet analysis of hydrologic and climatic time series, the following conclusions are made:

- The study of the slowest changing components in the time series obtained by multi-resolution decomposition of different time series indicate the existence of trends in the analyzed time series. The direction of trends agrees to a large extent with the results obtained from statistical trend analysis in the time domain in Chapter 3.

- By using the time integrated wavelet coefficients at different scales, the scales of dominant components were identified. The time periods corresponding to these scales were also found comparable to those found through conventional harmonic analysis of the analyzed time series in Chapter 4. Wavelet analysis confirmed the existence of such components at these scales.

- The study of the time-scale plots obtained by applying the continuous wavelet transform to the time series revealed interaction between components of different scales, as indicated by the change in the values of the wavelet coefficients over time for various components. These results are comparable to those observed in the time-frequency plots in Chapter 5.

The study of the distribution of details of the time series at different levels also offered some information about the structure of hydrologic and climatic time series. The following conclusions can be made based on the results of that analysis:

- Temperature time series seem to be the most consistent time series in this region. Large scale details for these time series agree to a large extent. Smaller scale details also agree in their statistical distribution.

- Flow time series are consistent over large and small scales. However, intermediate scales suffer from inconsistency between different time series. This may be related to the fact that while smaller scales represent added noise and larger scales may represent regional effects, intermediate scales are mostly affected by differences in the shapes and topography of watersheds producing these flow time series.

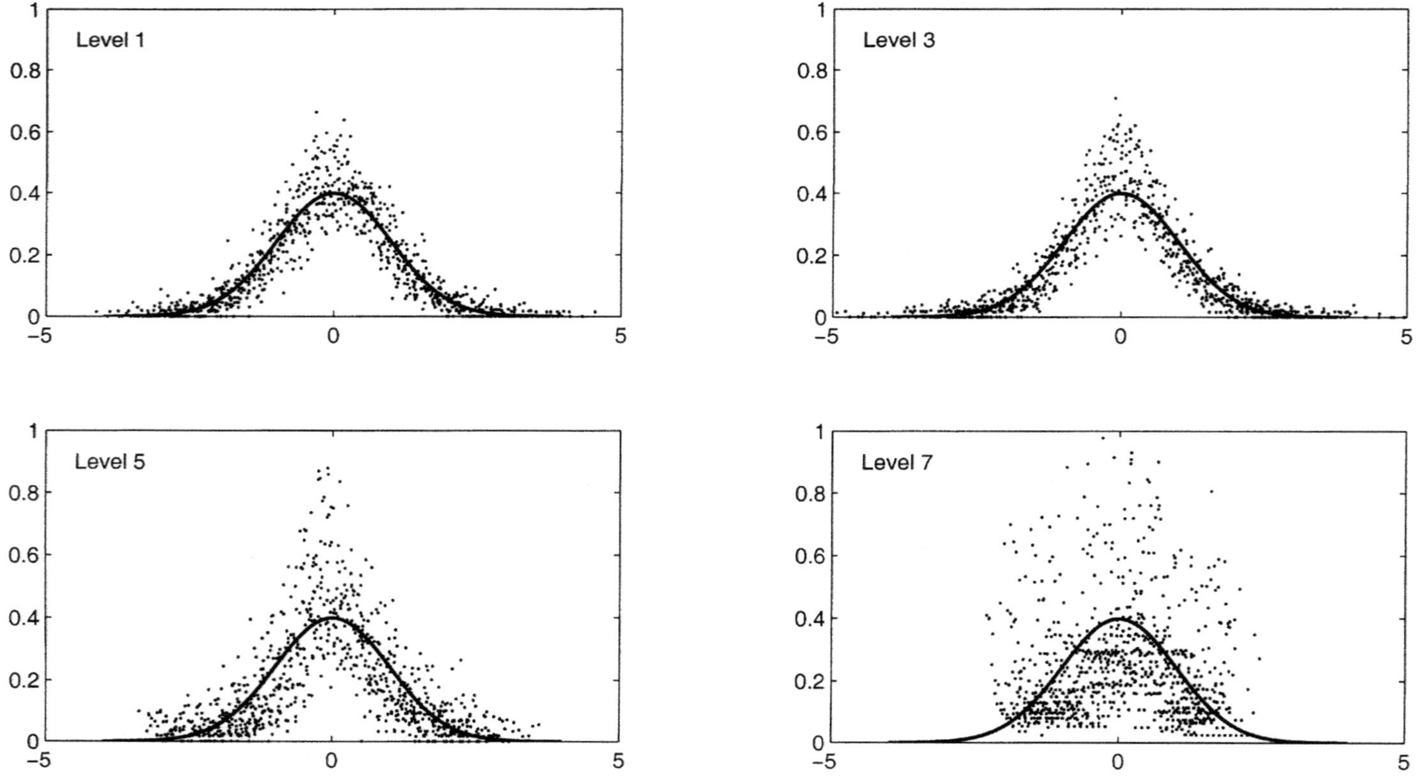

Figure 6.42. Histograms of the details of NOAA tree-ring time series at levels 1, 2, 3 and 4.

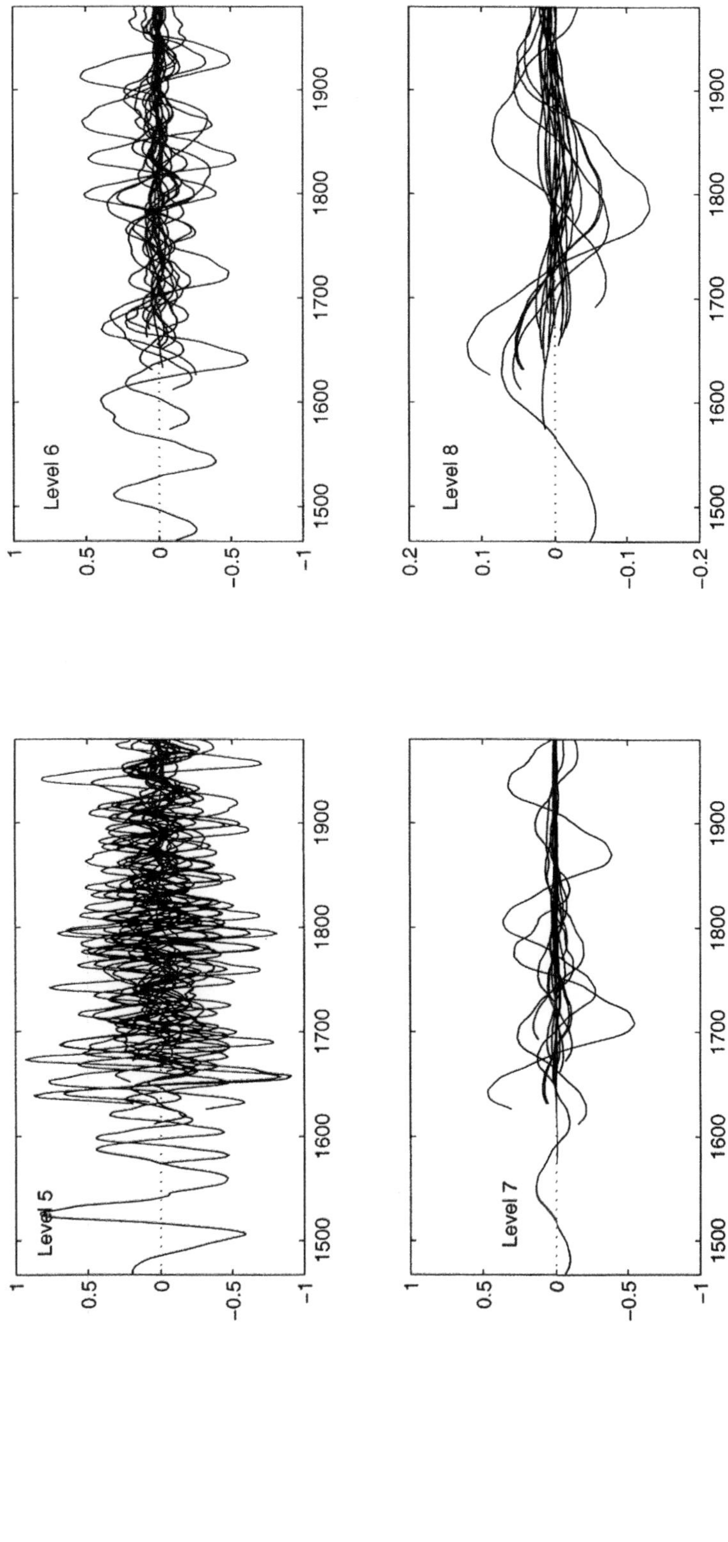

Figure 6.43. Superimposed details of NOAA tree-ring time series at levels 5 to 8.

- Rainfall time series are not in general similar at large scale. Small scale features, however, are similarly distributed. For rainfall time series this may be expected since they usually have less structure than flow time series.

- Tree-ring time series are inconsistent over all scales. This may be due to the nature of these time series which are influenced by temperature, rainfall, and other factors.

7. Segmentation of Non-Stationary Time Series

7.1. INTRODUCTION

In many practical situations, a model of piecewise-stationary time series with successive stationary segments is assumed for nonstationary series (Fig. 7.1). In order to establish this piecewise-stationary time series model, segmentation algorithms have been developed to detect segment boundaries and estimate the parameters characterizing each segment. Techniques of segmentation without requiring *a priori* spectral information have been developed based on autoregressive (AR) models. It is assumed in these techniques that statistical properties described by a set of AR parameters remain the same in each segment. If these algorithms yield a single series – the original series – then the series is stationary.

Several tests have been proposed for use in segmentation algorithms. de Souza and Thomson (1982) derived a test based on a likelihood ratio statistic, which is insensitive to changes in the signal energy but sensitive to changes in spectral shape. This test statistic has an asymptotic chi-square distribution with p degrees of freedom under the null hypothesis of no changes in the AR parameters. Appel and Brandt (1983) transformed a likelihood ratio to a measure of the information loss caused by the assumption that the null hypothesis is true. Under the null hypothesis of no changes in the AR parameters and variance, the threshold is empirically adjusted according to the particular application for which the segmentation procedure is designed. Because the threshold of the distance is empirically determined, it is difficult to analyze it analytically. Lovell and Boashash (1987) adopted the segmentation procedure of Appel

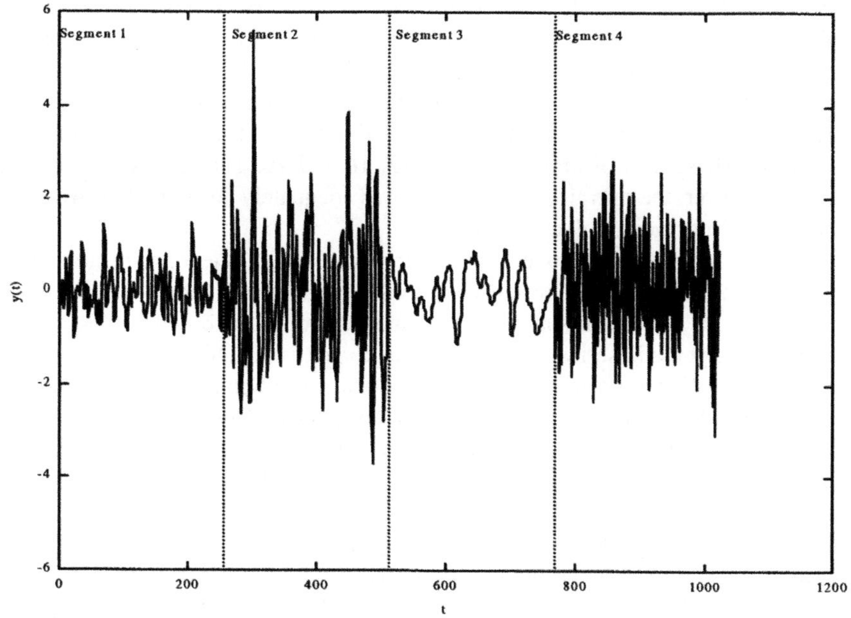

Figure 7.1. A piecewise-stationary time series with successive stationary segments.

and Brandt (1983), but replaced the test with the one of de Souza and Thomson (1982) instead. Since they maintain that variations in the signal energy are largely due to the measuring instrument, the segmentation algorithm of Lovell and Boashash (1987) is designed to determine segments according to changes in the spectral shape alone. Because the test derived by de Souza and Thomson (1982) is insensitive to changes in the signal energy, it is questionable whether the resulting segments from Lovell and Boashash's (1987) method are still stationary.

The algorithm used in Appel and Brandt's (1983) segmentation procedure consists of two stages: boundary detection and boundary optimization. In the first stage, a preliminary boundary is detected after the inherent change point; and the optimal boundary is then searched in the second stage. A disadvantage of this algorithm is that both boundary detection and optimization procedures are not symmetric with respect to time, which may give different results from the same time series in reverse order. Imberger and Ivey (1991) defined a distance derivative, a simple and symmetric segmentation technique suggested in part by personal communication with de Souza. The distance derivative is computed by taking the derivative of the test statistic derived by de Souza and Thomson (1982). A threshold is required for the segmentation algorithm so that stationary segments are determined where values of the distance derivative are below this threshold. However, the value of threshold is empirically adjusted rather than statistically determined. Besides, the distance derivative is based on the test statistic of de Souza and Thomson (1982) which is insensitive to changes in the signal energy.

There are three important parameters in these segmentation algorithms based on AR models: (1) the threshold for the test used in the algorithm; (2) the minimum starting segment length; and (3) the AR order p. For selecting the threshold, a tradeoff has to be considered between true and false detection rates. The minimum starting segment length is suggested by Appel and Brandt (1983) to be less than 70% of the length of the smallest segment and greater than $(p/3)^2$, in order to obtain small false detection rates. Imberger and Ivey (1991) compared estimated Batchelor spectra (defined in Section 2.2) with the AR spectra of order $p = 4$, 8, 12, 16 and 19. The results favored rather large lags; however, the overall spectral shape was adequately developed by AR models of order $p = 4$ (Imberger and Ivey, 1991).

Several tests based on AR models have been proposed for detecting spectral changes in time series, although the performances of these tests have not been evaluated. In this chapter, the results of comparison of performances of four tests based on AR models is presented first. Synthetic series are used for the comparison. The tests include the first test which is sensitive to changes in the spectral shape (de Souza and Thomson, 1982), the second test which is a derivative of the first test statistic (Imberger and Ivey, 1991), the third test which is designed to detect changes in the AR parameters (Davis, Huang and Yao, 1995), and the fourth test which is modified from Tsay (1988) to detect changes in the AR variance.

Tests based on AR models require specifying the AR order p which is unknown in observed data. In order to avoid the task of determining the AR order, a non-parametric test based on wavelet analysis is proposed. Based on the orthogonality and the property that at a given scale the correlation among wavelet coefficients is negligible (Wornell, 1996), the proposed test is derived for detecting changes in the wavelet variance. The performance of this non-parametric test based on wavelet analysis is compared with those based on AR models and the results are presented.

A modified segmentation algorithm consisting of three stages, including boundary detection and two stages of boundary optimization is presented. The first stage of this algorithm is a modified form of the boundary detection procedures of Appel and Brandt (1983). In the two stages of boundary optimization, a symmetric approach is modified from the segmentation algorithm of Imberger and Ivey (1991). In general, a preliminary boundary is detected in the first stage, the optimal ending boundary is determined in the second stage (boundary optimization one), and the optimal starting boundary is determined in the third stage (boundary optimization two).

This chapter is organized as follows: four tests based on AR models, including test 1 by de Souza and Thomson (1982), test 2 by Imberger and Ivey (1991), test 3 by Davis, Huang and Yao (1995) and test 4 modified from Tsay (1987), are defined in Section 7.2; a test based on wavelet analysis (test 5) is proposed in Section 7.3; a modified segmentation algorithm is presented in Section 7.4; two types of synthetic series are used in Section 7.5 to examine variations of the test statistics of tests 1, 3 and 4 based on AR models with respect to the AR order p; the performances of tests 1-5 are compared in Section 7.6 by using synthetic series consisting of multiple stationary segments and a comparison is made in Section 7.7 between the algorithms with and without boundary optimization, using synthetic series composed of multiple stationary segments and non-stationary segments.

7.2. TESTS BASED ON AR MODELS

7.2.1. Test 1 (de Souza and Thomson, 1982). Consider a series of random variables $\{Y_t\}$ satisfying the AR(p) model in Eq. 7.1:

$$Y_t = \varepsilon_t - a_0 - \sum_{i=1}^{p} a_i Y_{t-i}, \quad t = 1,2,\ldots,N. \tag{7.1}$$

Assume $a_0 = 0$, and $\{\varepsilon_t\}$ are independent and identically distributed (i.i.d.) Gaussian random variables with zero mean and variance σ^2. Let $\mathbf{a}$ be the p-dimensional column vector of AR(p) parameters: $[a_1, a_2,\ldots, a_p]'$, the least square estimates of the parameters of the AR model are

$$\hat{\mathbf{a}} = -\mathbf{D}^{-1}\mathbf{d}, \tag{7.2}$$

and

$$\hat{\sigma}^2 = \hat{\mathbf{a}}'\mathbf{C}\hat{\mathbf{a}}, \tag{7.3}$$

where $\mathbf{D}$ is a p by p submatrix of covariance matrix $\mathbf{C}$ obtained by deleting row and column zero, and $\mathbf{d}$ is the p-dimensional column vector equivalent to the first column of $\mathbf{D}$.

Let $\{Y_{Rt}\}$ and $\{Y_{Tt}\}$ denote two series of length N_R and N_T respectively. Let us assume that $\{Y_{Rt}\}$ is generated by an AR(p) model with $\mathbf{a} = \mathbf{a}_R$, $\sigma^2 = \sigma_R^2$, and $\{Y_{Tt}\}$, independent of $\{Y_{Rt}\}$, is generated by an AR(p) model with $\mathbf{a} = \mathbf{a}_T$, $\sigma^2 = \sigma_T^2$.

If σ_R^2 and σ_T^2, not necessarily the same, are known as *apriori*, the null hypothesis of no changes in the AR parameters is

$$\mathbf{H_0}: \mathbf{a}_R = \mathbf{a}_T.$$

The relevant likelihood ratio is

$$\exp\left[-\frac{1}{2}(\hat{\mathbf{a}}_R - \hat{\mathbf{a}}_T)'\left\{\sigma_R^2(N'_R\mathbf{D}_R)^{-1} + \sigma_T^2(N'_T\mathbf{D}_T)^{-1}\right\}^{-1}(\hat{\mathbf{a}}_R - \hat{\mathbf{a}}_T)\right], \tag{7.4}$$

where $N'_R = N_R - p$, $N'_T = N_T - p$, and $\mathbf{D}_R$ and $\mathbf{D}_T$ denote the submatrices of $\mathbf{C}_R$ and $\mathbf{C}_T$ obtained from $\{Y_{Rt}\}$ and $\{Y_{Tt}\}$, respectively. Under the null hypothesis, this statistic

$$(\hat{\mathbf{a}}_R - \hat{\mathbf{a}}_T)'\left\{\sigma_R^2(N'_R\mathbf{D}_R)^{-1} + \sigma_T^2(N'_T\mathbf{D}_T)^{-1}\right\}^{-1}(\hat{\mathbf{a}}_R - \hat{\mathbf{a}}_T) \tag{7.5}$$

has an asymptotic chi-squared distribution with p degrees of freedom. However, this test is of limited use in practice since the variances σ_R^2 and σ_T^2 are typically unknown, and the computation of the quadratic form

$$\left\{\sigma_R^2(N'_R\mathbf{D}_R)^{-1} + \sigma_T^2(N'_T\mathbf{D}_T)^{-1}\right\}^{-1} \tag{7.6}$$

may be infeasible, especially when singular matrices are encountered.

Two modifications are made to Eq. 7.5. The first modification is to replace σ_R^2 and σ_T^2 with the unbiased estimates

$$\sigma_R^{*2} = \frac{N'_R\hat{\sigma}_R^2}{N'_R - p}, \tag{7.7}$$

and

$$\sigma_T^{*2} = \frac{N'_T\hat{\sigma}_T^2}{N'_T - p}, \tag{7.8}$$

where $\hat{\sigma}_R^2$ and $\hat{\sigma}_T^2$ are the least squares estimates of σ_R^2 and σ_T^2. The second modification is to replace the quadratic form (Eq. 7.6) with

$$\frac{N'_R N'_T}{(N'_R + N'_T)^2}\left(\frac{N'_T \mathbf{D}_R}{\sigma_R^{*2}} + \frac{N'_R \mathbf{D}_T}{\sigma_T^{*2}}\right), \tag{7.9}$$

in the sense that the harmonic mean is replaced with the arithmetic mean of $\frac{\mathbf{D}_R}{\sigma_R^{*2}}$ and $\frac{\mathbf{D}_T}{\sigma_T^{*2}}$. These modifications yield

$$d_s = \frac{N'_R N'_T}{(N'_R + N'_T)^2}(\hat{\mathbf{a}}_R - \hat{\mathbf{a}}_T)'\left(\frac{N'_T \mathbf{D}_R}{\sigma_R^{*2}} + \frac{N'_R \mathbf{D}_T}{\sigma_T^{*2}}\right)(\hat{\mathbf{a}}_R - \hat{\mathbf{a}}_T), \tag{7.10}$$

where d_s has an asymptotic chi-squared distribution with p degrees of freedom under the null hypothesis. Consequently, the decision rule is

$$\begin{cases} d_s < \chi^2_\alpha(p), \text{accept Ho} \\ d_s \geq \chi^2_\alpha(p), \text{do not accept Ho} \end{cases}.$$

The threshold $\chi^2_\alpha(p)$ is chosen corresponding to a given level of significance α..

7.2.2. Test 2 (Imberger and Ivey, 1991). If $\{Y_{Rt}\}$ and $\{Y_{Tt}\}$ are two halves of a series; then, a simple alternative is defined as distance derivative by following the definition of test statistic d_s in Eq. 7.10

$$\text{dd/dt} = \frac{2d_s}{N_R \Delta}, \textit{if } N_R = N_T, \tag{7.11}$$

where Δ is the sampling time interval. A minimum value is specified for the distance derivative below which the record is assumed stationary. Unlike test 1, the threshold for the distance derivative is empirically selected.

7.2.3. Test 3 (Davis, Huang and Yao, 1995). Consider a series $\{Y_t\}$ specified in Eq. 7.1, and $\{\varepsilon_t\}$ is a fourth-order white noise series, that is, for $i \leq j \leq k \leq l$,

$$E(\varepsilon_i) = 0, \tag{7.12}$$

$$E(\varepsilon_i \varepsilon_j) = \begin{cases} \sigma^2, \textit{if } i = j, \\ 0, \textit{if } i < j, \end{cases} \tag{7.13}$$

$$E(\varepsilon_i \varepsilon_j \varepsilon_k) = \begin{cases} \mu_3, \textit{if } i = j = k, \\ 0, \textit{otherwise}, \end{cases} \tag{7.14}$$

$$E(\varepsilon_i \varepsilon_j \varepsilon_k \varepsilon_l) = \begin{cases} \mu_4, \textit{if } i = j = k = l, \\ \sigma^4, \textit{if } i = j < k = l, \\ 0, \text{otherwise.} \end{cases} \tag{7.15}$$

If $\{\varepsilon_t\}$ satisfies Eqs. 7.12 to 7.15 with $\sup_t E|\varepsilon_t|^{4+\delta} < \infty$ for some $0 < \delta \leq 1$, and $\{Y_t\}$ is strongly mixing with the mixing function $\rho(N) << N^{-(1+\varepsilon)(1+4/\delta)}$ for some $\varepsilon > 0$, the null hypothesis of no changes in the AR parameters after $t = k$ is

H$_0$: no change occurred in the AR parameters after $t = k$.

The log likelihood ratio, conditional on the first p observations, is given by

$$\Lambda_N(k) = -2\ln\left(\frac{L(N)}{L(k)L(N-k)}\right) = \min_a \sum_{t=p+1}^{N} \varepsilon_t^2 - \min_a \sum_{t=p+1}^{k} \varepsilon_t^2 - \min_a \sum_{t=k+1}^{N} \varepsilon_t^2 = Q_1 - Q_2 - Q_3, \tag{7.16}$$

where

$$Q_1 = Y_N' Y_N - Y_N' M_N (M_N' M_N)^{-1} M_N' Y_N', \tag{7.17}$$

$$Q_2 = Y_k' Y_k - Y_k' M_k (M_k' M_k)^{-1} M_k' Y_k', \tag{7.18}$$

$$Q_3 = \tilde{Y}_k' \tilde{Y}_k - \tilde{Y}_k' \tilde{M}_k (\tilde{M}_k' \tilde{M}_k)^{-1} \tilde{M}_k' \tilde{Y}_k', \tag{7.19}$$

$$Y_k = [Y_{p+1}, \ldots, Y_k]', \tag{7.20}$$

$$\tilde{Y}_k = [Y_{k+1}, \ldots, Y_N]', \tag{7.21}$$

$$M_k = \begin{bmatrix} 1 & Y_p & Y_{p-1} & \cdots & Y_1 \\ 1 & Y_{p+1} & Y_p & \cdots & Y_2 \\ \vdots & \vdots & \vdots & \ddots & \vdots \\ 1 & Y_{k-1} & Y_{k-2} & \cdots & Y_{k-p} \end{bmatrix}, \tag{7.22}$$

and

$$\tilde{M}_k = \begin{bmatrix} 1 & Y_k & Y_{k-1} & \cdots & Y_{k-p+1} \\ 1 & Y_{k+1} & Y_k & \cdots & Y_{k-p+2} \\ \vdots & \vdots & \vdots & \ddots & \vdots \\ 1 & Y_{N-1} & Y_{N-2} & \cdots & Y_{N-p} \end{bmatrix}. \tag{7.23}$$

Under the null hypothesis, Eq. 7.1 yields

$$Y_k = M_k a_{p+1} + E_k , \tag{7.24}$$

$$\tilde{Y}_k = \tilde{M}_k a_{p+1} + \tilde{E}_k , \tag{7.25}$$

where

$$a_{p+1} = [a_0, a_1, \dots, a_p]' , \tag{7.26}$$

$$E_k = [\varepsilon_{p+1}, \varepsilon_{p+2}, \dots, \varepsilon_k]' , \tag{7.27}$$

and

$$\tilde{E}_k = [\varepsilon_{k+1}, \varepsilon_{k+2}, \dots, \varepsilon_N]' . \tag{7.28}$$

By replacing Y_N, Y_k *and* $\tilde{Y}_k$ in Eqs. 7.17 to 7.19, the log likelihood ratio is rearranged as

$$\Lambda_N(k) = -S_N' P_N S_N + S_k' P_k S_k + (S_N - S_k)' \tilde{P}_k (S_N - S_k), \tag{7.29}$$

where

$$S_k = M_k' E_k , \tag{7.30}$$

$$P_k = (M_k' M_k)^{-1}, \tag{7.31}$$

and

$$\tilde{P}_k = (\tilde{M}_k' \tilde{M}_k)^{-1}. \tag{7.32}$$

Under the null hypothesis, the normalized log likelihood ratio in Eq. 7.33

$$\gamma_n = \frac{\sigma^{-2}\Lambda_N - b_N(p+1)}{a_N(p+1)}, \tag{7.33}$$

has the following distribution (Theorem 2.2 in Davis, Huang and Yao (1995)):

$$P[\gamma_n \le z] \to \exp\left(-2\exp\left(-\frac{z}{2}\right)\right), \tag{7.34}$$

where *P* denotes the accumulative probability,

$$b_N(p) = \frac{2\ln\ln(N) + \left(\frac{p}{2}\right)\ln\ln\ln(N) - \ln\Gamma\left(\frac{p}{2}\right)^2}{2\ln\ln(N)}, \tag{7.35}$$

$$a_N(p) = \sqrt{\frac{b_N(p)}{2\ln\ln(N)}}, \tag{7.36}$$

and Γ is the gamma function. The decision rule is

$$\begin{cases} z_{1-\frac{\alpha}{2}} < \gamma_n < z_{\frac{\alpha}{2}}, \text{ accept H}_0 \\ \gamma_n \le z_{1-\frac{\alpha}{2}} \text{ or } \gamma_n \ge z_{\frac{\alpha}{2}}, \text{ do not accept H}_0 \end{cases}.$$

The threshold is chosen corresponding to a given level of significance α.

It should be noted that this likelihood ratio test statistic is applied to detection of a change point in a single series.

7.2.4. Test 4. Tsay (1988) proposed a test for detecting changes in the AR variance. Assume a change occurs after time $t = k$ in the AR variance of a series $\{Y_t\}$ specified in Eq. 7.1 with $a_0 = 0$, such that

$$\varepsilon_t = \begin{cases} e_t, \text{ for } t < k \\ e_t(1+\omega_v), \text{ for } t \ge k \end{cases} \tag{7.37}$$

where $\{e_t\}$ are i.i.d. Gaussian with zero mean and variance σ^2, and ω_v is a constant. If **a** and k are known, the null hypothesis that there are no changes in the AR variance after $t = k$ is

H$_0$: $\omega_v = 0$.

The variance ratio of ε_t before and after the time k is given by r_v:

$$r_v = \frac{(k-1)\sum_{t=k}^{N}\varepsilon_t^2}{(N-k+1)\sum_{t=1}^{k-1}\varepsilon_t^2}. \tag{7.38}$$

The likelihood ratio test statistic r_v under the assumption of Gaussianity has a central F-distribution with degrees of freedom $(N + k - 1)$ and $(k - 1)$.

In order to apply the variance ratio test to two independent series, a simple modification is made in this study. Let $\{Y_{Rt}\}$ and $\{Y_{Tt}\}$ denote two independent series of length N_R and N_T, respectively. Assume that both series are generated from two AR(p) models with the same **a**, but different white-noise series $\{\varepsilon_{Rt}\}$ and $\{\varepsilon_{Tt}\}$ which are Gaussian distributed with zero mean and variance σ_R^2 and σ_T^2, respectively. Assuming **a** is known *a priori*, a new null hypothesis of no changes in the AR variance is:

$$\mathbf{H_{0n}}: \sigma_R^2 = \sigma_T^2.$$

The variance ratio is redefined as

$$r'_v = \frac{(N_R - p)\sum_{t=p+1}^{N_T} \varepsilon_{Tt}^2}{(N_T - p)\sum_{t=1}^{k-1} \varepsilon_{Rt}^2}. \tag{7.39}$$

Since **a** is unknown in practice, the least square estimates of **a** corresponding to $\{Y_{Rt}\}$ is used for both series under the null hypothesis and the same **a** assumption. Thus, the modified variance ratio is equivalent to

$$r'_V = \frac{\tilde{\hat{\sigma}}_T^2}{\hat{\sigma}_R^2}, \tag{7.40}$$

where

$$\hat{\sigma}_R^2 = \hat{a}'_R C_R \hat{a}_R, \tag{7.41}$$

$$\tilde{\hat{\sigma}}_T^2 = \hat{a}'_R C_T \hat{a}_R, \tag{7.42}$$

$$\hat{a}_R = -D_R^{-1} d_R. \tag{7.43}$$

This modified variance ratio r'_v has a central F-distribution with degrees of freedom $(N_T - p)$ and $(N_R - p)$. The decision rule is

$$\begin{cases} F_{1-\frac{\alpha}{2}}(N_T - p, N_R - p) < r'_v < F_{\frac{\alpha}{2}}(N_T - p, N_R - p), \text{ accept } H_{0n} \\ r'_v \le F_{1-\frac{\alpha}{2}}(N_T - p, N_R - p) \, or \, r'_v \ge F_{\frac{\alpha}{2}}(N_T - p, N_R - p), \text{ do not accept } H_{0n} \end{cases}.$$

The threshold is chosen corresponding to a given level of significance α.

7.3. A TEST BASED ON WAVELET ANALYSIS

Wavelet analysis is a rapidly developing interdisciplinary technique featuring time-frequency localization and multiresolution analysis (Daubechies, 1992; Chui, 1992; Kaiser, 1994). Wavelet based mathematical and application-oriented studies have been successfully conducted in a variety of fields. In the field of geophysics, various applications of wavelets have been proposed and discussed, including detection of atmospheric signals from noisy data, characterization of hydraulic conductivity, analysis of long memory geophysical processes, and other applications (Meneveau, 1991; Farge, 1992; Howell and Mahrt, 1994; Percival and Guttorp, 1994; Trevino and Andreas, 1996; and Farge et al., 1996). In this study, wavelet analysis is applied to detect change points in time series.

Long before the concept of wavelet transforms was formalized, Haar (1910) developed a decomposition method by using the Haar wavelet. There have been several families of wavelet bases developed in different applications in recent years. In statistical applications, two preferred properties are usually required in choosing a family of wavelet bases. The first is "orthogonality" which results in independence of wavelet coefficients; the second is "compact support" with which the wavelet functions are non-zero only on a certain interval of time τ. The basis functions $\phi_{j,k}(\tau)$ and $\psi_{j,k}(\tau)$ are generated from the scaling function (father wavelet) ϕ and the wavelet (mother wavelet) ψ with scale factor 2^j and translation parameter $2^j k$ as follows:

$$\phi_{j,k}(\tau) = 2^{-\frac{j}{2}} \phi\left(\frac{\tau - 2^j k}{2^j}\right), \tag{7.44}$$

$$\psi_{j,k}(\tau) = 2^{-\frac{j}{2}} \psi\left(\frac{\tau - 2^j k}{2^j}\right). \tag{7.45}$$

For the basis functions to satisfy these two properties, the following restrictions are posed on the scaling function (Kumar and Foufoula-Georgiou, 1994):

1. $\int \phi(\tau) d\tau = 1$, (7.46)

2. $\|\phi(\tau)\| = 1$, (7.47)

3. $\int \phi_{j,k}(\tau) \psi_{j',k'}(\tau) d\tau = 0$, (7.48)

4. $\int \phi_{j,k}(\tau) \phi_{j,k'}(\tau) d\tau = \delta_{nn'}$, (7.49)

5. $\phi(\tau) = \sum_n h(n) \phi(2\tau - n)$,

where $h(n)$ are called the scaling coefficients,

6. $\psi(\tau)=\sum_{n} g(n)\phi(2\tau-n)$, (7.50)

where $g(n)$ are coefficients derived from $h(n)$. (7.51)

For the particular case of Haar wavelet, $h(0) = h(1) = 1$, and $h(n) = 0$ for all other n; $g(1) = - g(0) = 1$, and $g(n) = 0$ for all other n.

A continuous time signal $y(\tau)$ may be decomposed into multiresolution components by using such an orthogonal wav
elet basis as

$$y(\tau)\approx\sum_{k} s_{J,k}\phi_{J,k}(\tau)+\sum_{k} d_{J,k}\psi_{J,k}(\tau)+\sum_{k} d_{J-1,k}\psi_{J-1,k}(\tau)+\cdots +\sum_{k} d_{1,k}\psi_{1,k}(\tau) \quad (7.52)$$

where J is the number of multiresolution components. Since these basis functions are orthogonal, the wavelet coefficients may be approximated by the wavelet transforms

$$s_{J,k}\approx\int\phi_{J,k}(\tau)y(\tau)d\tau, \quad (7.53)$$

$$d_{j,k}\approx\int\psi_{j,k}(\tau)y(\tau)d\tau, \qquad j=1,2,\ldots\ldots, J. \quad (7.54)$$

Two types of wavelet satisfying 'orthogonality' and 'compact support' are frequently used: the Haar and the Daubechies families of wavelets. In general, the higher order Daubechies wavelets are more compact in Fourier space, and better for representing larger scale variations. However, Howell and Mahrt (1994) suggest that the Haar wavelet is more efficient in capturing the sharp gradients associated with the transport physics. Since both the turbulence data and the Haar wavelet contain sharp boundaries, the turbulence data are efficiently decomposed with the Haar wavelet. Therefore, the Haar wavelet is applied to the wavelet variance analysis in this study. Here the scaling function of the Haar wavelet is given by Eq. 7.55

$$\phi(\tau)=\begin{cases}1, 0\le\tau<1\\ 0, \text{otherwise}\end{cases}, \quad (7.55)$$

and the corresponding wavelet is $\psi(\tau)=\begin{cases}1, 0\le\tau<\frac{1}{2}\\ -1, \frac{1}{2}\le\tau<1.\\ 0, \text{otherwise}\end{cases}$ (7.56)

From the orthogonality of wavelet bases, an analogue to Parseval's theorem in the wavelet domain is directly derived

$$\int y^2(\tau)d\tau = \sum_{j=1}^{J}\sum_{k} d_{j,k}^2 + \sum_{k} s_{J,k}^2 \quad . \tag{7.57}$$

Let $E(j)$ represent the wavelet variance at scale 2^j,

$$E(j) = \sum_{k} d_{j,k}^2 \quad . \tag{7.58}$$

It can be shown that (Hudgins, Mayer and Friehe, 1993)

$$E(j) = \frac{1}{2\pi}\int_{-\infty}^{\infty} S_y(\omega) S_{\psi j}(\omega) d\omega \,, \tag{7.59}$$

where $S_y(\omega)$ is the spectrum of $y(\tau)$, and $S_{\psi j}(\omega)$ is the spectrum of the wavelet at scale 2^j .

For most of the real data which are discrete and are of finite extent, it is necessary to introduce the discrete wavelet transform. Consider a discrete series $\{Y_t\}$ obtained by sampling $y(\tau)$ at a sampling interval Δ, so that $Y_t = y(t\Delta)$ then, the wavelet coefficients are calculated by

$$s_{J,k} \approx \sum_{t=0}^{N} \phi_{J,k}(t\Delta) Y_t \,, \tag{7.60}$$

$$d_{j,k} \approx \sum_{t=0}^{N} \psi_{j,k}(t\Delta) Y_t \,, \quad j = 1,2,......,J \,; \tag{7.61}$$

where k ranges from 1 to $N / (2^j)$.

Tewfik and Kim (1992) derived an asymptotic expression for the correlation among wavelet coefficients of power-law noise processes. Wornell (1996) concluded that the wavelet coefficients corresponding to distinct scales are uncorrelated, and at a given scale the correlation is extremely small. There is strong empirical evidence that the correlation among wavelet coefficients is rather small (Wornell, 1996; Subba Rao and Indukumar, 1996). Therefore, it is not unreasonable to neglect the correlation between wavelet coefficients at a given scale. Accordingly, a likelihood ratio test is proposed for detecting changes in the wavelet variance.

Consider the null hypothesis of no changes in the wavelet variances $E_R(j)$ and $E_T(j)$ computed from two independent series $\{Y_{Rt}\}$ and $\{Y_{Tt}\}$ of length N_R and N_T , respectively,

$\mathbf{H_0}$: $E_R(j) = E_T(j), \quad j = 1,2,\ldots,J;$

a likelihood ratio $r_w(j)$ is presented in Eq. 3.62:

$$r_W(j) = \frac{E_T(j)}{E_R(j)} \,, \quad j = 1,2,\ldots,J. \tag{7.62}$$

This wavelet variance ratio has a central F-distribution with $N_T/(2^j)$ and $N_R/(2^j)$ degrees of freedom under the null hypothesis. The decision rule is

$$\begin{cases} F_{1-\frac{\alpha}{2}}\left(\frac{N_T}{2^j},\frac{N_R}{2^j}\right) < r_w(j) < F_{\frac{\alpha}{2}}\left(\frac{N_T}{2^j},\frac{N_R}{2^j}\right), \; j=1,2,\ldots,J \quad \text{accept } H_0 \\ r_w(j) \le F_{1-\frac{\alpha}{2}}\left(\frac{N_T}{2^j},\frac{N_R}{2^j}\right) \; or \; r_w(j) \ge F_{\frac{\alpha}{2}}\left(\frac{N_T}{2^j},\frac{N_R}{2^j}\right), \; j\in\{1,2,\ldots,J\} \\ \qquad \text{do not accept } H_0. \end{cases}$$

The threshold is chosen corresponding to a given level of significance α.

7.4. SEGMENTATION ALGORITHM

A modified segmentation algorithm consisting of three stages is considered in this book. In general, a preliminary boundary is detected in the first stage, the optimal ending boundary is determined in the second stage (boundary optimization one), and the optimal starting boundary is determined in the third stage (boundary optimization two). The entire procedure is referred as the segmentation algorithm in the following discussion. The procedures of each stage are as follows:

1. The boundary detection procedures of Appel and Brandt (1983) are adopted in the first stage with a modification that the reference window (W_r) is fixed and has the same length of the test window (W_t). The test statistics derived from the AR parameters of the reference series $\{Y_{Rt}\}$ and the test series $\{Y_{Tt}\}$ are computed from data within the disjoined reference-test windows in Fig. 7.2. The reference series $\{Y_{Rt}\}$ and the test series $\{Y_{Tt}\}$ are displaced with the data of a single series within the reference window and the test window, respectively. Although stationary series might be of any arbitrary length, a minimum number of observations are required to compute acceptable statistical estimates. In this study, each window contains 64 observations; consequently, the minimum number of observations of a detected stationary segment is 128. While the reference window is fixed at a starting point (t_s) of the series, the test window starts next to the end of the reference window ($t_{t1} = t_{r2} + 1$), and slides along the series until the boundary (t_d) is detected. The sliding test window skips eight observations at each move. The preliminary ending boundary is detected at the observation where the null hypothesis of 'no change' fails. If the number of observations of the detected segment, $t_d - t_s + 1$, is not greater than 128, the segmentation algorithm will be reset to start at the next starting point which is next to the previous starting point ($t_s + 1$). This modified boundary detection procedures have the advantage that the AR parameters computed from the reference window need not be updated as the test window slides along the series. Only after the boundary is detected, the AR parameters are recomputed from the reference window which is moved to a new starting point.

2. In the first stage the preliminary boundary (t_d) is only detected after the real change point in the series; consequently, the real change point is still within the stopped test window in the first stage (between $t = (t_d - 64)$ and t_d). In order to optimize the boundary, a symmetric boundary optimization is applied in the second stage. Similar to the segmentation algorithm of Imberger and Ivey (1991), the test statistics are computed from data within the consecutive reference-test windows (between $t = t_{r1}$ and t_{t2}) which are sliding together along the series (Fig. 7.3). The consecutive reference-test windows are sliding at one observation each move. Each window contains 64 observations. The optimal ending boundary is determined at the local maximum (or minimum) value of the test statistic within the stopped test window in the first stage (between $t = (t_d - 64)$ and t_d).
3. Due to strong intermittency in the temperature gradient profiles, there are short stationary segments of less than 128 observations. It is not uncommon that these short segments are located between stationary segments of more than 128 observations. Consequently, there is a possibility that these short segments may be included in the beginning of a stationary segment determined from the first two stages. Therefore, it is necessary to optimize the starting boundary in the third stage. The optimal starting boundary is determined at the local maximum (or minimum) value of the test statistic within the reference window used in the first stage (between $t = t_s$ and $(t_s + 64)$ in Fig. 7.4). The test statistics are computed from data within the consecutive reference-test windows (between $t = t_{r1}$ and t_{t2}). The consecutive reference-test windows are sliding at one observation each move. After both the optimal starting and ending boundaries ($t = t_{so}$ and $t = t_{do}$) are determined, the stationary segment is identified between $t = t_{so}$ and t_{do} in Fig. 7.5. If the number of observations between the optimal starting and ending boundaries, $t_{do} - t_{so} + 1$, is not greater than 128, the segmentation algorithm will be reset to start at the next starting point which is next to the previous starting point in the series.

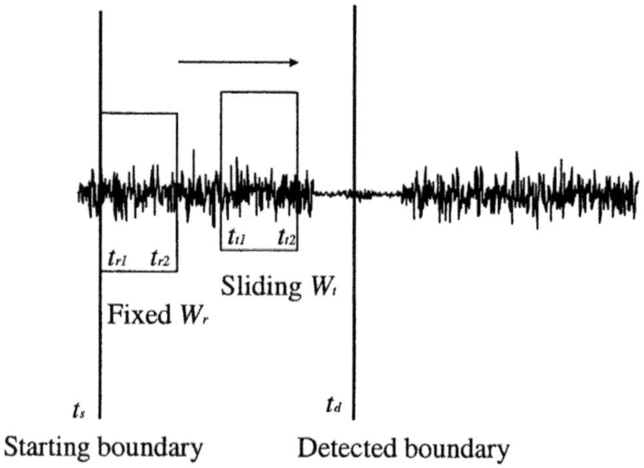

Figure 7.2. Stage 1: boundary detection.

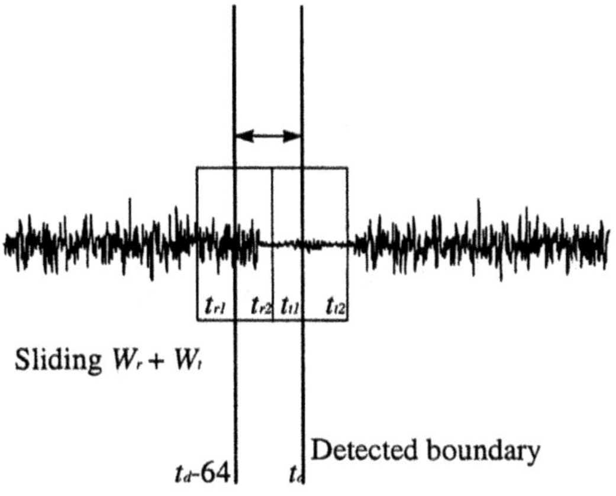

Figure 7.3. Stage 2: boundary optimization 1.

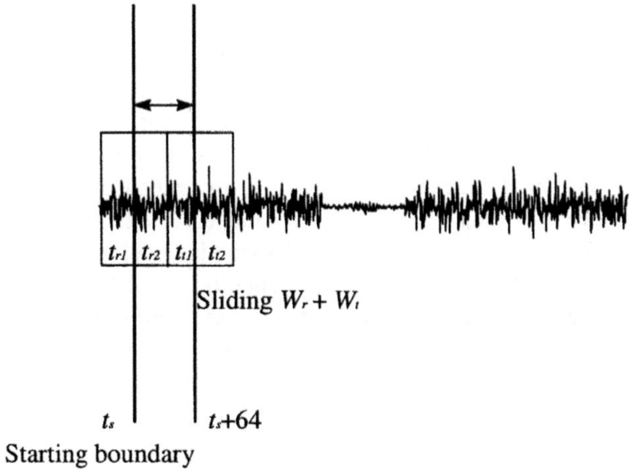

Figure 7.4. Stage 3: boundary optimization 2.

This segmentation algorithm performs in sequence so that the optimal boundaries of the first segment are determined before the next segmentation begins. A stationary segment is assumed to have been determined if the number of observations between the optimal starting and ending boundaries (including boundaries) is more than 128. After a stationary segment is determined, the algorithm will reset to begin at the observation next to the identified segment. The algorithm is depicted in Fig. 7.6.

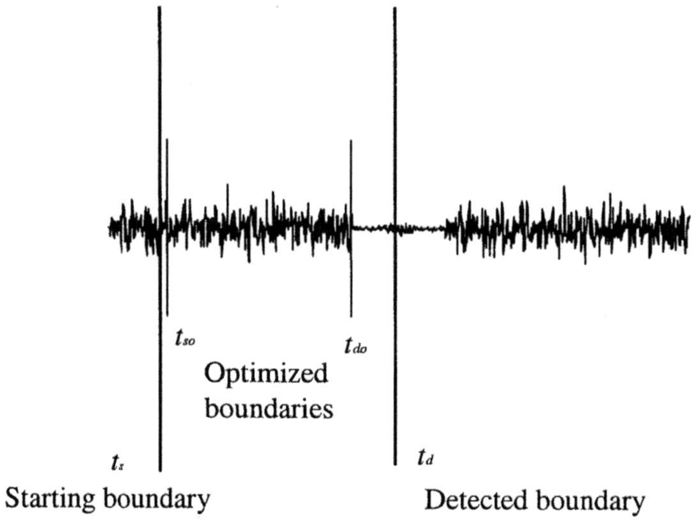

Figure 7.5. Determined segment.

7.5. VARIATIONS OF TEST STATISTICS WITH THE AR ORDER p

Imberger and Ivey (1991) observed that, based on their segmentation results, the AR spectrum, for a given order, has the poorest fit at low wave numbers compared to the estimated Batchelor spectra (Section 2.2). With increasing order, the overall fit improved at high wave numbers but only marginally at low wave numbers. A few examples of fitting estimated Batchelor spectra were examined by Imberger and Ivey (1991) with AR models of order p = 4, 8, 12, 16 and 19. Their results favored rather large order models. However, the overall spectral shape is developed by AR models of order p = 4. Imberger and Ivey's (1991) segmentation algorithm was intended for detecting major changes in the spectral shape; consequently, they decided not to optimize the order of the model, but rather to use the AR model of order $p = 4$.

In this study, synthetic series from two models (Models 2.1.1 and 2.1.11) are generated to examine the variation of test statistics d_s, r_n and r'_v (corresponding to tests 1, 3 and 4 defined in Section 7.2) with the AR orders p = 4, 8, 12 and 24. The first set of ten synthetic series are weakly stationary series generated by sampling from the Batchelor spectrum (Model 2.1.1), and the second set of ten synthetic series are generated by using a stationary AR(2) model (Model 2.1.11). Test statistics d_s and r'_v are computed from data within the disjoined reference-test windows in Fig. 7.7. Each window contains 64 observations. While the reference window is fixed at a starting point ($t = t_s$) of the series, the test window starts next to the end of the reference window ($t_{t1} = t_{r2} + 1$), and slides along the series to the end. The sliding test window skips eight observations at each move. Test statistic r_n is computed from data within the consecutive reference-test windows (between $t = t_{r1}$ and t_{t2}) which slide together along the series until the end (Fig. 7.8). The consecutive reference-test windows skip eight observations at each move. Each window contains 64 observations.

The results are discussed as follows:

1. The values of test statistic d_s in Fig. 7.9 are computed from the AR model of orders $p = 4$, 8, 12 and 24 for a synthetic series generated from Model 2.1.1. The results in Fig. 7.10 are for a synthetic series generated from Model 2.1.11. The straight lines in Figs. 7.9 and 7.10 represent the thresholds at the significance level $\alpha = 1\%$ corresponding to different AR orders. It is shown in both figures that the values of test statistic d_s computed from synthetic series of each model increase as the AR order p increases. The results for all the series generated from Models 2.1.1 and 2.1.11 are presented in Chen (1999), and are summarized in Table VII.1. According to the results in Table VII.1, nine out of ten series generated from Model 2.1.1 are identified as stationary, and eight out of ten series generated from Model 2.1.11 are identified as stationary. The higher the AR order p, the smaller the number of series that are identified as stationary. Especially when the AR order p = 24 is applied, only two stationary series out of ten generated from Model 2.1.1 are correctly identified. Consequently, test 1 based on a higher AR orders will lead to higher rate of false identification.

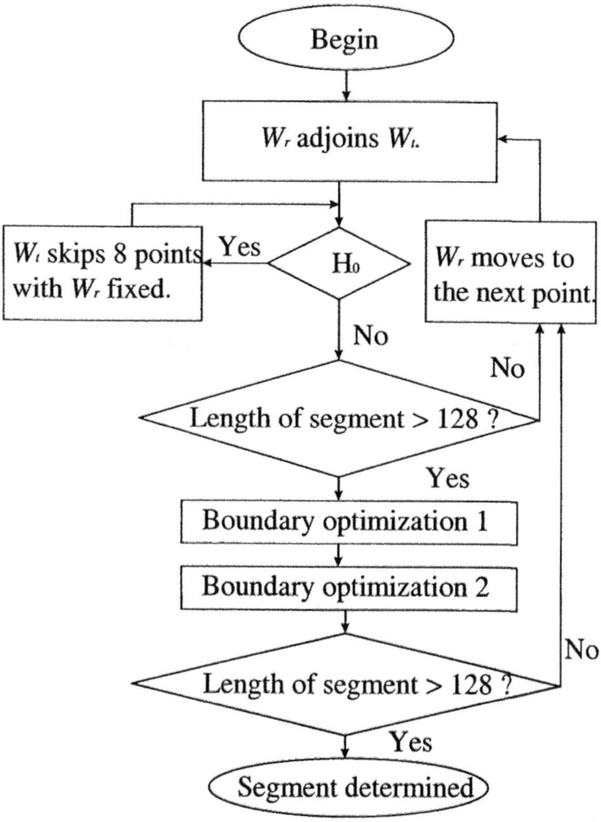

Figure 7.6. Algorithm of the adaptive sequential segmentation.

Table VII.1. Number of stationary series identified out of ten series.

Order	Series generated from Model 2.1.1			Series generated from Model 2.1.11		
P	Test 1	Test 3	Test 4	Test 1	Test 3	Test 4
4	9	0	7	8	0	7
8	9	0	4	7	0	6
12	8	10	3	8	10	1
24	2	0	0	0	0	0

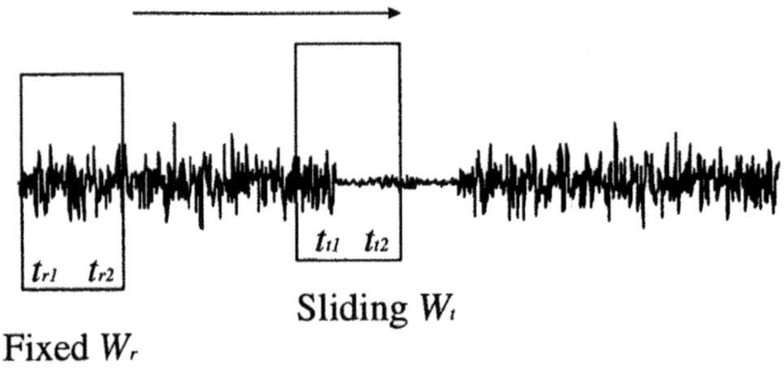

Figure 7.7. Disjoined reference-test windows for computing test statistics d_s and r'_v.

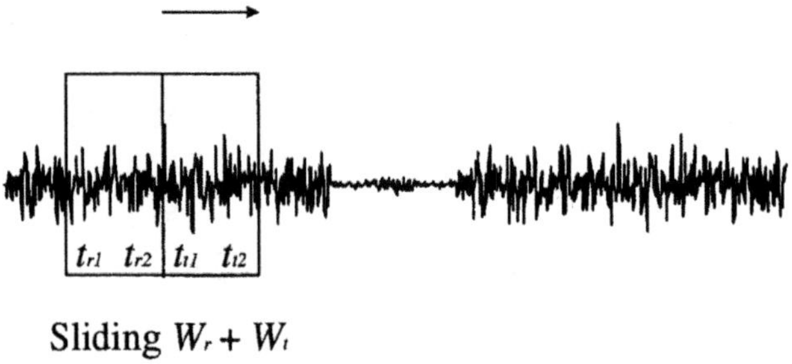

Figure 7.8. Consecutive reference-test windows for computing test statistic γ_n.

2. The values of test statistic γ_n in Fig. 7.11 are computed from the AR model of orders p = 4, 8, 12 and 24 for a synthetic series generated from Model 2.1.1. The results in Fig. 7.12 are for a synthetic series generated from Model 2.1.11. The straight lines in Figs. 7.11 and 7.12 represent the 99% confidence interval, which remains the same for test statistic r_n using different AR orders. The values of the test statistic r_n increase as the AR order p increases from 4 to 12, and then drop to the lowest level when the AR model of order p = 24 is used (Figs. 7.11 and 7.12). Except for the test statistic r_n based on AR model of order p = 12, the values of the

test statistic r_n based on the AR order p = 4, 8 and 24 are all outside the 99% confidence interval. Similar results for the other series generated from Models 2.1.1 and 2.1.11 are found in Chen (1999), which are summarized in Table VII.1. According to the results in Table VII.1, series generated from both Models 2.1.1 and 2.1.11 are identified as stationary by test 3 only when the AR order p = 12 is used. Since the second set of ten synthetic series are generated from an AR(2) model (Model 2.1.11), it does not conform to the accepted statistical criteria that AR models of order p = 12 should fit better than AR models of order p = 4. Consequently, segmentation results obtained by using test 3 are unreliable.

3. The values of test statistic r'_v in Fig. 7.13 are computed from the AR models of orders p = 4, 8, 12 and 24 for a synthetic series generated from Model 2.1.1. The results in Fig. 7.14 are for a synthetic series generated from Model 2.1.11. Four pairs of straight lines in Figs. 7.13 and 7.14 represent the 99% confidence intervals corresponding to test statistic r'_v using different AR orders. Similar to the results of test statistic d_s, the values of test statistic r'_v increase as the AR order p increases (Figs. 7.13 and 7.14). Especially when the AR model of order p = 24 is used, most of the values of test statistic r'_v are outside the 99% confidence interval. Results for the other series generated from Models 2.1.1 and 2.1.11 are found in Chen (1999). These results are summarized in Table VII.1. According to the results in Table VII.1, seven out of each set of ten series (Models 2.1.1 and 2.1.11) are identified as stationary. The higher the AR order p, the smaller the number of series that are identified as stationary. Especially when the AR model of order p = 24 is used, no stationary series is identified. Similar to test 1, test 4 based on higher AR orders will lead to higher rate of false identification.

These results show that those from test 3 are unreliable, and higher orders used in the AR models lead to higher values of test statistics d_s and r'_v which are computed from synthetic series generated from Models 2.1.1 and 2.1.11. According to the numbers of stationary series identified out of ten synthetic series generated from Model 2.1.1, the AR order p = 4 used in tests 1 and 4 has better performances than the other orders p = 8, 12 and 24 used. Consequently, the order p = 4 is applied to tests 1-4 in the segmentation algorithm discussed in this book.

7.6. SENSITIVITY OF TEST STATISTICS FOR DETECTING CHANGE POINTS

Four sets of synthetic series consisting of multiple segments from Models 2.1.2-2.1.5 are generated to examine the sensitivity to spectral changes on different tests, including tests 1, 2, 3 and 4 (Section 3.2), and test 5 (Section 3.3) which is based on wavelet analysis. Tests 1, 2, 3 and 4 are based on the AR order p = 4. Test 5 is based on the wavelet variances corresponding to the scale parameter j = 1 and 2 for which the degrees of freedom are 32 and 16 (Eq. 7.62), respectively.

A change-point detection algorithm without boundary optimization is applied to tests 1, 4 and 5. This algorithm is executed in sequence so that the first change point is detected before the next detection begins. Test statistics are computed from data within the disjoined reference-test windows in Fig. 7.15. The sliding test window skips eight observations at each move. If the number of observations between two change points (t_d $-t_s$ +1) is not greater than 128, the detection algorithm is reset to start at $t = t_s +1$ (Fig.

7.15). After a change point is detected, the algorithm is reset to begin at the next observation ($t = t_d + 1$ in Fig. 7.15). All the "no change" null hypotheses are tested at the significance level $\alpha = 1\%$. This change-point detection algorithm is depicted in Fig. 7.16.

For tests 2 and 3, the distance derivative *dd/dt* (Section 7.2.2) and test statistic γ_n are computed from data within the consecutive reference-test windows (Fig. 7.8). The consecutive reference-test windows slide together along the series and skip eight observations at each move. The change points are detected where the values of distance derivative are greater than the threshold 400, and test 3 fails at the significance level $\alpha =$ 1%, respectively.

The first and second sets of synthetic series (Models 2.1.2 and 2.1.3) are generated according to gradually increasing variations in χ_T. Each of the ten synthetic series from Model 2.1.2 consists of ten segments for which χ_T increases from $1*10^{-7}$ to $10*10^{-7}$ $(\text{degC})^2/\text{sec}$. It is shown in Fig. 7.17 that average variance corresponding to each segment increases as χ_T increases. Each of the ten synthetic series from Model 2.1.3 consists of three segments for which χ_T increases from $(10/3)*10^{-7}$ to $10*10^{-7}$ $(\text{degC})^2/\text{sec}$. Average variance corresponding to each segment is shown in Fig. 7.18. The third and fourth sets of synthetic series (Models 2.1.4 and 2.1.5) are generated according to gradually increasing variations in ε. Each of the ten synthetic series from Model 2.1.4 consists of ten segments for which ε increases from $1*10^{-7}$ to $10*10^{-7}$ m^2/sec^3. It is shown in Fig. 7.19 that average variance corresponding to each segment remains the same as ε increases. Each of the ten synthetic series from Model 2.1.5 consists of three segments for which ε increases from $(10/3)*10^{-7}$ to $10*10^{-7}$ m^2/sec^3. Average variance corresponding to each segment is shown in Fig. 7.20.

7.6.1. Detection Results for Synthetic Series from Model 2.1.2. The first ten synthetic series (Model 2.1.2) are generated according to variations in χ_T from 10^{-7} to 10^{-6} $(\text{degC})^2/\text{sec}$ (Fig. 7.17). Each series contains ten segments. The segmentation results for two series are given in Figs. 7.21 and 7.22. It is shown in these results that no stationary segment is identified by test 3, and no change point is detected by using a threshold 400 for test 2. Change points detected by using tests 1, 4 and 5 are shown in Table VII.2. If the detected change point is earlier than the designed change point, it is a false detection. If the detected change point is later than the designed change point and the difference between the detected change point and the designed change point is less than 64 (still within the test window), the change point is considered to be detected in time; otherwise, it is a late detection if the difference between the detected change point and the designed change point is greater than 64 but before the next change point. Both in-time and late detection are classified as true detection. The results in Table VII.2 show that only one change point is detected in time by using test 1, while 20 out of 90 change points are detected in time by using test 4, and 16 out of 90 by using test 5. Considering true detection, 7 change points are detected by using test 1, while 40 out of 90 change points are detected by using test 4, and 34 out of 90 by using test 5.

The sensitivity of these tests to the magnitude of variation in χ_T are analyzed by comparing the first-time detected variations in χ_T. The lower the magnitude at which the variation in χ_T is detected, the more sensitive is the test. The first column in Table VII.3 indicates the variation in χ_T. The results of tests 1, 4 and 5 in Table VII.3 show that the first-time detection in ten out of ten series is at the smallest variation in x_T (from $1*10^{-7}$ to $2*10^{-7}$ $(degC)^2/sec$) by using test 4. In six out of ten series the first-time detection is at the smallest variation in x_T by using test 5, and in four out of ten series is at the second smallest variation (from $1*10^{-7}$ to $3*10^{-7}$ $(degC)^2/sec$). Test 1 does not detect any change point in six out of ten series.

Table VII.2. Change point detection results.
($\varepsilon = 10^{-6}$ m^2/sec^3 and $x_T = 1*10^{-7} \sim 10*10^{-7}$ $(degC)^2/sec$).

Detection	Test 1	Test 4	Test 5
In time	1	20	16
Late	6	20	18
False detection	0	4	5
True detection	7/90	40/90	34/90

7.6.2. Detection Results for Synthetic Series from Model 2.1.3. The second set of ten synthetic series (Model 2.1.3) are generated according to variations in x_T from $(10/3)*10^{-7}$ to 10^{-6} $(degC)^2/sec$ (Fig. 7.18). Each series contains three segments. The segmentation results for two series are given in Figs. 7.23 and 7.24. Similar to the results in Section 7.6.1, these results show that no stationary segment is identified by using test 3, and no change point is detected by using a threshold 400 for test 2. The results in Table VII.4 show that only one change point is detected in time by using test 1, while 10 out of 20 change points are detected in time by using test 4, and 7 out of 20 by using test 5. Considering true detection, 4 change points are detected by using test 1, while 18 out of 20 change points are detected by using test 4, and 18 out of 20 by using test 5.

The sensitivity analyses in Table VII.5 show that tests 4 and 5 detect the first-time variation in each series at the smallest variation in x_T (from $(10/3)*10^{-7}$ to $(20/3)*10^{-7}$ $(degC)^2/sec$), and test 1 does not detect any change point in six out of ten series.

Table VII.3. Number of the first-time detected variation (out of ten series)
($\varepsilon = 10^{-6}$ m^2/sec^3 and $x_T = 1*10^{-7} \sim 10*10^{-7}$ $(degC)^2/sec$).

Variation	Test 1	Test 4	Test 5
$x_T = 2*10^{-7}$		10	6
$3*10^{-7}$			4
$4*10^{-7}$	1		
$5*10^{-7}$	2		
$6*10^{-7}$			
$7*10^{-7}$			
$8*10^{-7}$	1		
$9*10^{-7}$			
$1*10^{-6}$			

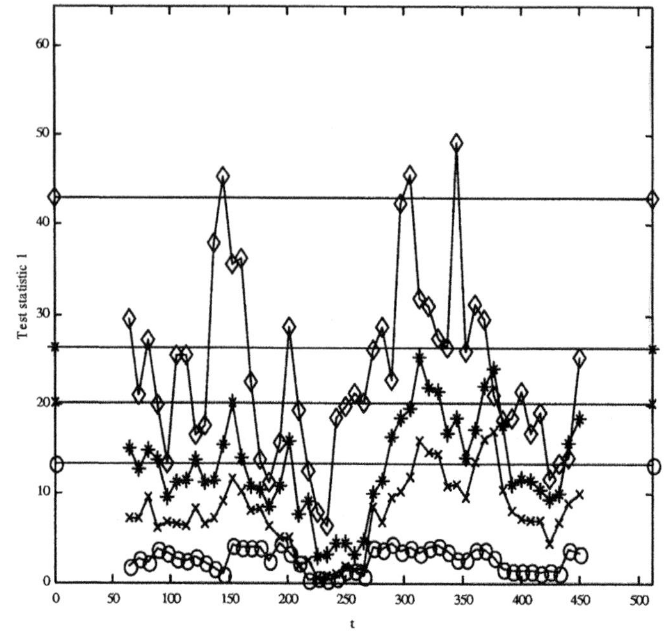

Figure 7.9. Variation of test statistic d_s with AR order for a series from model 2.1.1.

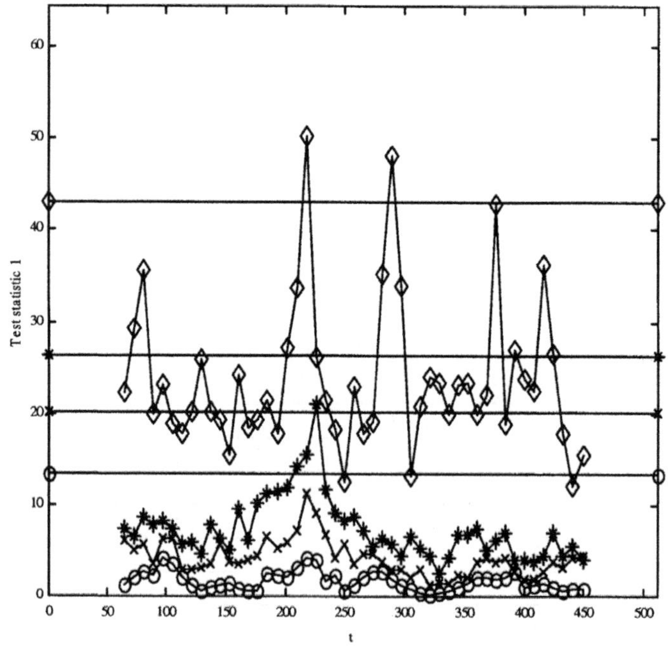

Figure 7.10. Variation of test statistic d_s with AR order for a series from model 2.1.11. Straight lines represent the thresholds at the significance level $\alpha = 1\%$.
o: AR(4); x: AR(8); *: AR(12); ◊: AR(24).

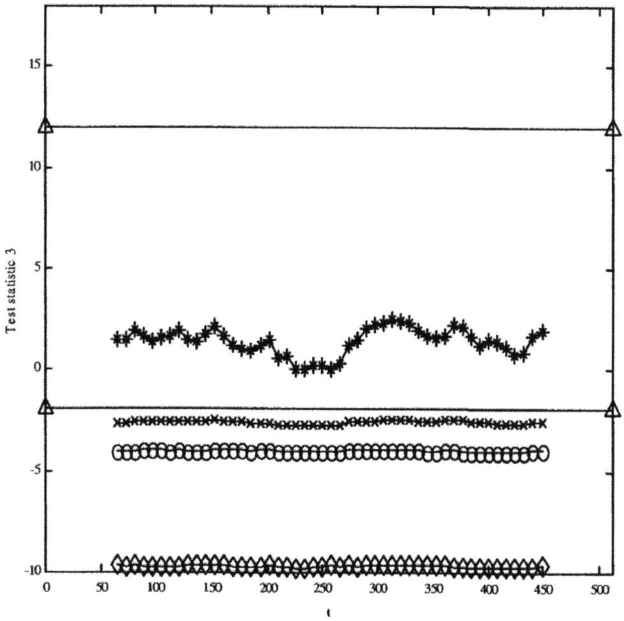

Figure 7.11. Variation of test statistic r_n with AR order for a series from Model 2.1.1.

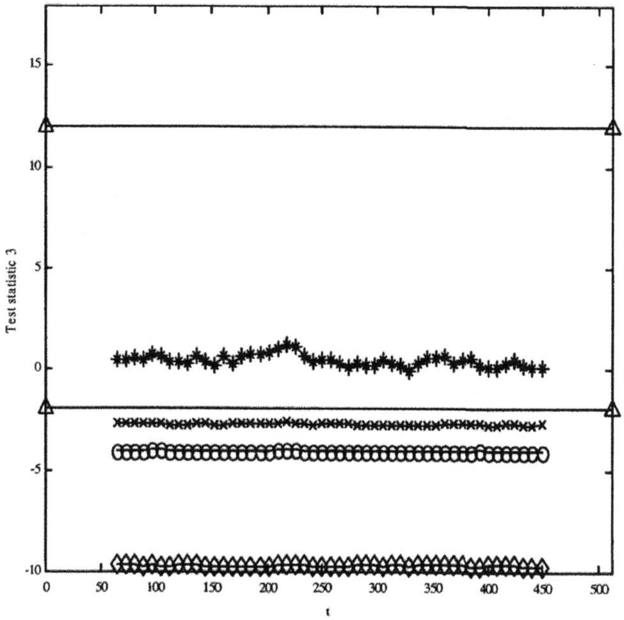

Figure 7.12. Variation of test statistic r_n with AR order for a series from model 2.1.11.
Straight lines represent the 95% confidence interval.
o: AR(4); x: AR(8); *: AR(12); ◊: AR(24).

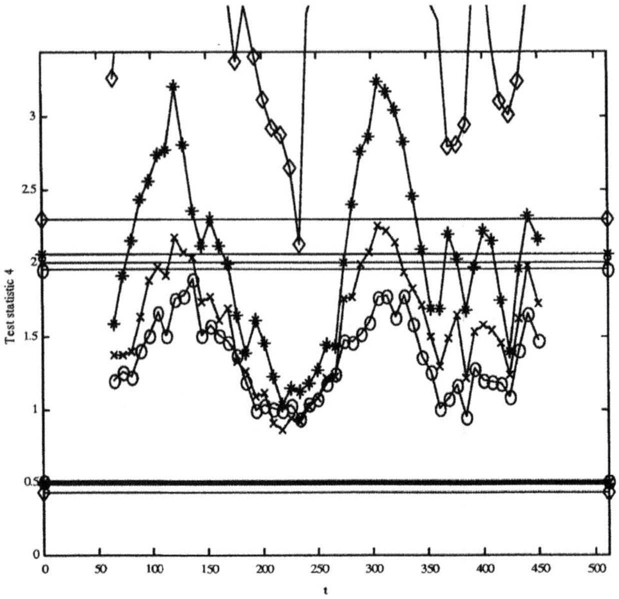

Figure 7.13. Variation of test statistic r'_{ν} with AR order for a series from model 2.1.1.

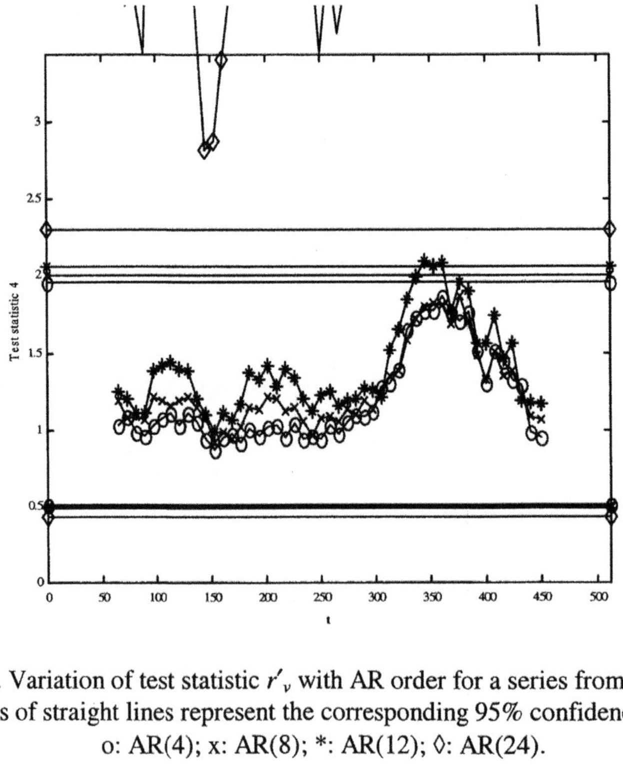

Figure 7.14. Variation of test statistic r'_{ν} with AR order for a series from model 2.1.11. Four pairs of straight lines represent the corresponding 95% confidence intervals. o: AR(4); x: AR(8); *: AR(12); ◊: AR(24).

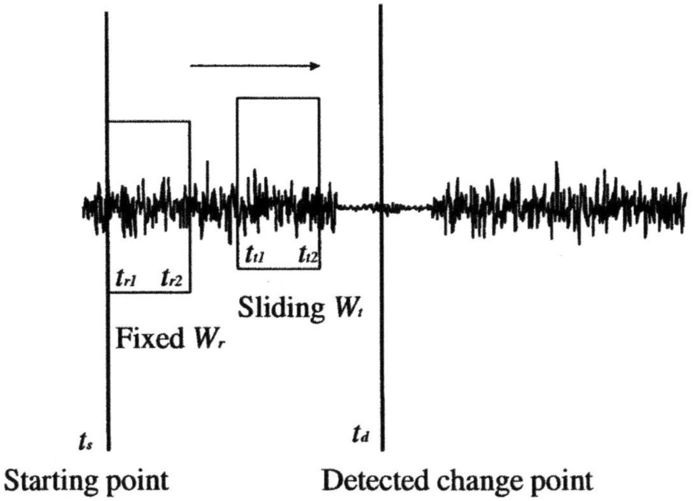

Figure 7.15. Disjoined reference-test windows in the change-point detection algorithm.

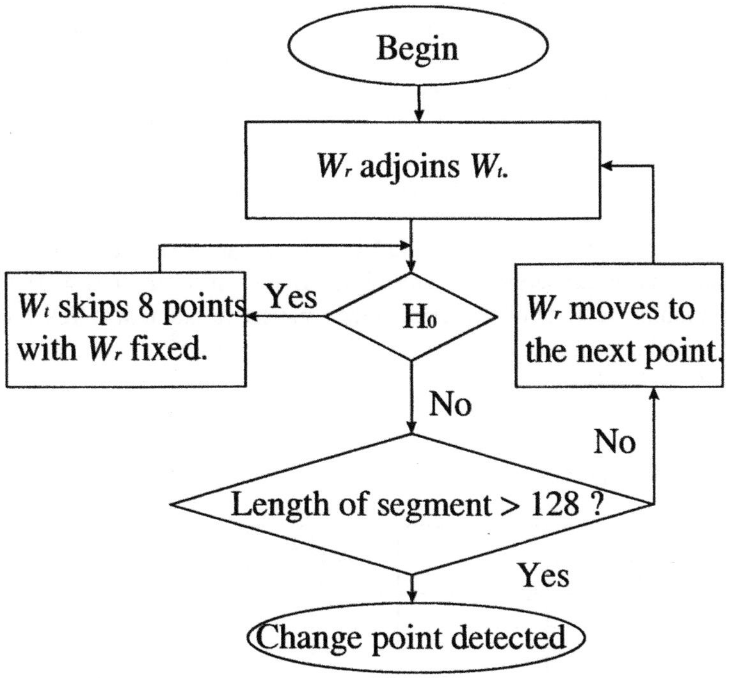

Figure 7.16. Change-point detection algorithm without boundary optimization.

Table VII.4. Change point detection results
($\varepsilon = 10^{-6}$ m^2/sec^3 and $x_T = (10/3) * 10^{-7} \sim 10* 10^{-7}$ (degC)^2/sec).

Detection	Test 1	Test 4	Test 5
In time	1	10	7
Late	3	8	11
False detection	5	10	12
True detection	4/20	18/20	18/20

Table VII.5. Number of the first-time detected variation (out of ten series)
($\varepsilon = 10^{-6}$ m^2/sec^3 and $x_T = (10/3) * 10^{-7} \sim 10* 10^{-7}$ (degC)^2/sec).

Variation	Test 1	Test 4	Test 5
$x_T = (20/3)*10^{-7}$	3	10	10
$1*10^{-6}$	1		

7.6.3. Detection Results for Synthetic Series from Model 2.1.4. The next ten synthetic series (Model 2.1.4) are generated according to variations in ε from 10^{-7} to 10^{-6} m^2/sec^3 (Fig. 7.19). Each series contains ten segments. The segmentation results for two series are given in Figs. 7.25 and 7.26. It is shown in these figures that no stationary segment is identified by test 3, and no change point is detected by using a threshold 400 for test 2. Results of change-point detection in Table VII.6 show that test 5 which correctly detects 18 change points out of 90 is less sensitive to changes in ε, than test 4 which correctly detects 44 out of 90 change points, and test 1 which correctly detects 37 out of 90.

The results of sensitivity analyses in Table VII.7 show that test 1 detects the first-time variation in nine out of ten series at the smallest variation in ε, and test 4 detects the first-time variation in eight out of ten series at the smallest variation in ε. Test 5 detects the first-time variation in six out of ten series before $\varepsilon = 4*10^{-7}$ m^2/sec^3, and in ten out of ten before $\varepsilon = 6*10^{-7}$ m^2/sec^3.

Table VII.6. Change point detection results
($\varepsilon = 1* 10^{-7} \sim 10* 10^{-7}$ m^2/sec^3 and $x_T = 10^{-6}$ (degC)^2/sec).

Detection	Test 1	Test 4	Test 5
In time	19	28	4
Late	18	16	14
False detection	3	3	1
True detection	37/90	44/90	18/90

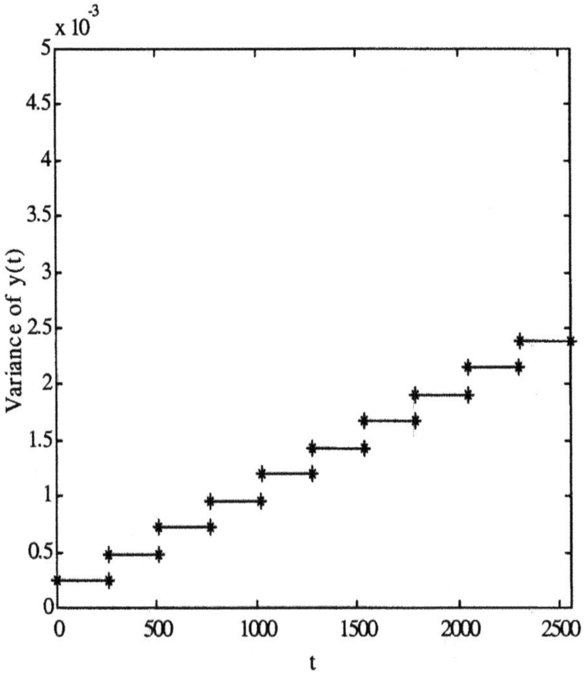

Figure 7.17. Variance of synthetic segments from Model 2.1.2.

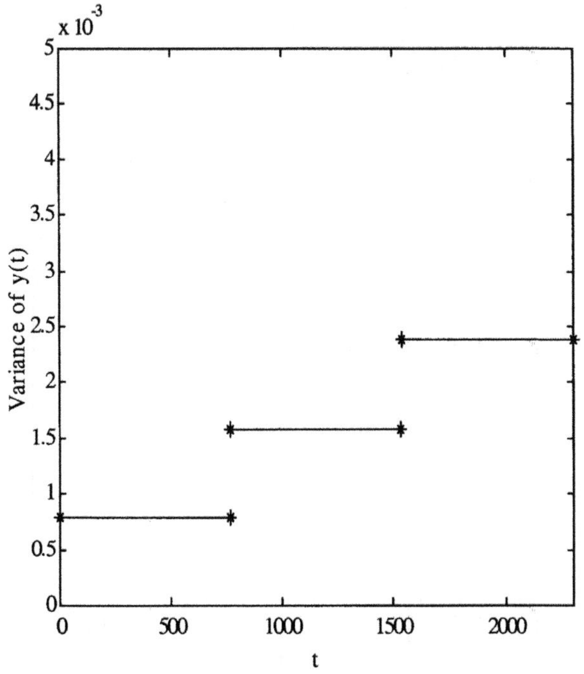

Figure 7.18. Variance of synthetic segments from Model 2.1.3.

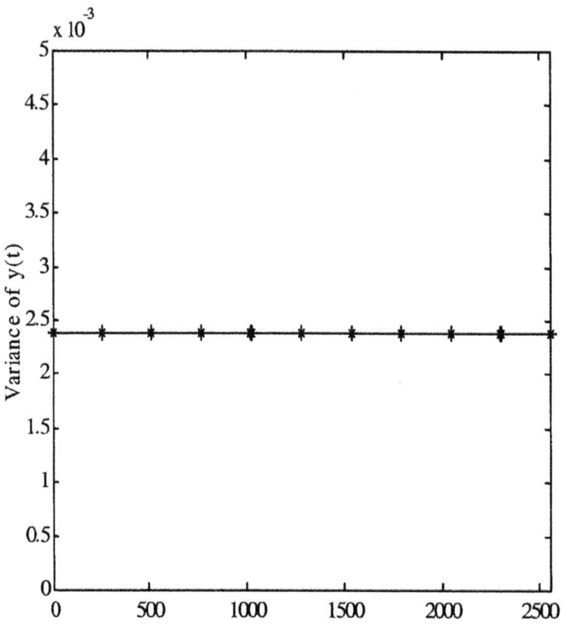

Figure 7.19. Variance of synthetic segments from Model 2.1.4.

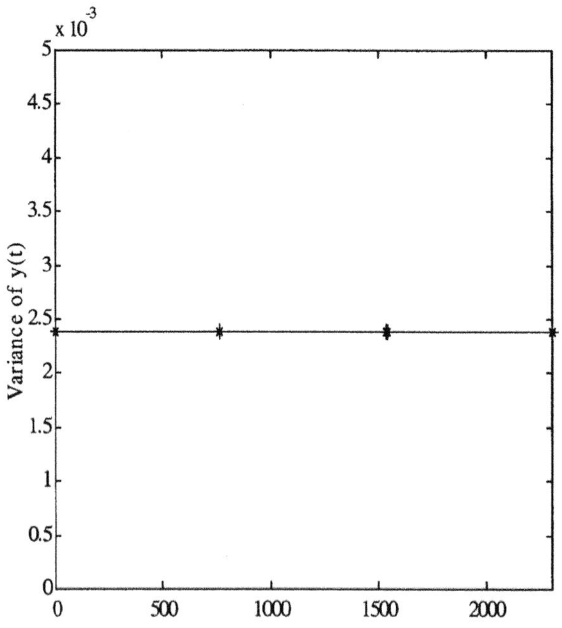

Figure 7.20. Variance of synthetic segments from model 2.1.5.

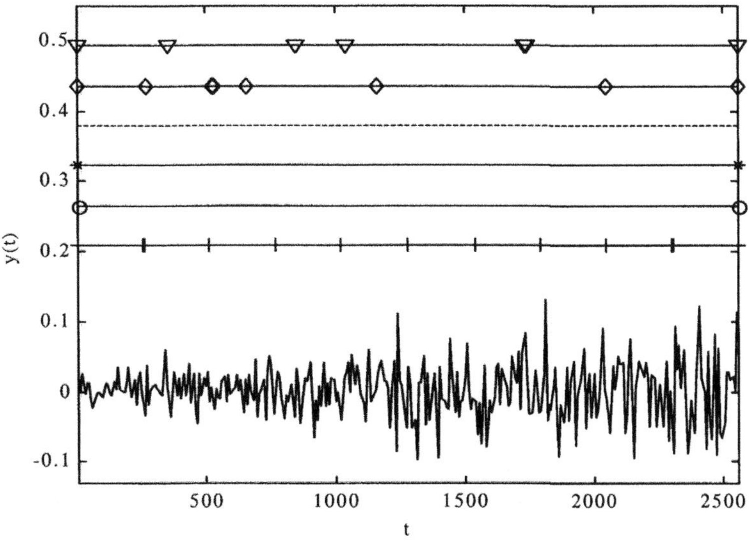

Figure 7.21. Segmentation results for a series from Model 2.1.2.
∇: Test 5; ◊: Test 4; - -: Test 3 (no segments);
*: Test 1; O: Test 2; +: Synthetic segments.

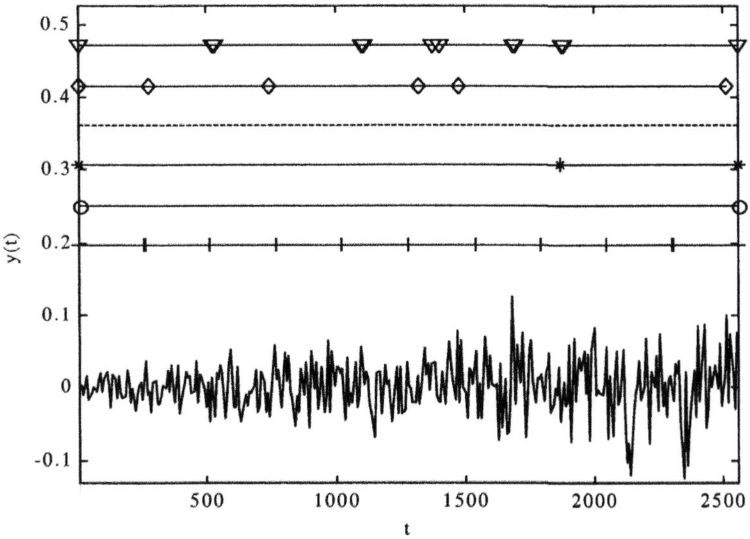

Figure 7.22. Segmentation results for a series from Model 2.1.2.
∇: Test 5; ◊: Test 4; - -: Test 3 (no segments);
*: Test 1; O: Test 2; +: Synthetic segments.

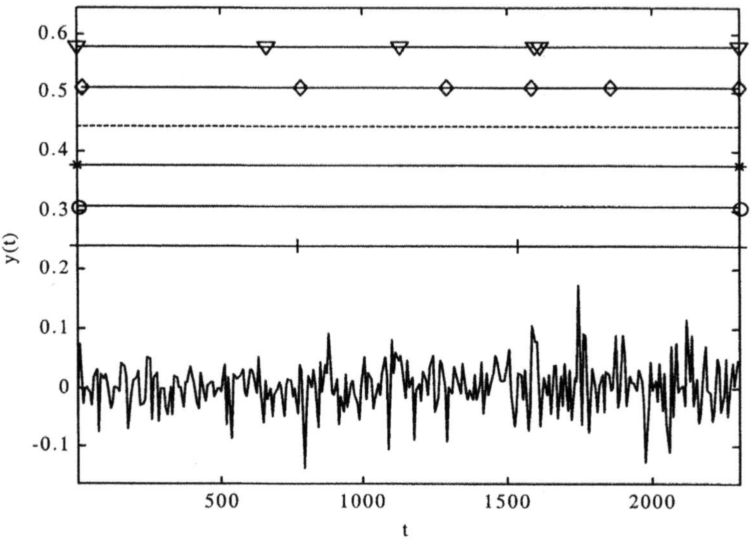

Figure 7.23. Segmentation results for a series from Model 2.1.3.
∇: Test 5; ◊: Test 4; - -: Test 3 (no segments);
*: Test 1; O: Test 2; +: Synthetic segments.

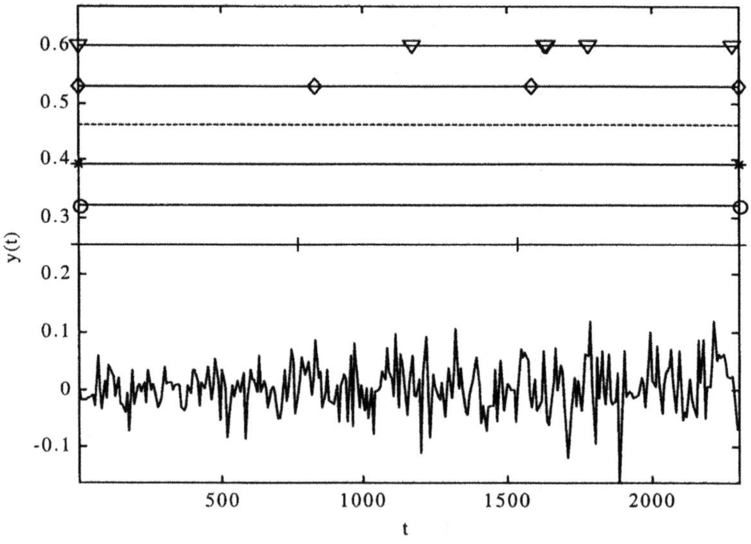

Figure 7.24. Segmentation results for a series from Model 2.1.3.
∇: Test 5; ◊: Test 4; - -: Test 3 (no segments);
*: Test 1; O: Test 2; +: Synthetic segments.

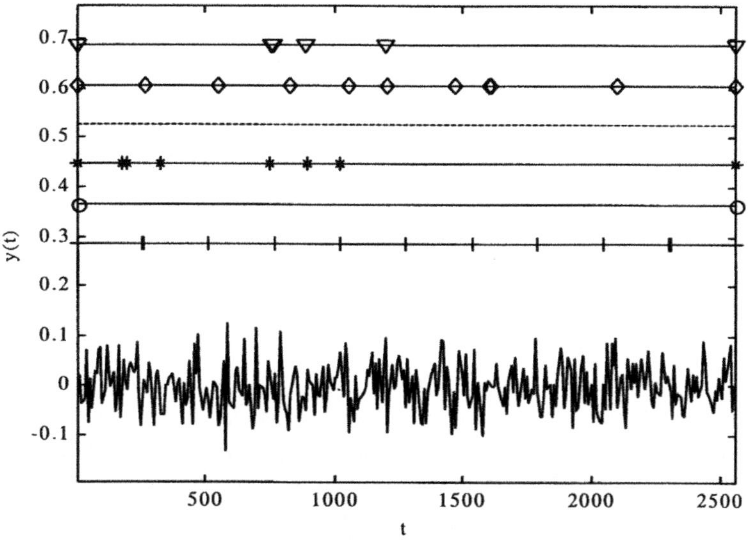

Figure 7.25. Segmentation results for a series from Model 2.1.4.
∇: Test 5; ◊: Test 4; - -: Test 3 (no segments);
∗: Test 1; O: Test 2; +: Synthetic segments.

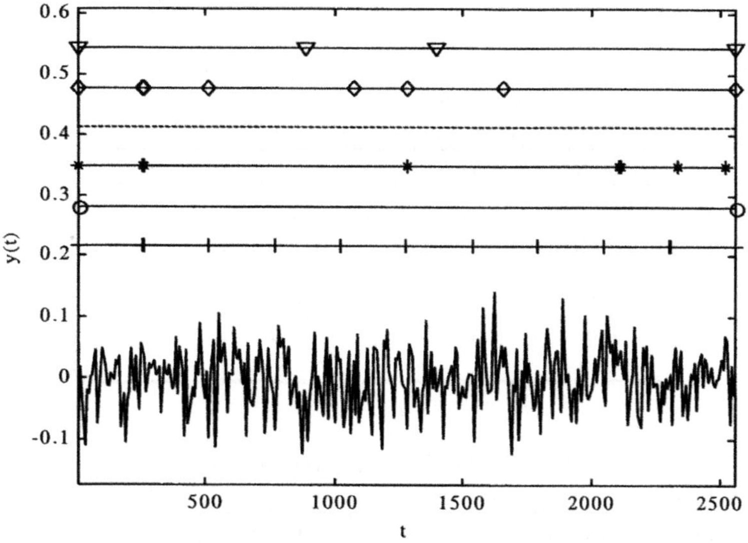

Figure 7.26. Segmentation results for a series from Model 2.1.4.
∇: Test 5; ◊: Test 4; - -: Test 3 (no segments);
∗: Test 1; O: Test 2; +: Synthetic segments.

Table VII.7. Number of the first-time detected variation (out of ten series) ($\varepsilon = 1 * 10^{-7} \sim 10 * 10^{-7}$ m^2/sec^3 and $x_T = 10^{-6}$ $(degC)^2/sec$).

Variation	Test 1	Test 4	Test 5
$\varepsilon = 2*10^{-7}$	9	8	1
$3*10^{-7}$	1	1	5
$4*10^{-7}$		1	2
$5*10^{-7}$			2
$6*10^{-7}$			
$7*10^{-7}$			
$8*10^{-7}$			
$9*10^{-7}$			
$1*10^{-6}$			

7.6.4. Detection Results for Synthetic Series from Model 2.1.5. The last set of ten synthetic series (Model 2.1.5) are generated according to variations in ε from $(10/3)*10^{-7}$ to 10^{-6} m^2/sec^3 (Fig. 7.20). Each series contains three segments. The segmentation results for two series are given in Figs. 7.27 and 7.28. According to these results, no stationary segment is identified by test 3, and no change point is detected by using a threshold 400 for test 2. In Table VII.8 it is shown that test 4 which correctly detects 17 out of 20 is more sensitive to changes in ε than test 1 which correctly detects 10 out of 20 change points, and test 5 which detects 10 out of 20.

The results of sensitivity analyses in Table VII.9 show that the first-time detection in nine out of ten series is at the first variation in ε (from $(10/3)*10^{-7}$ to $(20/3)*10^{-7}$ m^2/sec^3) by using test 1, in ten out of ten series by using test 4, and in seven of ten by using test 5.

Table VII.8. Change point detection results ($\varepsilon = (10/3) * 10^{-7} \sim 10 * 10^{-7}$ m^2/sec^3 and $x_T = 10^{-6}$ $(degC)^2/sec$).

Detection	Test 1	Test 4	Test 5
In time	6	10	2
Late	4	7	8
False detection	11	7	8
True detection	10/20	17/20	10/20

Table VII.9. Number of the first-time detected variation (out of ten series) ($\varepsilon = (10/3) * 10^{-7} \sim 10 * 10^{-7}$ m^2/sec^3 and $x_T = 10^{-6}$ $(degC)^2/sec$).

Variation	Test 1	Test 4	Test 5
$\varepsilon = (20/3)*10^{-7}$	9	10	7
$1*10^{-6}$			1

7.6.5. Conclusions on Performances of Tests 1-5. The results presented in this section indicate that under an acceptable specification of the AR order $p = 4$ and the significance level $\alpha = 1\%$: (1) no stationary segment is identified in these synthetic series by test 3 which is unreliable according to the results in this section; (2) test 2 does not detect any change point in these synthetic series consisting of multiple segments at the arbitrary threshold 400; (3) test 1 is only sensitive to changes of ε in these synthetic series; (4) test 4 is sensitive to changes of both ε and x_T; (5) test 5 is sensitive to changes of x_T, but less sensitive to changes of ε. Because tests based on AR models are parametric approaches, they may have better performances than non-parametric approaches (wavelet analysis) when their parameters are appropriately specified. Therefore, it is reasonable that test 5 is not as sensitive to changes in the synthetic series as test 4. It should be noted that two of the three important parameters in the segmentation algorithm, the significance level and the minimum starting segment length, are fixed in this study. Variations of these two parameters will have influences on performances of different tests to different degrees. It is possible that at some significant level some tests will have better performances, or by increasing the minimum starting segment length these tests will become more sensitive or less.

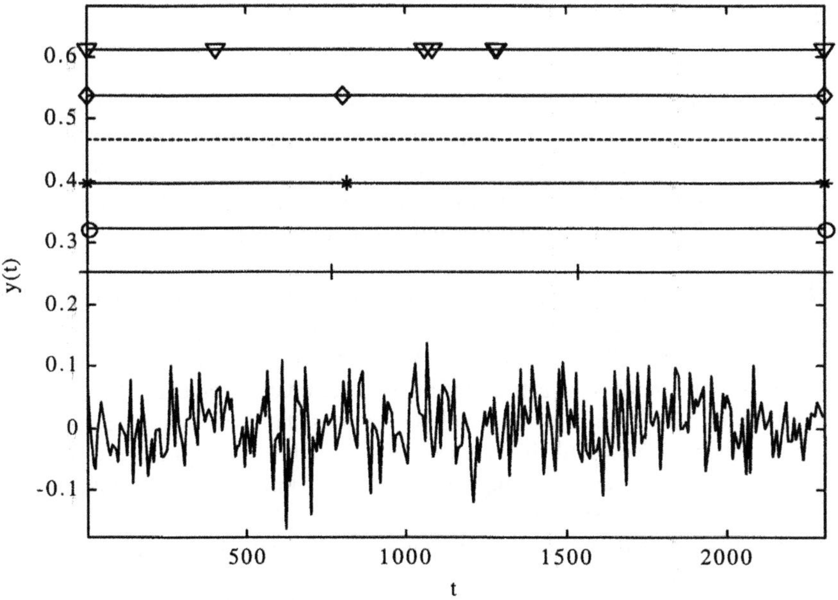

Figure 7.27. Segmentation results for a series from Model 2.1.5.
∇: Test 5; ◊: Test 4; - -: Test 3 (no segments);
*: Test 1; O: Test 2; +: Synthetic segments.

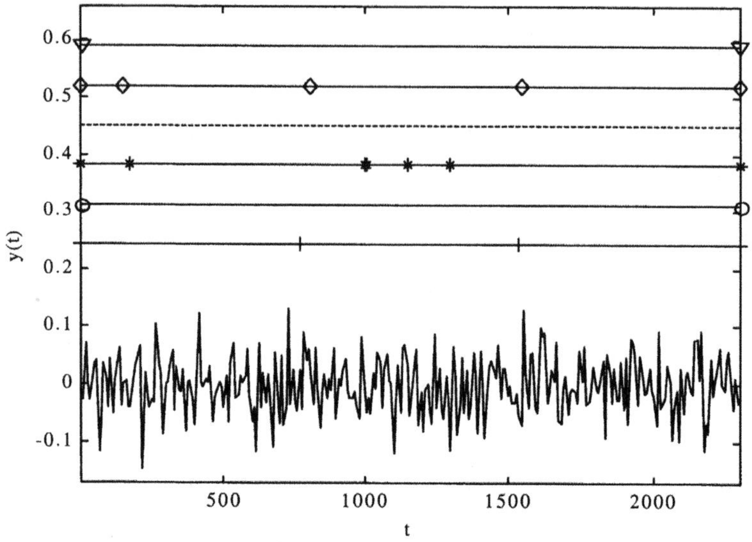

Figure 7.28. Segmentation results for a series from Model 2.1.5.
∇: Test 5; ◊: Test 4; - -: Test 3 (no segments);
*: Test 1; O: Test 2; +: Synthetic segments.

7.7. PERFORMANCES OF ALGORITHMS WITH AND WITHOUT BOUNDARY OPTIMIZATION

The segmentation algorithm (defined in Section 7.4) with two stages of boundary optimization is compared to the change-point detection algorithm without boundary optimization (defined in Section 7.6), in order to examine the performances of boundary optimization. The former algorithm is referred to as algorithm 1, and the latter is referred to as algorithm 2 in the following discussion. Two sets of synthetic series (Models 2.1.6 and 2.1.7) are generated for analysis in this section.

7.7.1. Detection of Non-Stationary Segment. The number of observations for a stationary segment to be identified is greater than 128 in this study; consequently, any segment with less than 128 observations should not be identified as a stationary segment, and should produce a gap between two stationary segments. Ten synthetic series from Model 2.1.6 are generated. Each series consists of two stationary segments with a "non-stationary segment" in between. Each stationary segment contains 512 observations, and the non-stationary segment consists of three short segments of white noise. Each of the short segments of white noise contains 64 observations. The variance of each segment is shown in Fig. 7.29. The boundaries of the non-stationary segment are identified by algorithms 1 and 2, respectively.

The results of boundary detection for two series from Model 2.1.6 are shown in Figs. 7.30 and 7.31 and summarized in Tables VII.10 and VII.11. The values in Tables VII.10 and VII.11 represent the differences between the detected locations and the designed boundaries. It is shown in Table VII.10 that in most of the synthetic series the ending boundary of the first stationary segment is detected at later locations by algorithm 2. In general, algorithm 1 (with boundary optimization) detects the ending boundary of the first stationary segment at closer locations than algorithm 2 (without boundary optimization). However, algorithm 1 using test 5 detects the ending boundary of the first stationary segment in two series out of ten at farther locations. In Table VII.11, algorithm 1 using tests 4 or 5 detect the starting boundary of the second stationary segment at closer location, while algorithm 2 using test 1 detects the starting boundary of the second stationary segment at closer locations. It should be noted that algorithm 2 using test 4 or 5 detects the starting boundary of the second stationary segment at earlier locations in most of the synthetic series (Table VII.11). Apparently, a part of the non-stationary segment (consisting of short segments) is included in the beginning of the second stationary segment determined by algorithm 2, which justifies the third stage of the segmentation algorithm (algorithm 1) to optimize the starting boundary. In general, algorithm 1 using test 5 detects both starting and ending boundaries at closer locations than the others, except in two series out of ten the ending boundary of the first stationary segment is detected at farther but earlier locations.

It is shown in Table VII.10 that the non-stationary segment (a 'gap' between stationary segments determined in Fig. 7.31) is also better identified by using the algorithms with boundary optimization. Algorithm 1 using test 5 identifies the non-stationary segment in five out of ten series, which is the best among these three tests.

Table VII.10. Number of detected non-stationary segments in ten series.

Test 1		Test 4		Test 5	
Algorithm 1	Algorithm 2	Algorithm 1	Algorithm 2	Algorithm 1	Algorithm 2
2	0	3	1	5	0

7.7.2. Detection of Multi-Segment Series. A set of ten synthetic series (Model 2.1.7) are generated according to increasing variations in both ε and x_T. Average variance corresponding to each segment is shown in Figure 7.32. Each series contains four segments. The number of observations in each segment is 512.

The boundaries of each segment are detected by using algorithms 1 and 2, respectively. The boundary detection results for two series from Model 2.1.7 are given in Figs. 7.33 and 7.34. The results are summarized in Tables VII.11 and VII.12. Similar to previous results in Section 7.7.1, it is shown in Tables VII.11 and VII.12 that the boundaries of the first two segments are better detected in most of the synthetic series by using algorithm 1. The positive values of difference between the detected and designed boundaries indicate that there is an evident detection time delay caused by using algorithm 2. In general, algorithm 1 using tests 4 or 5 detects both starting and ending boundaries at closer locations than the algorithm using test 1. However, there are a significant number of false detection of boundaries in the last two segments where the level of x_T is only at 1% and 0.1% of the level corresponding to the first segment (Figs. 7.33 and 7.34). It is also shown in Figs. 7.33 and 7.34 that test 5 has much lower false detection rate than tests 1 and 4.

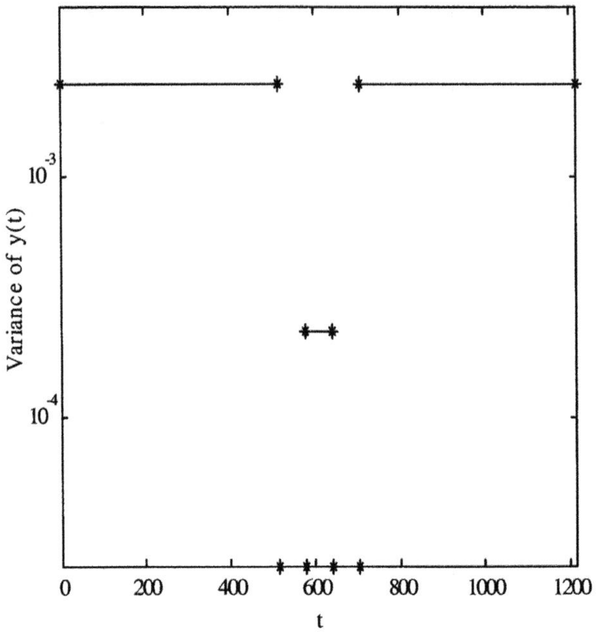

Figure 7.29. Variance of synthetic segments from Model 2.1.6.

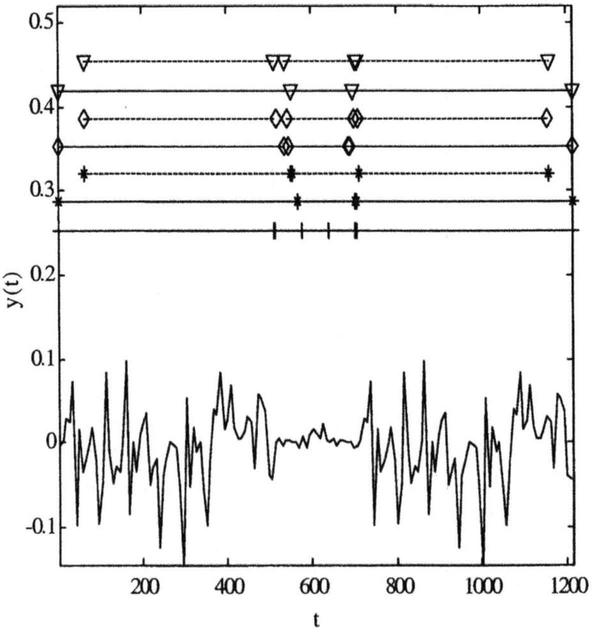

Figure 7.30. Segmentation results for a series from Model 2.1.6.
(∇: Test 5; ◊: Test 4; *: Test 1; +: Synthetic segments.)
(- -: Algorithm 1; —: Algorithm 2.)

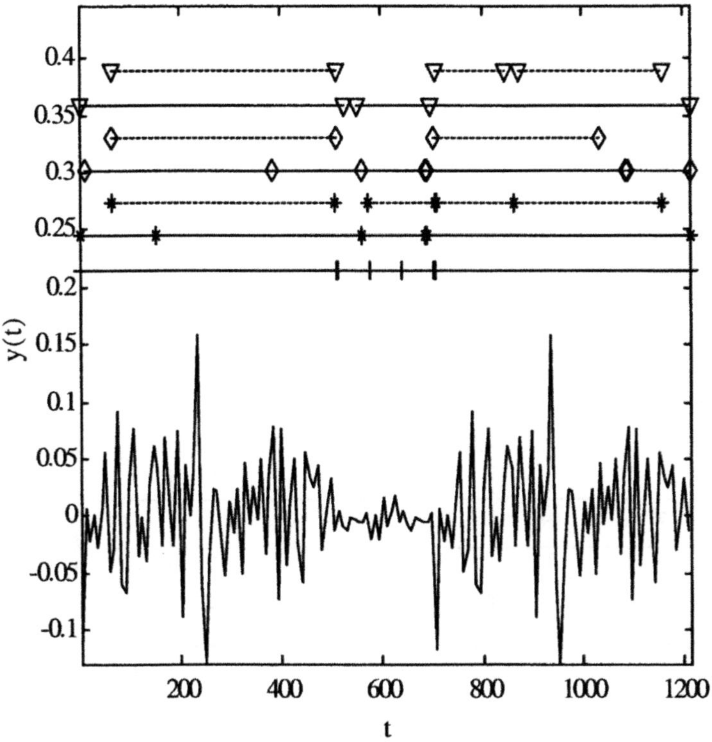

Figure 7.31. Segmentation results for a series from model 2.1.6.
(∇: Test 5; ◊: Test 4; *: Test 1; +: Synthetic segments.)
(- -: Algorithm 1; —: Algorithm 2.)

7.8. CONCLUSIONS ABOUT THE SEGMENTATION ALGORITHM.

Two disadvantages are often encountered in using the segmentation algorithm without boundary optimization (algorithm 2): the first is detection time delays between the detected and designed ending boundaries of a stationary segment; the second is a part of preceding short segments included in the beginning of the stationary segment identified next to them. The results presented in this section indicate that, in general, the segmentation algorithm with boundary optimization detects both starting and ending boundaries at closer locations. Based on the overall performance and non-stationary segment (gap) identification, the segmentation algorithm using test 5 is better than the algorithm using test 1 or 4. It should be noted that both ε and χ_T of the sampled Batchelor spectrum vary at change points in the synthetic series used in this chapter. Consequently, tests 1, 4 and 5 are all sensitive to these change points, according to the results in Section 7.6. It is possible that the segmentation algorithm using test 1 or 4 has better performance if the synthetic series are generated from other models.

Table VII.11. Detected location from the ending boundary of the first stationary segment.

Series	Test 1		Test 4		Test 5	
Algorithm	1	2	1	2	1	2
1	4	48	25	32	0	56
2	20	56	-2	32	4	32
3	-7	72	2	16	7	40
4	7	56	0	40	2	48
5	-13	40	-2	16	0	24
6	-8	40	22	24	-7	32
7	57	56	5	24	0	24
8	107	64	-3	32	2	40
9	25	56	-6	24	0	32
10	34	56	5	32	1	32

Table VII.12. Detected location from the starting boundary of the second stationary segment.

Series	Test 1		Test 4		Test 5	
Algorithm	1	2	1	2	1	2
1	4	48	27	32	0	56
2	23	56	12	32	18	32
3	-6	72	3	16	8	40
4	9	56	2	40	3	48
5	-12	40	1	16	1	24
6	-1	40	26	24	1	32
7	58	56	5	24	5	24
8	110	64	1	32	34	40
9	26	56	26	32	2	32
10	36	56	5	32	8	32

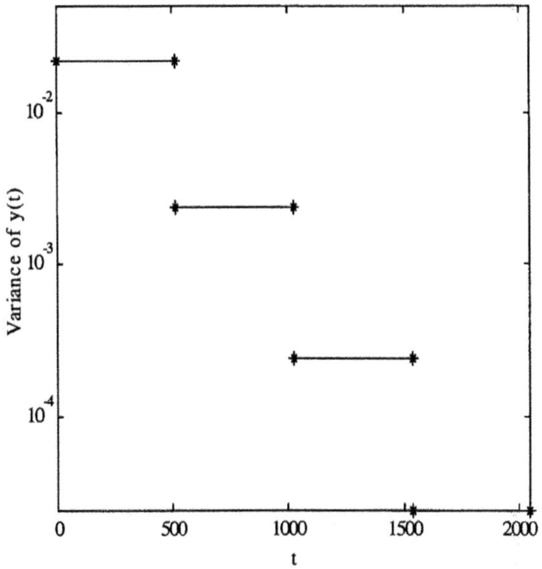

Figure 7.32. Variance of synthetic segments from model 2.1.7.

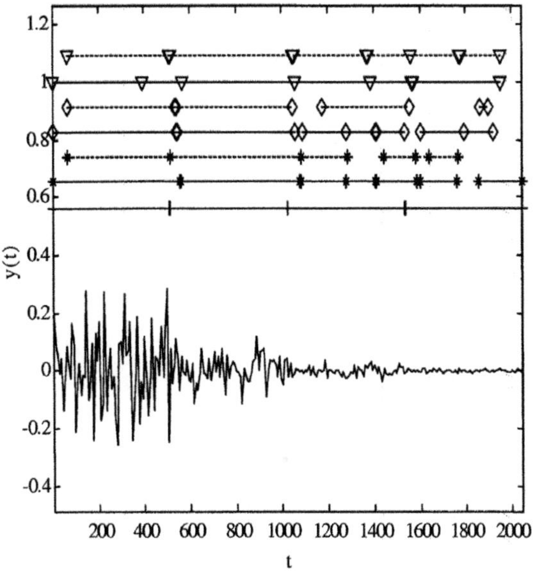

Figure 7.33. Segmentation results for a series from model 2.1.7.
(∇: Test 5; ◊: Test 4; ∗: Test 1; +: Synthetic segments.)
(- -: Algorithm 1; —: Algorithm 2.)

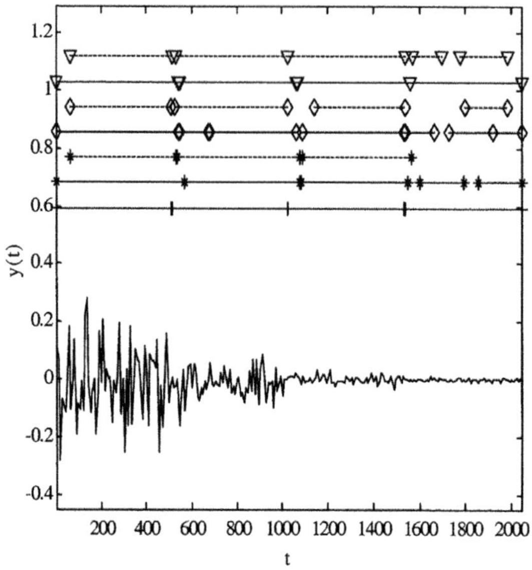

Figure 7.34. Segmentation results for a series from model 2.1.7.
(∇: Test 5; ◊: Test 4; ∗: Test 1; +: Synthetic segments.)
(- -: Algorithm 1; —: Algorithm 2.)

8. Estimation of Turbulent Kinetic Energy Dissipation

8.1. INTRODUCTION

Several studies have suggested that small-scale fluid motion, of scales smaller than a centimeter, influences the kinetics of biological and chemical particles in aquatic environments (O'Melia and Tiller, 1993; Denman and Gargett, 1995; Hondzo and Lyn, 1999). The direct interaction of fluid motion and particles exists if there is an overlap in length scales or gradients produced by the fluid flow and particles in an aquatic environment. One of the key intrinsic fluid flow parameters is the kinetic energy dissipation rate ε. This parameter provides an estimate for the smallest length and time scales relevant to interaction among small scale fluid motion, biological, and chemical particles.

The kinetic energy dissipation rate ε is obtainable through direct and indirect methods. A direct method uses a gradient of the velocity microstructure, and an indirect method is based on a temperature gradient microstructure. While the field measurements of velocity microstructure have been a technological challenge, the field measurements of temperature gradient microstructure are a well-established procedure (Imberger and Ivey, 1991). In an indirect method (Dillon and Caldwell, 1980; Oakey, 1982), the kinetic energy dissipation rate ε is estimated from the Batchelor spectrum (defined in Section 2.2.1). The form of the Batchelor spectrum is determined at higher wave numbers of the temperature gradient spectrum, where viscous and diffusive effects are important. For fluids with large Prandtl number Pr ($\nu >> D_T$, ν is the kinematic viscosity, and D_T is the diffusivity of heat), such as water (Pr ≈ 10 at 7 °C, and $Pr \approx 6$ at 25 °C), the Batchelor spectrum contains two subranges (Fig. 8.1): the viscous-convective subrange and the viscous-diffusive subrange (Newberger and Caldwell, 1981). The estimation procedure assumes homogeneous and isotropic turbulence at small scales.

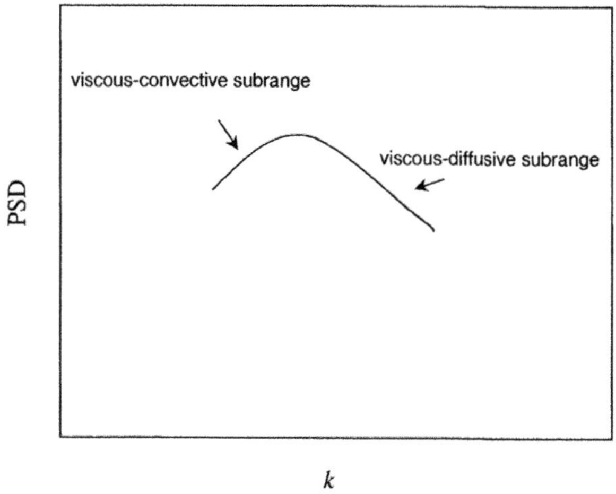

Figure 8.1. Two subranges of the Batchelor spectrum.
(k: wave number; PSD: power spectrum density)

Profiles of temperature gradient microstructure used in this book are measured by using the SCAMP (Self-Contained Autonomous Microstructure Profiler) profiling instrument manufactured by Precision Measurement Engineering of Encinitas, CA. The temperature sensor of SCAMP is assumed to have a first-order response (Fozdar, Parker and Imberger, 1985). For an applied step input sequence, the output is a first-order exponential sequence. In order to reconstruct the input signal from the sensor output, an inverse transfer function of the assumed first-order response is applied. Unfortunately, the inverse transfer function amplifies the noise at high frequencies. Thus, it is necessary to reduce the high-frequency noise by refiltering the reconstructed signal. The entire procedure of reconstruction and refiltering is referred to as signal correction. As a result, the corrected signal from SCAMP has a 3-dB cutoff frequency at 30 Hz (Fig. 8.2).

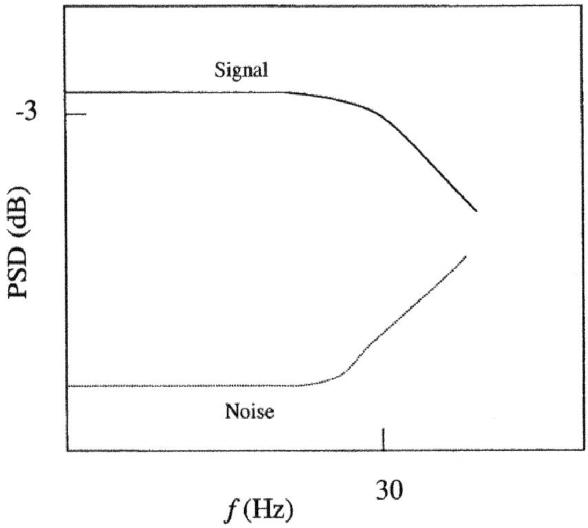

Figure 8.2. Corrected signal and amplified noise.

Different algorithms for obtaining ε from fitting the Batchelor spectrum have been used by researchers. Key issues in fitting the Batchelor spectrum were discussed by Luketina and Imberger (1998), including: (1) the appropriate portion of the estimated spectrum to use; (2) the dissipation of temperature variance, x_T, be determined from the estimated spectrum; and (3) an appropriate error function to use in determining the optimum fit. The entire procedure of fitting the Batchelor spectrum is referred to as Batchelor curve fitting in this chapter.

Dillon and Caldwell (1980) considered only fully resolved spectra for Batchelor curve fitting. A spectrum was defined as 'resolved' if it fell to a certain percentage of the spectral peak and still above the noise level before the wave number equivalent to 30 Hz which corresponds to the 3-dB cutoff frequency of the corrected response of the sensor. Dillon and Caldwell's (1980) method calculates x_T by integrating the resolved spectrum, and finds ε by performing a non-linear weighted least square fitting to the Batchelor spectrum.

Luketina and Imberger (1998) found the best fit by a trial-and-error approach instead of non-linear curve fitting algorithms which often find only the best local rather than global fit. In this method, x_T is estimated by the sum of the integration of the estimated spectrum between two intersections of the theoretical Batchelor spectrum, and the integration of the theoretical Batchelor spectrum outside these two intersections. Different portions of the estimated spectrum are fitted between the lowest wave number of the estimated spectrum and the noise wave number. The noise wave number is determined at the intersection of the Batchelor spectrum and the noise spectrum which has a positive slope at higher wave numbers of the estimated spectrum. However, the definition of the noise wave number by Luketina and Imberger (1998) is not applicable to the estimated spectra in which the noise spectrum is not obvious or the slope at higher wave numbers is not positive. A complicated error function is used in this method, which was defined by Luketina (1987) as the difference between the theoretical Batchelor and estimated spectra in the slope of -1 in log-log space. The error function is empirically adjusted by using a bias factor so as to accept a larger portion of the spectrum. A difficulty in applying this error function may occur when the estimated spectrum does not meet the fitted curve in the slope of -1; consequently, an infinite value of the difference is encountered by the error function.

Instead of directly delimiting estimated spectra after the noise wave number, a noise removal scheme was tested by Sherman and Davis (1995). A spectrum averaged over 76 profiles is used in this method to model the shape of the noise floor. Since this approach requires multiple profiles of measurements, it is not applicable to single-profile records. Ruddick, Anis and Thompson (1999) considered the instrumental noise spectrum as a component of the theoretical spectrum. The noise model was made from a parametric fit based on the SCAMP sensor circuit, and a noise parameter selected according to observations of real data. A disadvantage common to these two approaches is that the attenuation of the signal (Fig. 8.2), excluding the amplified noise, due to the correction of the sensor response at higher frequencies is not considered. If the attenuated signal is still above the amplified noise level at higher frequencies, the Batchelor spectrum will be underestimated by using this method.

An essential problem encountered but not discussed by previous Batchelor curve fitting methods is due to the spectral bias in the periodogram and the smoothing window effect. The periodogram is a conventional spectral estimator which suffers sidelobe energy leakage (bias) in frequency bands where the true spectrum is zero. The application of a smoothing window may reduce the variance of periodogram, but at the cost of exaggerating the bias. Because the Batchelor spectrum has a high dynamic range (defined as the difference between the highest and lowest spectral values in log), the bias will cause the lower spectral values to increase. Consequently, the accuracy in previous Batchelor curve fitting methods is affected. Also, the observed noise level may vary with signals at different vertical depths, which questions that a noise model of fixed spectral values should be assumed in both Sherman and Davis's (1995) and Ruddick, Anis and Thompson's (1999) methods.

A new spectral estimation method called multitaper spectral estimation was introduced by Thomson (1982), using multiple orthogonal tapers to achieve a better tradeoff between the bias and variance of the spectrum (see Sec. 4.3). Temperature gradient spectra computed from conventional and multitaper spectral estimation methods, including the periodogram, a Hamming-window smoothed spectral estimator, and the eigenvalue-weighted multitaper spectral estimator are compared in this chapter. Secondly a modified Batchelor curve fitting method is tested. Only fully resolved spectra are considered in this method. ε and x_T are obtained by performing a trial-and-error approach. A comparison is made from the results of Batchelor curve fitting based on three error functions.

The contents of this chapter are arranged as follows: multitaper spectral estimation (Thomson, 1982) is introduced in Section 8.2; a modified Batchelor curve fitting method and three error functions are defined in Section 8.3; synthetic series and twelve segments selected from a measured temperature gradient profile are used in Section 8.4 for a comparison of spectral estimation methods; results of the modified Batchelor curve fitting method performed on synthetic series with three levels of noise contamination are presented in Section 8.5, and a comparison is made based on three error functions. Conclusions about Batchelor curve fitting method are presented in Section 8.6.

8.2. MULTITAPER SPECTRAL ESTIMATION

The periodogram is an asymptotically unbiased estimator of the true spectrum. The expected value of periodogram is equivalent to the true spectral density convolved with Fejer's kernel. That the periodogram is asymptotically unbiased does not mean that its bias is negligible for any particular sample size. Percival and Walden (1993) compared bias in the spectra of three specific stationary processes with dynamic ranges of 0, 14 and 65 dB. The dynamic range is defined by

$$10\left[\log_{10}\left(\frac{\max_f S(f)}{\min_f S(f)}\right)\right] \qquad [\text{dB}],$$

where S is the spectrum and f is the frequency. They showed that the bias is almost 20 dB in the periodogram for processes with high dynamic range of 65 dB even for the sample size as large as 1024. Due to the energy leakage from the sidelobes of Fejer's kernel, the periodogram is badly biased in the frequency bands where the true spectral value is small compared to the rest of the spectrum.

One of the commonly used techniques to reduce the bias is tapering, which suppresses the sidelobes of Fejer's kernel in the convolution with a sequence of real-valued constants called a data taper. However, the sidelobes are suppressed at the expense of a wider central lobe. Therefore, there is an inherent tradeoff between reducing bias due to the sidelobes and introducing the bias due to the width of the central lobe (Percival and Walden, 1993). Moreover, the variance of the tapered spectral estimator increases as the central lobe becomes wider. While the bias of the periodogram can be considerably reduced by using an appropriate data taper, the variance of the periodogram becomes another problem. Since the periodogram is an inconsistent estimator, it does not help to reduce the variance by increasing the sample size. A conventional method to reduce the variance is to smooth the periodogram across frequencies. The variance decreases as the bandwidth of the smoothing window increases. However, it is shown that the larger the bandwidth, the greater the bias in Percival and Walden (1993).

Thomson (1982) introduced multitaper spectral estimation by using multiple orthogonal tapers to achieve a better tradeoff between the bias and variance for spectral estimators. The simplest form of a multitaper spectral estimator is the average of K direct spectral estimators is the kth-order eigenspectrum of a stationary process

$$\hat{S}^{(mt)}(f) = \frac{1}{K}\sum_{k=0}^{K-1} \hat{S}_k^{(mt)}(f), \tag{8.1}$$

where

$$\hat{S}_k^{(mt)}(f) = \Delta\left|\sum_{t=1}^{N} h_{t,k} Y_t e^{-i2\pi ft\Delta}\right|^2, \tag{8.2}$$

$\{Y_t\}$ with sample size N, Δ is the sampling interval, and $\{h_{t,k}\}$ is the data taper for the kth-order eigenspectrum. A set of K orthogonal data tapers is given by portions of the discrete prolate spheroidal sequences (dpss) with parameter W. The resolution bandwidth is defined as $2W$, and K is chosen to be less than $2NW\Delta$. A reasonable choice of W, usually less than 4, must take into account the tradeoff between decreasing the variance and smearing the resolution. Due to the bias properties of the eigenspectra, the higher-order eigenspectra have greater bias and must be downweighted. Therefore, an eigenvalue-weighted multitaper spectral estimator (Thomson, 1982; and Percival and Walden , 1993) is used in this study

$$\bar{S}^{(mt)}(f) = \frac{\sum_{k=0}^{K-1} \lambda_k \hat{S}_k^{(mt)}(f)}{\sum_{k=0}^{K-1} \lambda_k}, \tag{8.3}$$

where λ_k is the kth-order eigenvalue of the discrete prolate spheroidal wave functions (dpswf). The eigenvalue-weighted multitaper spectral estimator is chi-square distributed with $2K$ degrees of freedom.

8.3. BATCHELOR CURVE FITTING

Once a stationary segment of the temperature gradient record is determined, an estimated spectrum is computed from the segment for Batchelor curve fitting. Before fitting the Batchelor spectrum, a resolved-spectrum screening procedure for Batchelor curve fitting is adopted from Dillon and Caldwell's (1980) method to select the estimated spectra which have two distinct subranges with the cutoff wave number k_c less than the wave number equivalent to 30 Hz. The cutoff wave number k_c was defined by Caldwell et al. (1980) as the wave number at which the spectrum has decreased to 3.5% of its peak value. The theoretical Batchelor spectrum falls to this percentage at wave number k equivalent to k_B (the Batchelor wave number defined in Eq. 2.1).

Three error functions are tested for Batchelor curve fitting. The first is defined by the square of the spectral difference between the theoretical Batchelor and estimated spectra in log-log space,

$$E_1 = \sum_k \left(\ln\left(S_{est}(k)\right) - \ln\left(S_B(k)\right)\right)^2 , \tag{8.4}$$

where S_{est} and S_B denote the estimated and the theoretical Batchelor spectra (computed from Eq. 2.1), respectively. The second error function is defined by a chi-square distributed function (Bevington and Robinson, 1969),

$$E_2 = \sum_k \left(\frac{S_{est}(k) - S_B(k)}{S_B(k)} \right)^2 , \tag{8.5}$$

and the last one is a weighted chi-square distributed function,

$$E_3 = \sum_k S_B(k) \left(\frac{S_{est}(k) - S_B(k)}{S_B(k)} \right)^2 , \tag{8.6}$$

where the weights are the values of the Batchelor spectrum itself.

Prior to the first set of trials, x_T is calculated by integrating the entire estimated spectrum. The error function is calculated with k ranging from the first non-zero wave number ordinate to the wave number equivalent to 30 Hz (3-dB cutoff frequency). At each set of trials, 101 geometrically distributed values of ε are tried, ranging from 10^{-12} to $2.47*10^{-7}$ m^2/sec^3. The upper bound $2.47*10^{-7}$ m^2/sec^3 of ε is chosen so that the corresponding Batchelor wave number k_B is equivalent to 30 Hz. If the value of ε is greater than $2.47*10^{-7}$ m^2/sec, then the estimated spectrum is not resolved. The best fit of the first set of trials is determined at the value of ε corresponding to the minimum of the error function.

After the best fit of the first set of trials is found, x_T is re-estimated by summing two parts: the first is from the integration of the estimated spectrum over the range between the smallest (k_1) and the largest wave numbers (k_2) at which the values of the theoretical Batchelor spectrum are within the 95% confidence interval (Fig. 8.3), and the second is from the integration of the theoretical Batchelor spectrum outside this range. The error function is re-calculated with k ranging between k_1 and k_2 (Fig. 8.3). After the best fit of the second set of trials is found, a new pair of ε and x_T are obtained. If the difference in log-log space between two consecutive sets of trials is less than 0.01 for both ε and x_T, the trial-and-error procedures stop and the optimum fit is obtained. Otherwise, a new set of trials will begin.

A diagnostics procedure is used to examine the quality of fitting. If the theoretical Batchelor spectrum has more than 20% of the range between k_1 and k_2 outside the 95% confidence interval, or if the range between k_1 and k_2 is less than 80 cycle/m (suggested in Luketina and Imberger, 1998), the fitting result is considered as 'bad fit'. The procedure of the modified Batchelor curve fitting method is depicted in Fig. 8.4.

8.4. COMPARISON OF SPECTRAL ESTIMATION METHODS

Three spectral estimators are compared in this study, including the periodogram, a Hamming-window smoothed spectral estimator, and the eigenvalue-weighted multitaper spectral estimator (Eq. 8.3). The Hamming window is a lag window defined by Blackman and Tukey (1958), which minimizes the magnitude in the first sidelobe. The lag of Hamming window is chosen to be *N*/2, and the corresponding frequency resolution is 0.64*(4/*N*) (see Eq. 12.2.64 in Proakis and Manolakis, 1996).

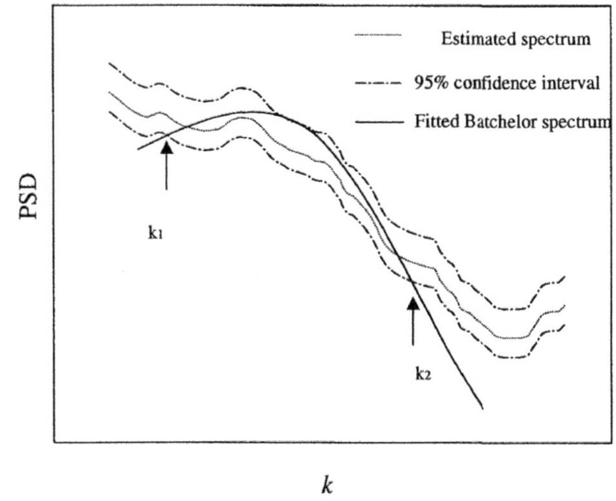

Figure 8.3. The smallest (k_1) and the largest wave numbers (k_2) at which the values of the theoretical Batchelor spectrum are within the 95% confidence interval.

The parameter *N***W* (*W* is defined in Section 8.2) for the eigenvalue-weighted multitaper spectral estimator is specified as 2 so that the resolution bandwidth 4/*N* is slightly greater than that of the Hamming-window smoothed spectral estimator. The eigenvalue-weighted multitaper spectral estimator (*NW* = 2) is chi-square distributed with 8 degrees of freedom.

Ten synthetic series (data length *N* is 256) are generated from Model 2.1.9, according to the Batchelor spectrum of $\varepsilon = 10^{-7}$ m^2/sec^3 and $\chi_T = 10^{-7}$ (degC)^2/sec with the noise level $\sigma_{wn}^2 = 4*10^{-7}$ (degC/m)^2/(cyc/m). The noise level $\sigma_{wn}^2 = 4*10^{-7}$ (degC/m)^2/(cyc/m) is equivalent to 10-dB signal-to-noise ratio at the wave number equivalent to 30 Hz (3-dB cutoff frequency). Spectra computed from these three spectral estimation methods for two series are shown in Figs. 8.5 and 8.6. These results show that the multitaper spectra have less bias than the Hamming-window smoothed spectra at high frequencies, except for two series. Although the Hamming-window smoothed spectrum in Fig. 8.6 has lower spectral values at high frequencies than the multitaper spectrum, these low values at high frequencies in the Hamming-window smoothed spectrum show greater bias, and are underestimated compared to the designed spectrum.

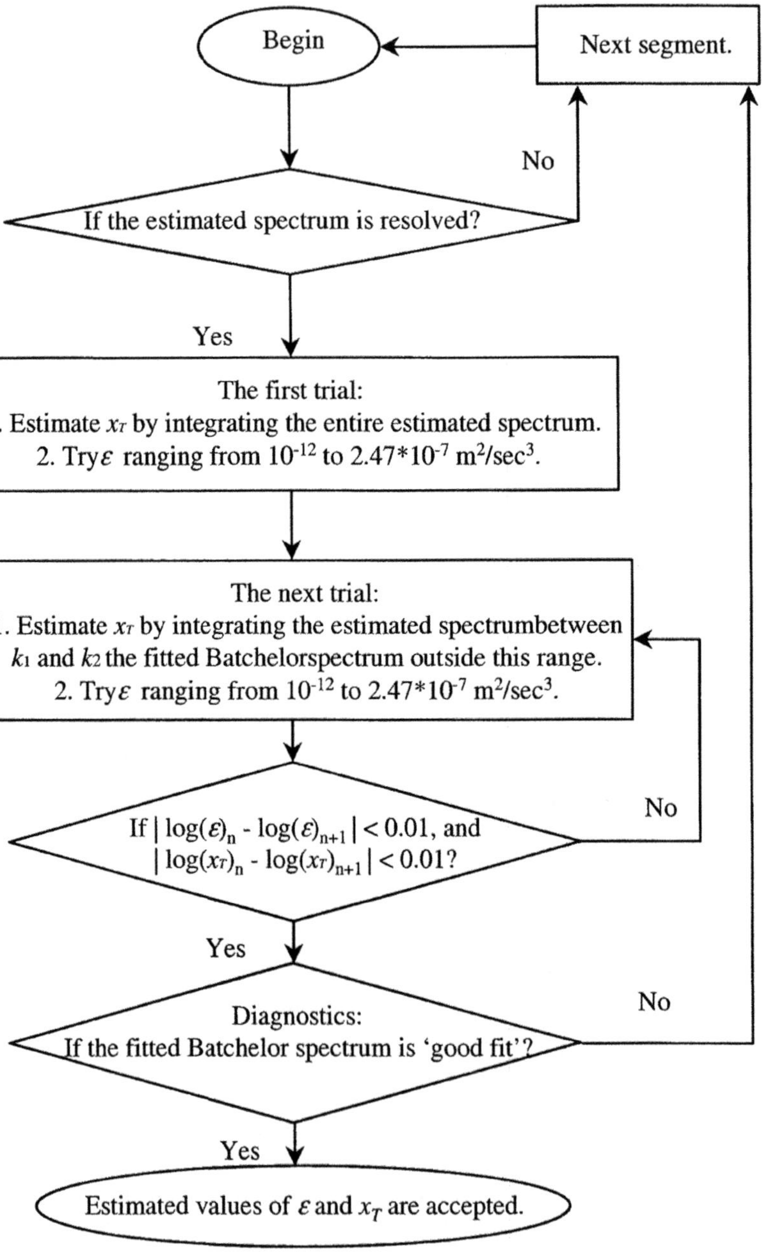

Figure 8.4. The modified Batchelor curve fitting method.

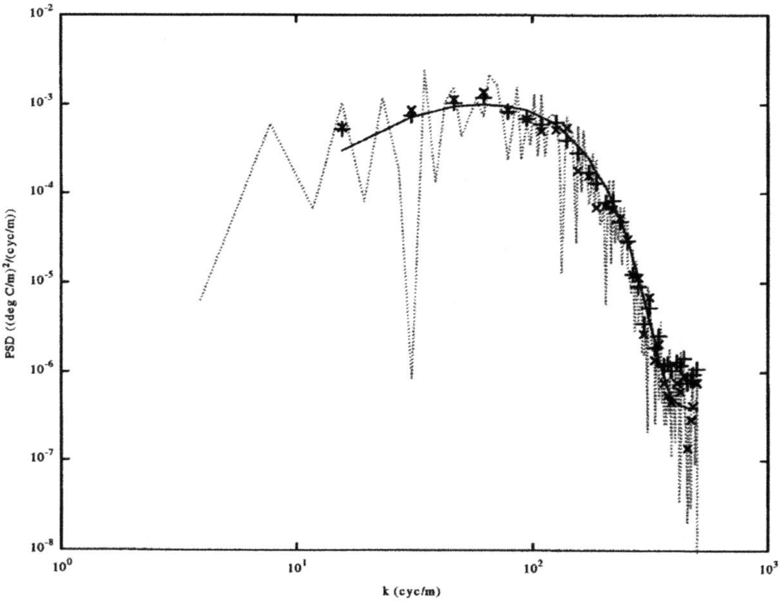

Figure 8.5. Spectra computed from three methods for a series from model 2.1.9.
—: Designed spectrum; - - -: Periodogram;
x: Multitaper spectrum; +: Hamming-window smoothed spectrum.

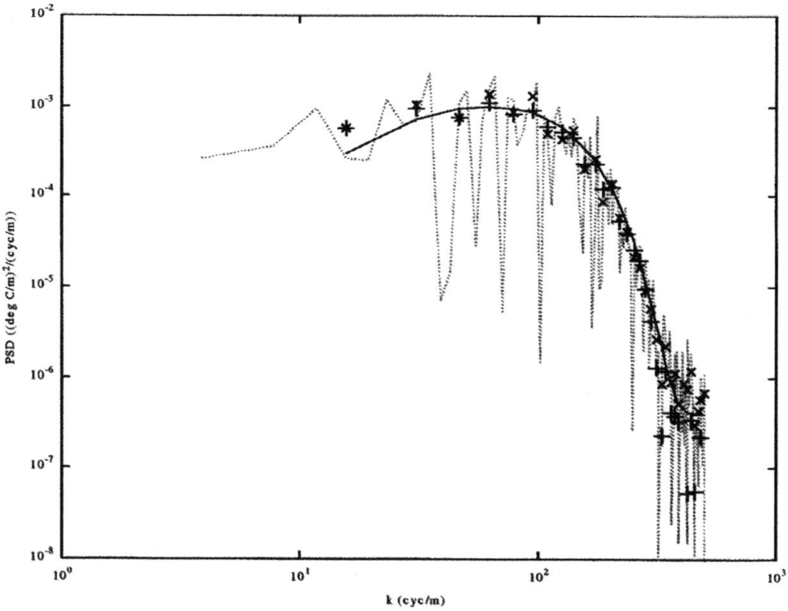

Figure 8.6. Spectra computed from three methods for a series from model 2.1.9.
—: Designed spectrum; - - -: Periodogram;
x: Multitaper spectrum; +: Hamming-window smoothed spectrum.

Twelve arbitrary segments selected from a temperature gradient profile (given in Chapter 2) are used to compare the performances of these three spectral estimation methods. Each segment contains 128 data points. Results for two segments are shown in Figs. 8.7 and 8.8. These results show that the periodogram has greater variations than the other two estimators, and the multitaper spectrum has lower spectral values at high frequencies than the Hamming-window smoothed spectrum, except for two segments (one in Fig. 8.7) lower spectral values are shown at high frequencies in the Hamming-window smoothed spectrum. There is the possibility that these lower values in the Hamming-window smoothed spectrum are underestimated due to local spectral variations within the bandwidth of the Hamming window.

8.5. BATCHELOR CURVE FITTING TO SYNTHETIC SERIES

The accuracy of Batchelor curve fitting and dissipation estimation is tested on estimated spectra computed from the synthetic series for which the theoretical Batchelor spectra are known. Since the noise level may vary in the estimated spectra due to spectral bias, it is necessary to examine the sensitivity of Batchelor curve fitting by using spectra contaminated with different levels of noise. It should be noted that even if estimated spectra are free of noise, there is the possibility that fitting errors result from the procedures of Batchelor curve fitting.

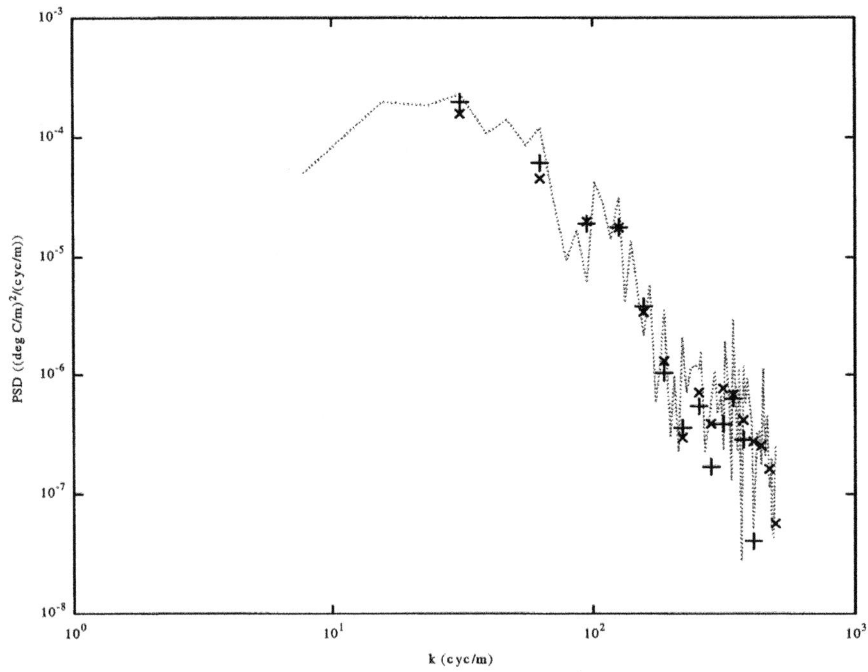

Figure 8.7. Spectra for a selected segment from profile 1210 (data 4129-4256).
- - -: Periodogram; x: Multitaper spectrum; +: Hamming-window smoothed spectrum.

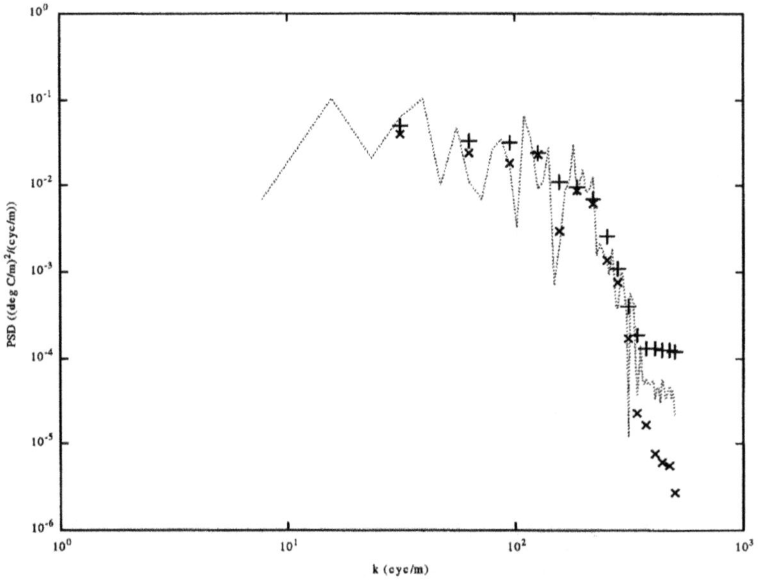

Figure 8.8. Spectra for a selected segment from profile 1210 (data 7001-7128). - - -: Periodogram; x: Multitaper spectrum; +: Hamming-window smoothed spectrum.

Three sets of synthetic series (data length N is 256) are generated from Models 2.1.8-2.1.10, based on the Batchelor spectrum of $\varepsilon = 10^{-7}$ m^2/sec^3 and $\chi_T = 10^{-7}$ (degC)^2/sec with three different levels of noise $\sigma_{wn}^2 = 0$, $4{*}10^{-7}$ and $4{*}10^{-5}$ (degC/m)^2/(cyc/m). The noise level $\sigma_{wn}^2 = 4{*}10^{-7}$ (degC/m)^2/(cyc/m) is equivalent to 10-dB signal-to-noise ratio at the wave number equivalent to 30 Hz (3-dB cutoff frequency), and the noise level $\sigma_{wn}^2 = 4{*}10^{-5}$ (degC/m)^2/(cyc/m) is equivalent to 10-dB signal-to-noise ratio. The eigenvalue-weighted multitaper spectral estimator (NW = 2) is used to compute estimated spectra for synthetic series. The estimated results of ε and χ_T corresponding to three error functions (Eqs. 8.4, 8.5 and 8.6) are compared for each set of synthetic series.

8.5.1. Batchelor Curve Fitting Using the First Error Function. Results of Batchelor curve fitting using the first error function (Eq. 8.4) for two series of each model are shown in Figs. 8.9-8.14. The complete results are summarized in Tables VIII.1 and VIII.2. These results show that the average errors of estimated χ_T are within 10% for series of these three models (Table VIII.2). The average errors of estimated ε are within 10% for series of Models 2.1.8 and 2.1.9 (Table VIII.1). For series from Model 2.1.10, the average of estimated values of ε is overestimated up to 1.8 times of the designed value ($\varepsilon = 10^{-7}$ m^2/sec^3) (Table VIII.1). These highly overestimated values are due to the bias caused by the high noise level ($\sigma_{wn}^2 = 4{*}10^{-5}$ (degC/m)^2/(cyc/m)) used in Model 2.1.10. The results for the series from Model 2.1.10 also show that the cutoff wave number k_c at which the spectrum has decreased to 3.5% of its peak value is obviously larger than the wave number 300 cycle/m equivalent to 30 Hz (Figs. 8.13 and 8.14). Therefore, all the estimated spectra of series of Model 2.1.10 do not pass the resolved-spectrum screening procedure defined in Section 8.3.

Therefore, all the estimated spectra of series of Model 2.1.10 do not pass the resolved-spectrum screening procedure defined in Section 8.3.

Table VIII.1. Estimated dissipation rate ε [m^2/sec^3] for ten synthetic series by using the first error function.

Model	2.1.8	2.1.9	2.1.10
1	$8.9095*10^{-8}$	$8.9095*10^{-8}$	$1.2523*10^{-7}$
2	$1.0563*10^{-7}$	$1.0563*10^{-7}$	$1.7603*10^{-7}$
3	$8.9095*10^{-8}$	$1.0563*10^{-7}$	$1.7603*10^{-7}$
4	$1.2523*10^{-7}$	$1.2523*10^{-7}$	$2.0870*10^{-7}$
5	$1.0563*10^{-7}$	$1.0563*10^{-7}$	$1.2523*10^{-7}$
6	$1.2523*10^{-7}$	$1.2523*10^{-7}$	$1.4848*10^{-7}$
7	$1.0563*10^{-7}$	$1.0563*10^{-7}$	$1.4848*10^{-7}$
8	$1.0563*10^{-7}$	$1.0563*10^{-7}$	$1.7603*10^{-7}$
9	$1.2523*10^{-7}$	$1.2523*10^{-7}$	$2.4743*10^{-7}$
10	$1.0563*10^{-7}$	$1.0563*10^{-7}$	$2.4743*10^{-7}$
Average	$1.0820*10^{-7}$	$1.0986*10^{-7}$	$1.7791*10^{-7}$
Designed	10^{-7}	10^{-7}	10^{-7}

Table VIII.2. Estimated dissipation of temperature variance x_T [$(degC)^2/sec$] for ten synthetic series by using the first error function.

Model	2.1.8	2.1.9	2.1.10
1	$1.0101*10^{-7}$	$1.0117*10^{-7}$	$1.1057*10^{-7}$
2	$1.0746*10^{-7}$	$1.0712*10^{-7}$	$1.1191*10^{-7}$
3	$1.0687*10^{-7}$	$1.0677*10^{-7}$	$1.1573*10^{-7}$
4	$8.5110*10^{-8}$	$8.6025*10^{-8}$	$1.0227*10^{-7}$
5	$9.9817*10^{-8}$	$9.9438*10^{-8}$	$1.0385*10^{-7}$
6	$8.9124*10^{-8}$	$8.9537*10^{-8}$	$1.0031*10^{-7}$
7	$9.5820*10^{-8}$	$9.6746*10^{-8}$	$1.1295*10^{-7}$
8	$9.2465*10^{-8}$	$9.2950*10^{-8}$	$1.0508*10^{-7}$
9	$7.8025*10^{-8}$	$7.8608*10^{-8}$	$9.0333*10^{-8}$
10	$8.8814*10^{-8}$	$8.8346*10^{-8}$	$9.2064*10^{-8}$
Average	$9.4453*10^{-8}$	$9.4672*10^{-8}$	$1.0451*10^{-7}$
Designed	10^{-7}	10^{-7}	10^{-7}

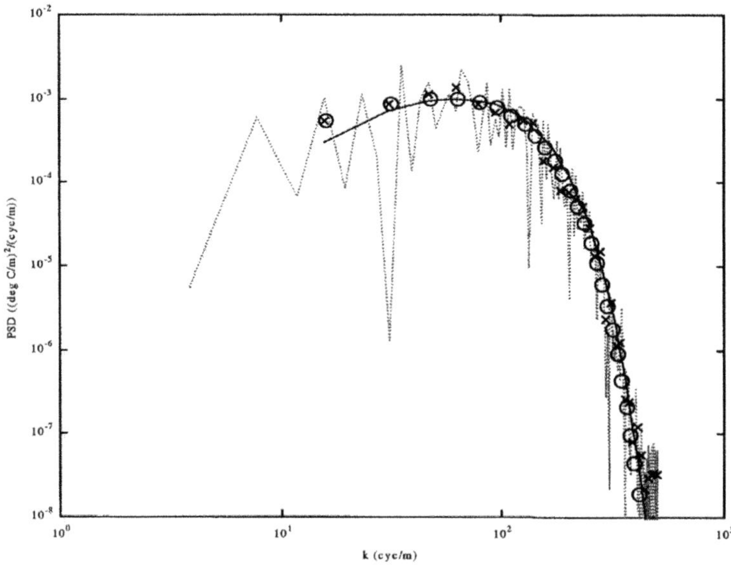

Figure 8.9. Batchelor curve fitting to the multitaper spectrum of a series from model 2.1.8, using the first error function.

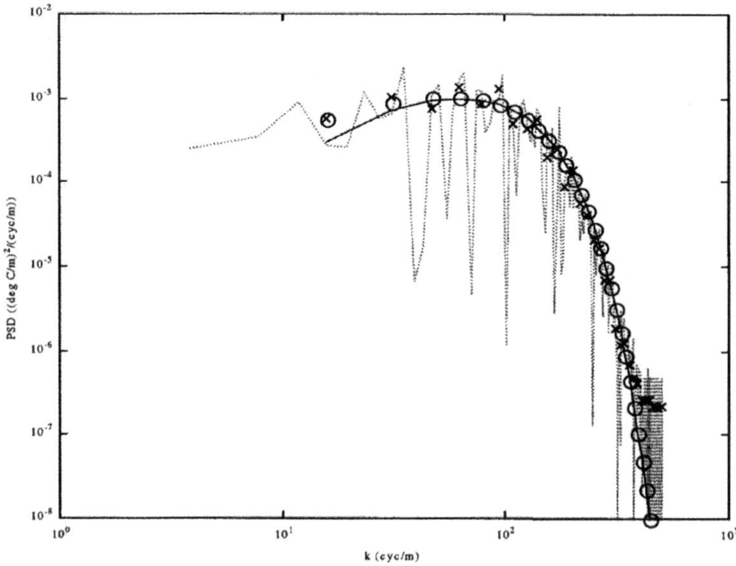

Figure 8.10. Batchelor curve fitting to the multitaper spectrum of a series from model 2.1.8, using the first error function.
—: Designed Batchelor spectrum; - - -: Periodogram;
x: Multitaper spectrum; o: Fitted Batchelor spectrum.

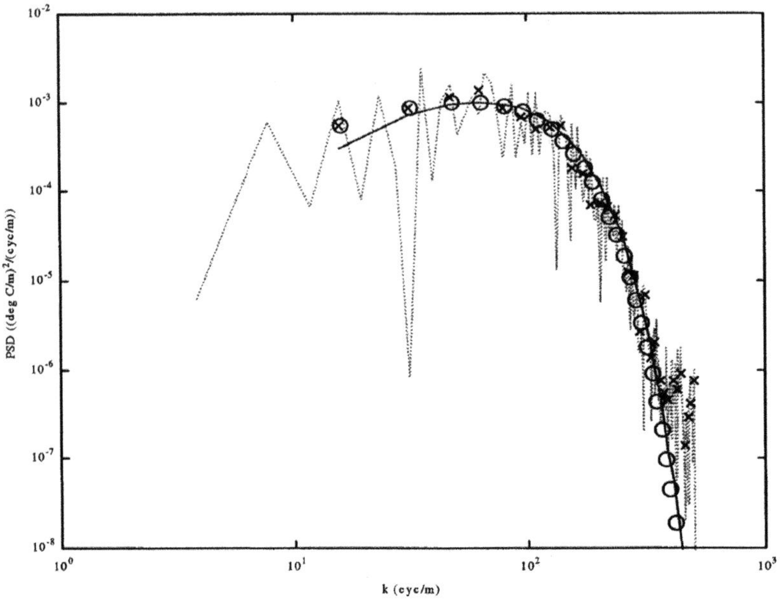

Figure 8.11. Batchelor curve fitting to the multitaper spectrum of a series from model 2.1.9, using the first error function.

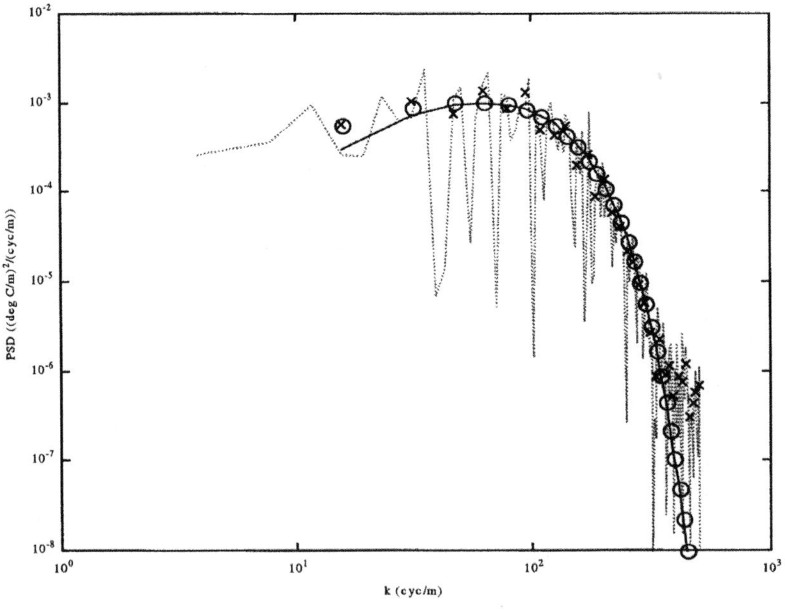

Figure 8.12. Batchelor curve fitting to the multitaper spectrum of a series from model 2.1.9, using the first error function.

—: Designed Batchelor spectrum; - - -: Periodogram;
x: Multitaper spectrum; o: Fitted Batchelor spectrum.

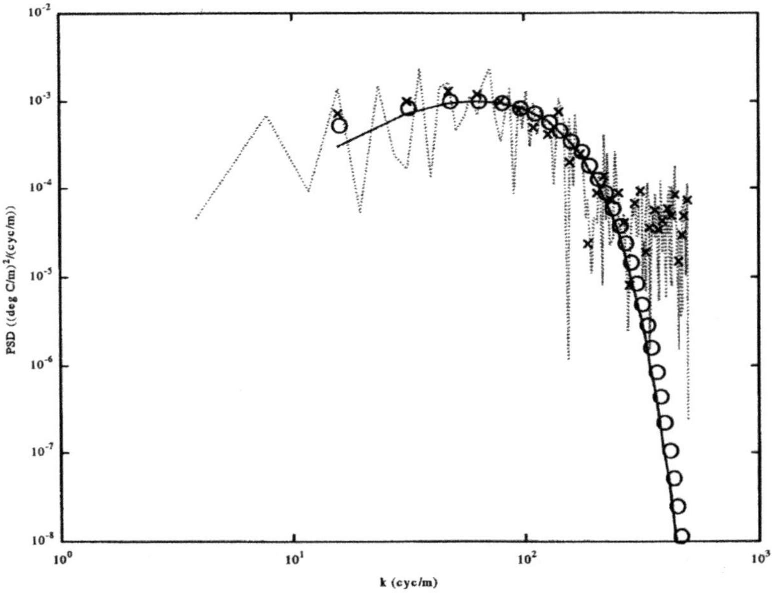

Figure 8.13. Batchelor curve fitting to the multitaper spectrum of a series from model 2.1.10, using the first error function.

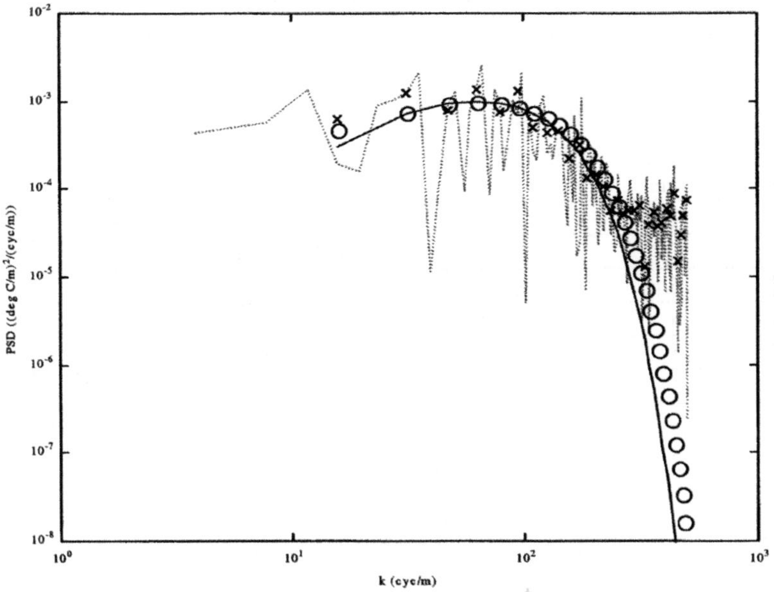

Figure 8.14. Batchelor curve fitting to the multitaper spectrum of a series from model 2.1.10, using the first error function.
—: Designed Batchelor spectrum; - - -: Periodogram;
x: Multitaper spectrum; o: Fitted Batchelor spectrum.

VIII.4. These results show that the average errors of estimated x_T are within 10% for series from these three models (Table VIII.4). The average errors of estimated ε are greater than 20%, and up to 2.4 times for series of Model 2.1.10 (Table VIII.3). Similar to the results in Section 8.5.1, the estimated spectra for series from Model 2.1.10 do not pass the resolved-spectrum screening procedure (Figs. 8.19 and 8.20).

Table VIII.3. Estimated dissipation rate ε [m^2/sec^3] for ten synthetic series by using the second error function.

Model	2.1.8	2.1.9	2.1.10
1	$1.0563*10^{-7}$	$1.0563*10^{-7}$	$2.4743*10^{-7}$
2	$1.0563*10^{-7}$	$1.0563*10^{-7}$	$2.4743*10^{-7}$
3	$1.0563*10^{-7}$	$1.0563*10^{-7}$	$2.4743*10^{-7}$
4	$1.2523*10^{-7}$	$1.2523*10^{-7}$	$2.0870*10^{-7}$
5	$1.4848*10^{-7}$	$1.4848*10^{-7}$	$2.4743*10^{-7}$
6	$1.4848*10^{-7}$	$1.4848*10^{-7}$	$2.4743*10^{-7}$
7	$1.0563*10^{-7}$	$1.0563*10^{-7}$	$2.4743*10^{-7}$
8	$1.2523*10^{-7}$	$1.2523*10^{-7}$	$2.4743*10^{-7}$
9	$1.2523*10^{-7}$	$1.2523*10^{-7}$	$2.4743*10^{-7}$
10	$1.2523*10^{-7}$	$1.2523*10^{-7}$	$2.4743*10^{-7}$
Average	$1.2204*10^{-7}$	$1.2204*10^{-7}$	$2.4356*10^{-7}$
Designed	10^{-7}	10^{-7}	10^{-7}

Table VIII.4. Estimated dissipation of temperature variance x_T [$(degC)^2/sec$] for ten synthetic series by using the second error function.

Model	2.1.8	2.1.9	2.1.10
1	$1.0078*10^{-7}$	$1.0094*10^{-7}$	$1.1080*10^{-7}$
2	$1.0746*10^{-7}$	$1.0712*10^{-7}$	$1.1245*10^{-7}$
3	$1.0663*10^{-7}$	$1.0677*10^{-7}$	$1.1610*10^{-7}$
4	$8.5110*10^{-8}$	$8.6025*10^{-8}$	$1.0227*10^{-7}$
5	$9.9563*10^{-8}$	$9.9187*10^{-8}$	$1.0428*10^{-7}$
6	$8.9005*10^{-8}$	$8.9418*10^{-8}$	$1.0115*10^{-7}$
7	$9.5820*10^{-8}$	$9.6746*10^{-8}$	$1.1349*10^{-7}$
8	$9.2293*10^{-8}$	$9.2777*10^{-8}$	$1.0575*10^{-7}$
9	$7.8025*10^{-8}$	$7.8608*10^{-8}$	$9.0333*10^{-8}$
10	$8.8649*10^{-8}$	$8.8181*10^{-8}$	$9.2064*10^{-8}$
Average	$9.4334*10^{-8}$	$9.4577*10^{-8}$	$1.0487*10^{-7}$
Designed	10^{-7}	10^{-7}	10^{-7}

8.5.3. Batchelor Curve Fitting Using the Third Error Function. Results of Batchelor curve fitting using the third error function (Eq. 8.6) for two series of each model are shown in Figs. 8.21-8.26. The complete results are summarized in Tables VIII.5 and VIII.6. These results show that the average errors of estimated of x_T are within 10% for series of these three models (Table VIII.6). The average errors of estimated ε are less than 3% (Table VIII.5), except for series of Model 2.1.10, the average of estimated values of ε is overestimated at 1.4 times the designed value (Table VIII.5). The estimated spectra for series of Model 2.1.10 do not pass the resolved-spectrum screening procedure defined (Figs. 8.25 and 8.26).

Table VIII.5. Estimated dissipation rate ε [m^2/sec^3] for ten synthetic series by using the third error function.

Model	2.1.8	2.1.9	2.1.10
1	$8.9095*10^{-8}$	$8.9095*10^{-8}$	$1.2523*10^{-7}$
2	$8.9095*10^{-8}$	$8.9095*10^{-8}$	$1.0563*10^{-7}$
3	$1.0563*10^{-7}$	$1.0563*10^{-7}$	$1.2523*10^{-7}$
4	$1.0563*10^{-7}$	$1.0563*10^{-7}$	$1.4848*10^{-7}$
5	$1.0563*10^{-7}$	$1.0563*10^{-7}$	$1.2523*10^{-7}$
6	$8.9095*10^{-8}$	$8.9095*10^{-8}$	$1.2523*10^{-7}$
7	$8.9095*10^{-8}$	$8.9095*10^{-8}$	$1.0563*10^{-7}$
8	$1.2523*10^{-7}$	$1.2523*10^{-7}$	$1.7603*10^{-7}$
9	$1.2523*10^{-7}$	$1.2523*10^{-7}$	$1.7603*10^{-7}$
10	$1.0563*10^{-7}$	$1.0563*10^{-7}$	$2.0870*10^{-7}$
Average	$1.0294*10^{-7}$	$1.0294*10^{-7}$	$1.4214*10^{-7}$
Designed	10^{-7}	10^{-7}	10^{-7}

Table VIII.6. Estimated dissipation of temperature variance x_T [(degC)^2/sec] for ten synthetic series by using the third error function.

Model	2.1.8	2.1.9	2.1.10
1	$1.0101*10^{-7}$	$1.0117*10^{-7}$	$1.1057*10^{-7}$
2	$1.0771*10^{-7}$	$1.0737*10^{-7}$	$1.1068*10^{-7}$
3	$1.0663*10^{-7}$	$1.0677*10^{-7}$	$1.1539*10^{-7}$
4	$8.5268*10^{-8}$	$8.6185*10^{-8}$	$1.0217*10^{-7}$
5	$9.9817*10^{-8}$	$9.9438*10^{-8}$	$1.0385*10^{-7}$
6	$8.9497*10^{-8}$	$8.9912*10^{-8}$	$1.0030*10^{-7}$
7	$9.6041*10^{-8}$	$9.6969*10^{-8}$	$1.1197*10^{-7}$
8	$9.2293*10^{-8}$	$9.2777*10^{-8}$	$1.0508*10^{-7}$
9	$7.8034*10^{-8}$	$7.8619*10^{-8}$	$9.0804*10^{-8}$
10	$8.8814*10^{-8}$	$8.8346*10^{-8}$	$9.1862*10^{-8}$
Average	$9.4512*10^{-8}$	$9.4756*10^{-8}$	$1.0427*10^{-7}$
Designed	10^{-7}	10^{-7}	10^{-7}

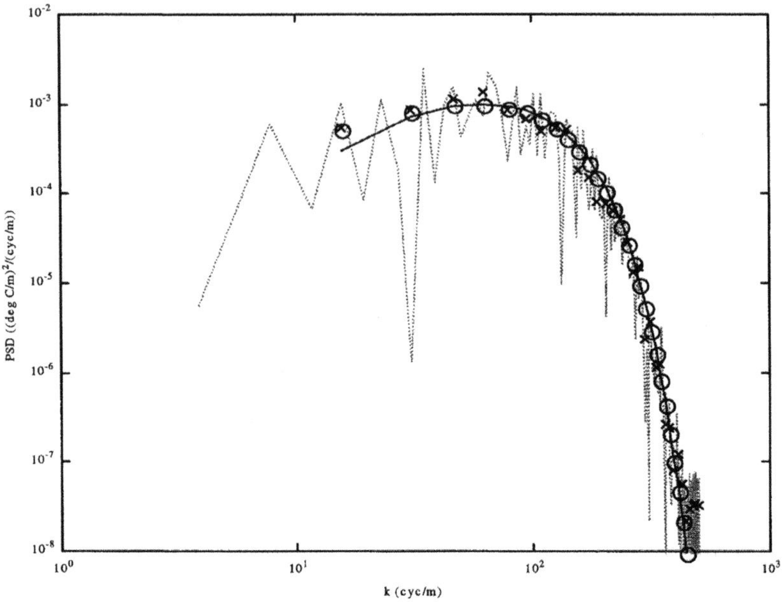

Figure 8.15. Batchelor curve fitting to the multitaper spectrum of a series from model 2.1.8, using the second error function.

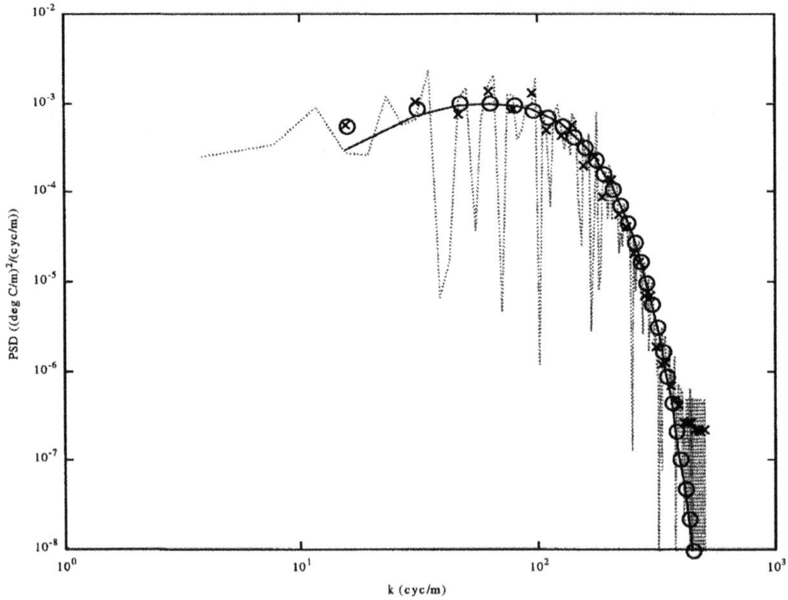

Figure 8.16. Batchelor curve fitting to the multitaper spectrum of a series from model 2.1.8, using the second error function.
—: Designed Batchelor spectrum; - - -: Periodogram;
x: Multitaper spectrum; o: Fitted Batchelor spectrum.

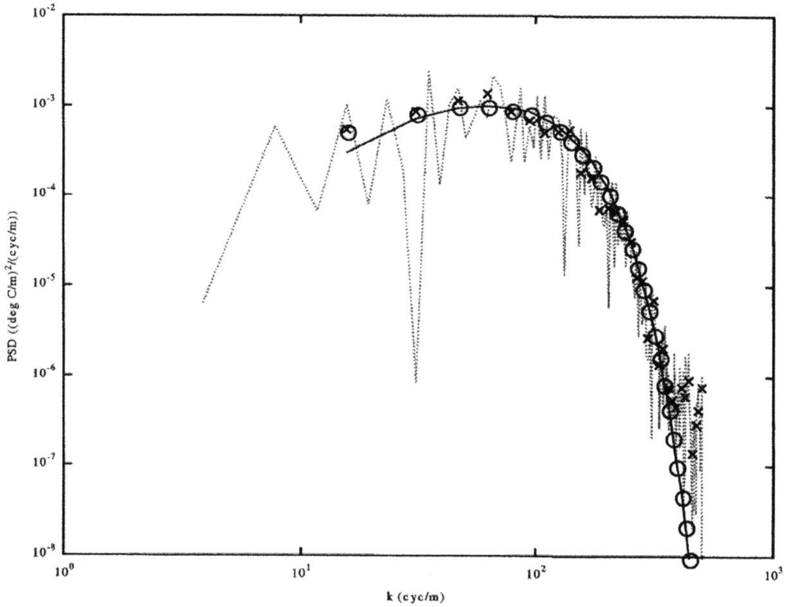

Figure 8.17. Batchelor curve fitting to the multitaper spectrum of a series from model 2.1.9, using the second error function.

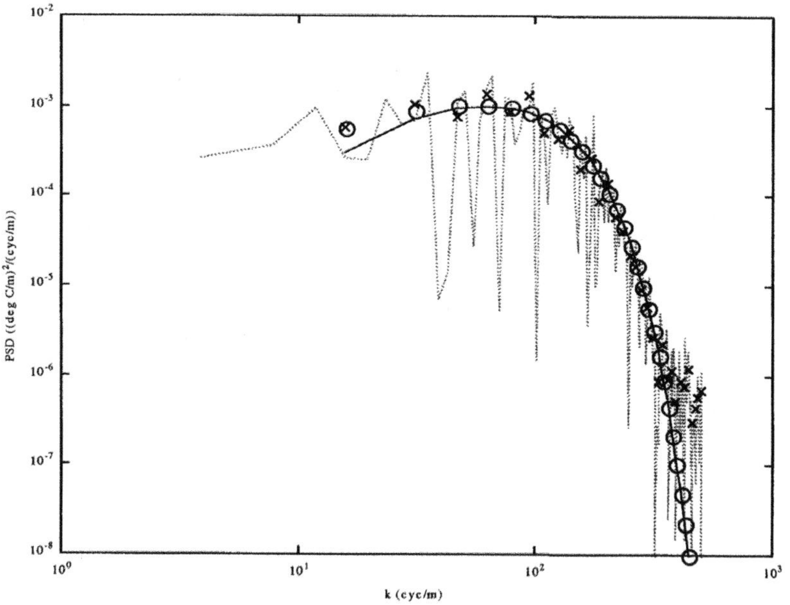

Figure 8.18. Batchelor curve fitting to the multitaper spectrum of a series from model 2.1.9, using the second error function.
—: Designed Batchelor spectrum; - - -: Periodogram;
x: Multitaper spectrum; o: Fitted Batchelor spectrum.

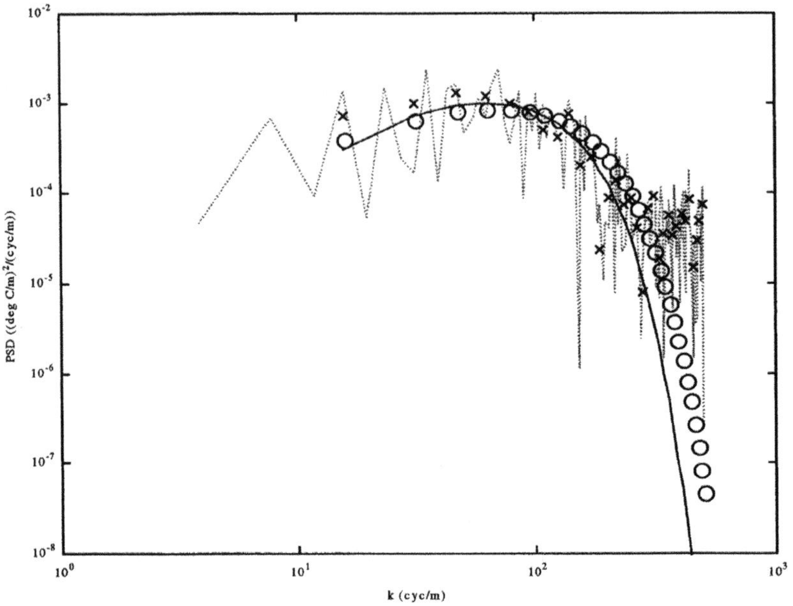

Figure 8.19. Batchelor curve fitting to the multitaper spectrum of a series from model 2.1.10, using the second error function.

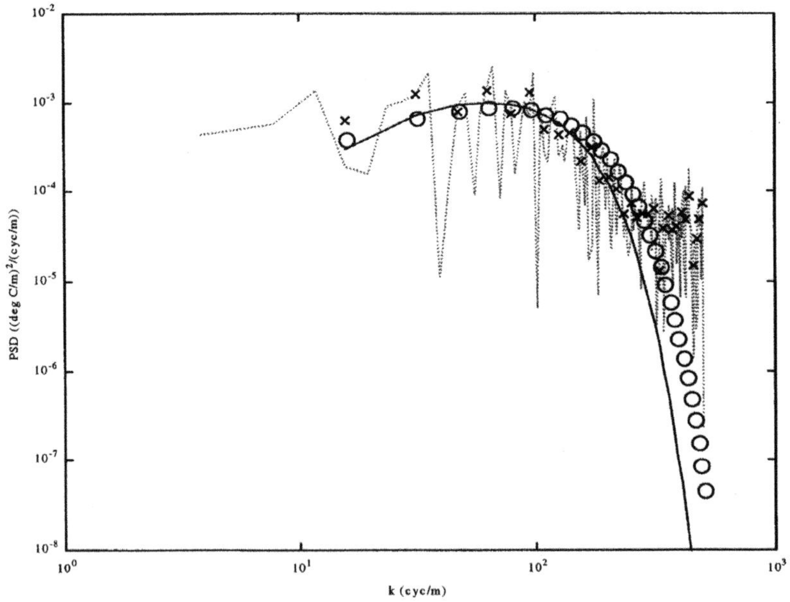

Figure 8.20. Batchelor curve fitting to the multitaper spectrum of a series from model 2.1.10, using the second error function.
—: Designed Batchelor spectrum; - - -: Periodogram;
x: Multitaper spectrum; o: Fitted Batchelor spectrum.

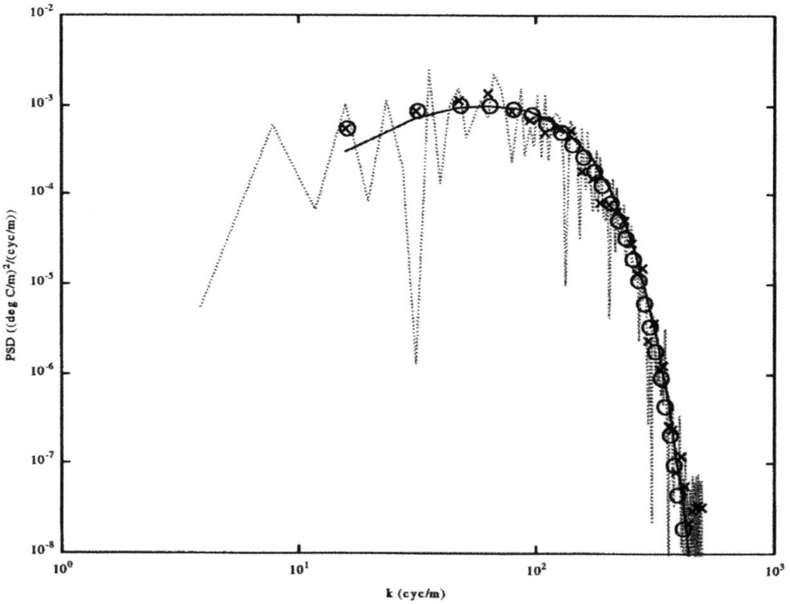

Figure 8.21. Batchelor curve fitting to the multitaper spectrum of a series from model 2.1.8, using the third error function.

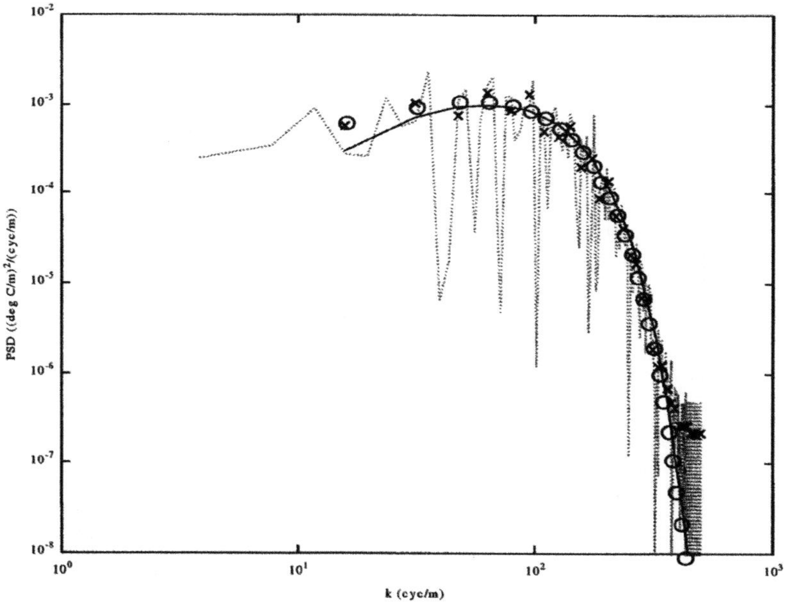

Figure 8.22. Batchelor curve fitting to the multitaper spectrum of a series from model 2.1.8, using the third error function.
—: Designed Batchelor spectrum; - - -: Periodogram;
x: Multitaper spectrum; o: Fitted Batchelor spectrum.

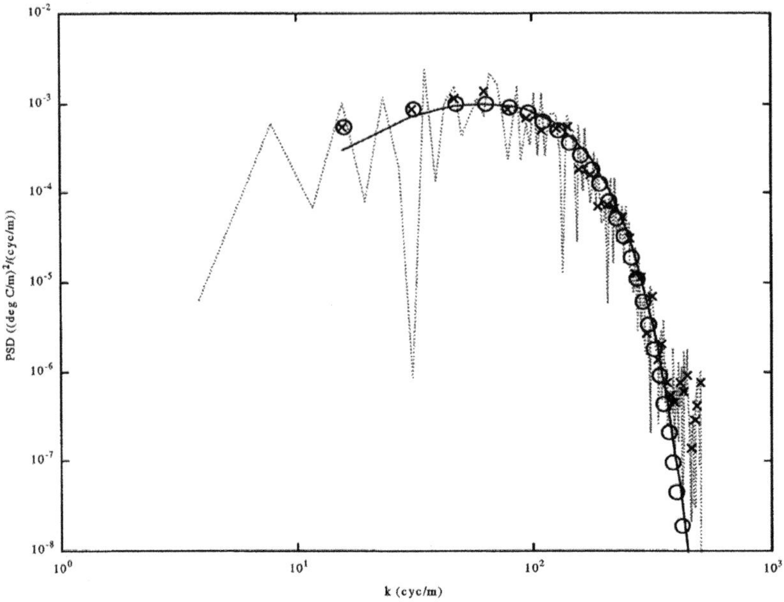

Figure 8.23. Batchelor curve fitting to the multitaper spectrum of a series from model 2.1.9, using the third error function.

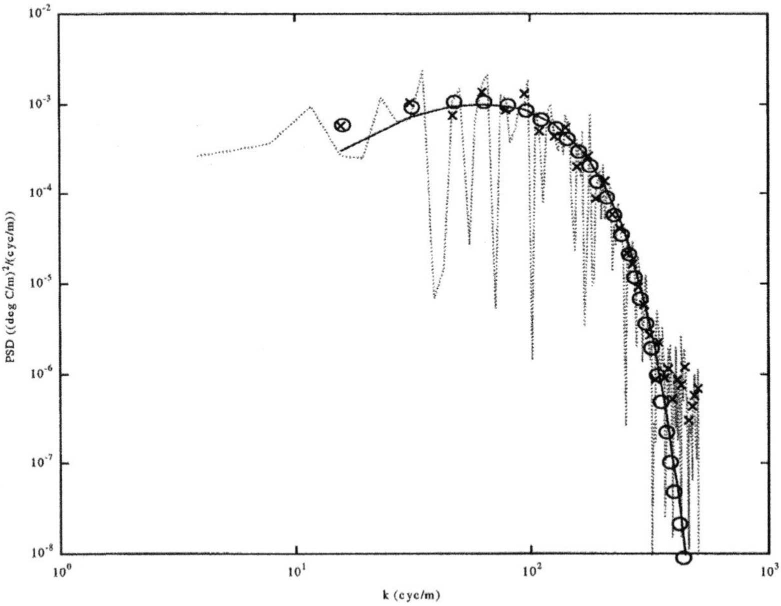

Figure 8.24. Batchelor curve fitting to the multitaper spectrum of a series from model 2.1.9, using the third error function.
—: Designed Batchelor spectrum; - - -: Periodogram;
x: Multitaper spectrum; o: Fitted Batchelor spectrum.

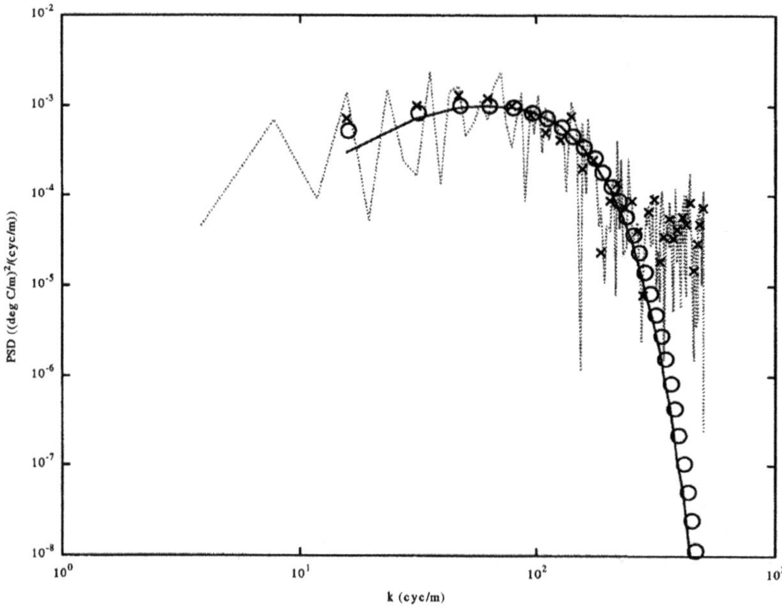

Figure 8.25. Batchelor curve fitting to the multitaper spectrum of a series from model 2.1.10, using the third error function.

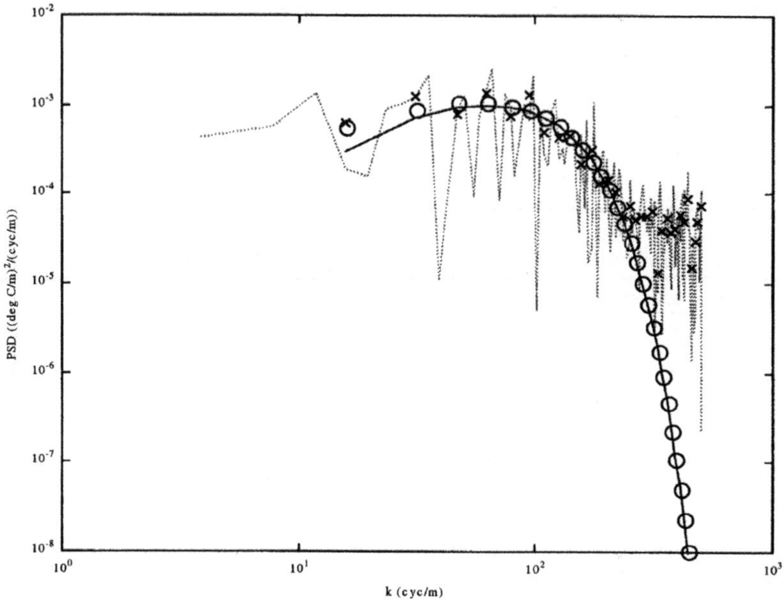

Figure 8.26. Batchelor curve fitting to the multitaper spectrum of a series from model 2.1.10, using the third error function.
—: Designed Batchelor spectrum; - - -: Periodogram;
x: Multitaper spectrum; o: Fitted Batchelor spectrum.

8.6. CONCLUSIONS ON BATCHELOR CURVE FITTING

The average values of estimated ε and x_T for synthetic series of different models are shown in Tables VIII.7 - VIII.9. These results show that, in general, both estimated values of ε and x_T increase as the noise level increases. All the average estimated errors for x_T are less than 10 % of the designed values. It is also shown that the estimated values of ε and x_T using the third error function (weighted chi-square distributed error function in Eq. 8.6) are the closest to the designed values for series of these three models.

Due to the property that the Batchelor spectrum has a high dynamic range, the higher spectral values at low frequencies are more reliable than the lower spectral values at high frequencies. The third error function is a chi-square distributed error function which is weighted with the spectral values of the theoretical Batchelor spectrum. Because the bias contributed to the chi-square errors at high frequencies in the third error function are downweighted significantly, the third error function has the best performance in estimating ε and x_T. Therefore, the third error function is applied to observed data for Batchelor curve fitting in Chapter 9. However, it should be noted that the weighting factor chosen for the third error function in this study does not warrant the optimal fitting results. It is possible that other choices of weighting factor could achieve better fitting results.

Table VIII.7. Estimated ε and x_T from estimated spectra for model 2.1.8.

Error function	1	2	3
Average ε [m^2/sec^3]	$1.0820*10^{-7}$	$1.2204*10^{-7}$	$1.0294*10^{-7}$
Average x_T [$(degC)^2/sec$]	$9.4453*10^{-8}$	$9.4334*10^{-8}$	$9.4512*10^{-8}$
Designed	10^{-7}	10^{-7}	10^{-7}

Table VIII.8. Estimated ε and x_T from estimated spectra for model 2.1.9.

Error function	1	2	3
Average ε [m^2/sec^3]	$1.0986*10^{-7}$	$1.2204*10^{-7}$	$1.0294*10^{-7}$
Average x_T [$(degC)^2/sec$]	$9.4672*10^{-8}$	$9.4577*10^{-8}$	$9.4756*10^{-8}$
Designed	10^{-7}	10^{-7}	10^{-7}

Table VIII.9. Estimated ε and x_T from estimated spectra for model 2.1.10.

Error function	1	2	3
Average ε [m^2/sec^3]	$1.7791*10^{-7}$	$2.4356*10^{-7}$	$1.4214*10^{-7}$
Average x_T [$(degC)^2/sec$]	$1.0451*10^{-7}$	$1.0487*10^{-7}$	$1.0427*10^{-7}$
Designed	10^{-7}	10^{-7}	10^{-7}

9. Segmentation of Observed Data

9.1. INTRODUCTION

The kinetic energy dissipation rate ε (defined in Chapter 2) is estimated by the Batchelor curve fitting method discussed in Chapter 8. The estimation procedure assumes homogeneous and isotropic turbulence at small scales. Therefore, the procedure requires partitioning the non-stationary temperature gradient series into stationary segments. In this chapter, the ten profiles of temperature gradient series measured in three lakes (discussed in Chapter 2) are analyzed and the results reported. Stationary segments are determined by using the segmentation algorithm defined in Chapter 7. The kinetic energy dissipation rate ε of each stationary segment is estimated by using the Batchelor curve fitting method described in Chapter 8.

Non-stationarity is also encountered in hydrologic and climatic data. A standardization procedure defined in Chapter 2 is often used to transform monthly series into a series with zero mean and unit standard deviation. However, this standardization procedure does not assure stationarity in the transformed series. The second objective of this chapter is to examine if the standardized monthly series are stationary. The segmentation algorithm defined in Chapter 7 is applied so that non-stationary series will be identified and partitioned into stationary segments. Four sets of hydrologic data are analyzed, including streamflow, temperature, precipitation and Palmer's drought severity index (PDSI) series (discussed in Chapter 2). It should be noted that the tests used in the segmentation algorithm are based on the assumption that the constant mean of the series is zero. However, trends and periodicities are often detected in hydrologic and climatic data (Hamed, 1997; Bhattacharya, 1997; Chen and Rao, 1998). Because the standardization is a procedure which removes the periodicity of twelve months, trends or other periodicities than twelve months may still remain in the standardized monthly series. In order to at least assure no trends in the series, they are detrended by using the first differences of standardized monthly series. In the second part (Section 9.3) of this chapter, both the differenced standardized monthly series and standardized monthly series are analyzed, and the results of segmentation are compared.

9.2. TEMPERATURE GRADIENT PROFILES

Ten temperature gradient profiles (Chapter 2) measured in the Calhoun Lake, the Waconia Lake and Christmas Lake in Minneapolis, Minnesota are analyzed in this chapter. Each profile is partitioned into stationary segments by the segmentation algorithm using tests 1, 2, 4 and 5 defined in Chapter 7. The Batchelor curve fitting method described in Chapter 8 is applied to stationary segments partitioned from each profile. These results show that some spectra are unresolved or do not have two sub-ranges (Figs. 9.1 and 9.2), and some of the fitted spectra are 'bad fits' (Figs. 9.3 and 9.4). The spectra in Figs. 9.5 and 9.6 are 'good fits'. An index, Fit(I), in these figures indicates 'good fit', 'bad fit' or 'unresolved', if the integer, I, is equal to, less or greater than zero.

9.2.1. Ratios of Unresolved, Bad-Fit and Good-Fit Segments. Estimated values of ε and χ_T (defined in Chapter 2) corresponding to stationary segments of three profiles are shown in Figs. 9.7-9.9 and Figs. 9.10-9.12, respectively. ε is the kinetic energy dissipation rate [m^2/sec^3], and x_T is the dissipation of temperature variance due to turbulence [$(degC)^2/sec$]. Each of these three profiles is taken from a different lake. Each of these figures includes seven subplots: (a) temperature; (b) temperature gradient; (c) distance derivative; (d) estimated values of ε (or x_T) from segments partitioned by the algorithm using test 1; (e) estimated values of ε (or x_T) from segments partitioned by the algorithm using test 2; (f) estimated values of ε (or x_T) from segments partitioned by the algorithm using test 4; and (g) estimated values of ε (or x_T) from segments partitioned by the algorithm using test 5. The segments of which estimated spectra are unresolved or do not have two spectral sub-ranges are marked with 'x'. A question mark (?) above the value of estimated ε or x_T indicates 'bad fit'. The ratios of stationary segments corresponding to unresolved, bad-fit and good-fit spectra obtained from each profile are shown in Table IX.1. The denominator indicates the number of stationary segments of each profile, and the numerator indicates the number of stationary segments which have unresolved, bad-fit or good-fit spectra. For example, Profile 1111 is partitioned into 61 stationary segments by the segmentation algorithm

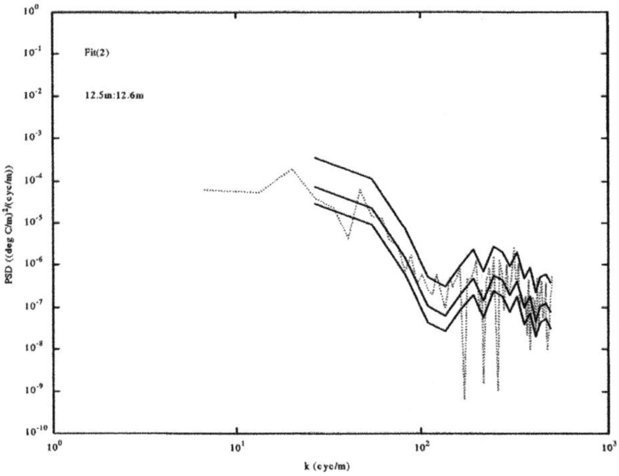

Figure 9.1. An unresolved spectrum of segment at depth between 12.5 and 12.6 m of profile 1111.

using test1, out of which 36, 13 and 12 stationary segments have unresolved, bad-fit and good-fit spectra. In general, the numbers of segments partitioned by the algorithm using tests 1, 4 and 5 are about the same order and are greater than the number of those obtained by using test 2. These results are similar to the results for synthetic series in Chapter 7. The high numbers of segments partitioned by the algorithm using tests 1, 4 and 5 indicate strong intermittency in these profiles of temperature gradient.

Profiles taken from different lakes have different ratios of unresolved, 'bad fit' and good-fit segments (Table IX.1). In spite of using different tests, the profiles taken from the Calhoun lake and the Waconia lake have more than 50% of identified stationary segments of which the measured spectra are unresolved and 'bad fit', while most of the profiles taken from the Christmas lake have less than 50% of those of which the measured spectra are unresolved and 'bad fit'.

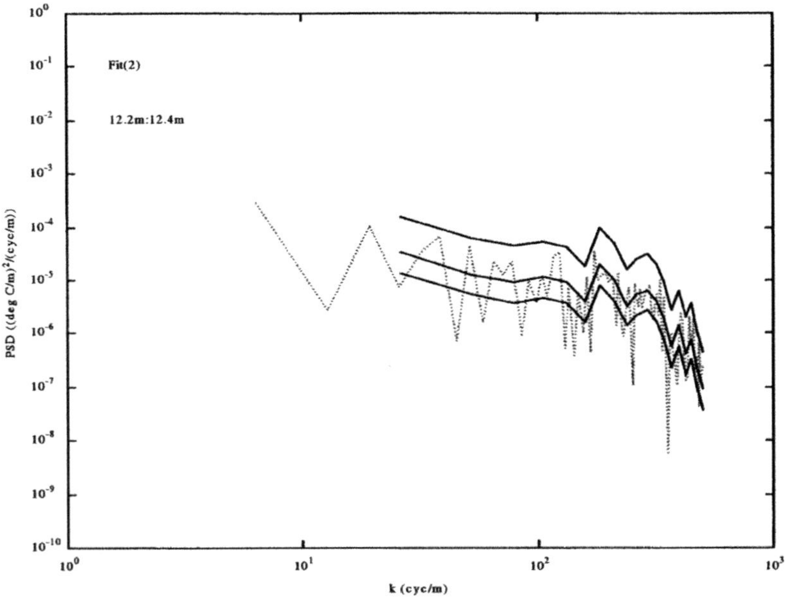

Figure 9.2. An unresolved spectrum of segment at depth between 12.2 and 12.4 m of profile 1228.
- - -: Periodogram; —: Multitaper spectrum and 95% C.I.
Fit(I > 0): Unresolved spectrum; Fit(I < 0): 'Bad fit'; Fit(I = 0): 'Good fit'.

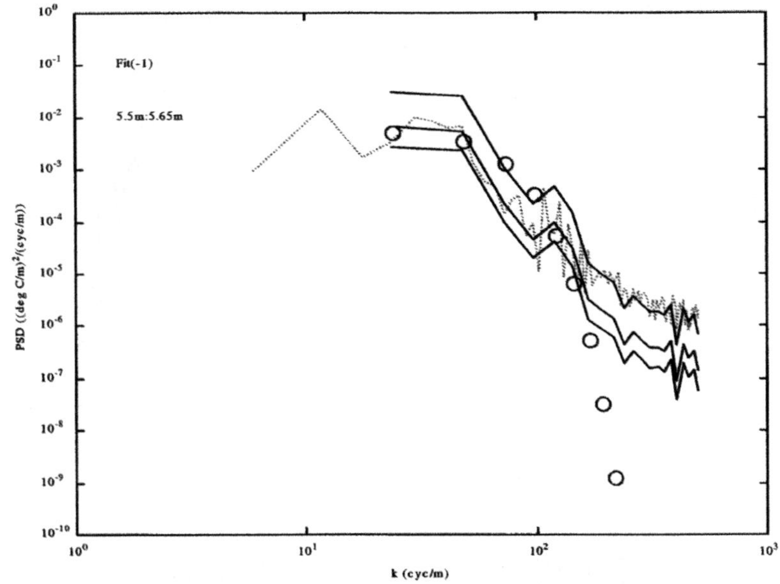

($\varepsilon = 3.5083 * 10^{-9}$ m^2/sec^3; $\chi_T = 2.3692 * 10^{-7}$ (degC)^2/sec)
Figure 9.3. A 'bad fit' for a segment at depth between 5.5 and 5.6 m of profile 1111.

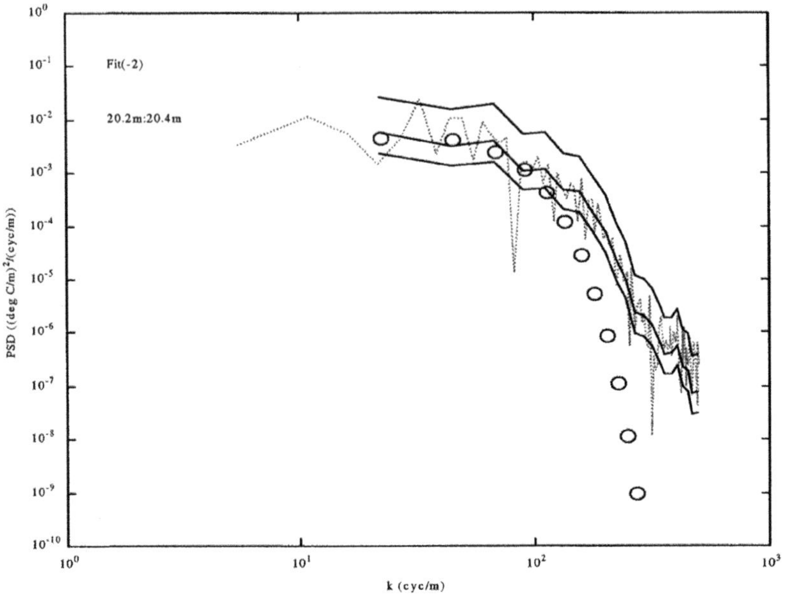

($\varepsilon = 8.2180*10^{-9}$ m^2/sec^3; $\chi_T = 2.6215*10^{-7}$ (degC)^2/sec)

Figure 9.4. A 'bad fit' for a segment at depth between 20.2 and 20.4 m of profile 1228. - - -: Periodogram; —: Multitaper spectrum and 95% C.I.; o: Fitted Batchelor spectrum.

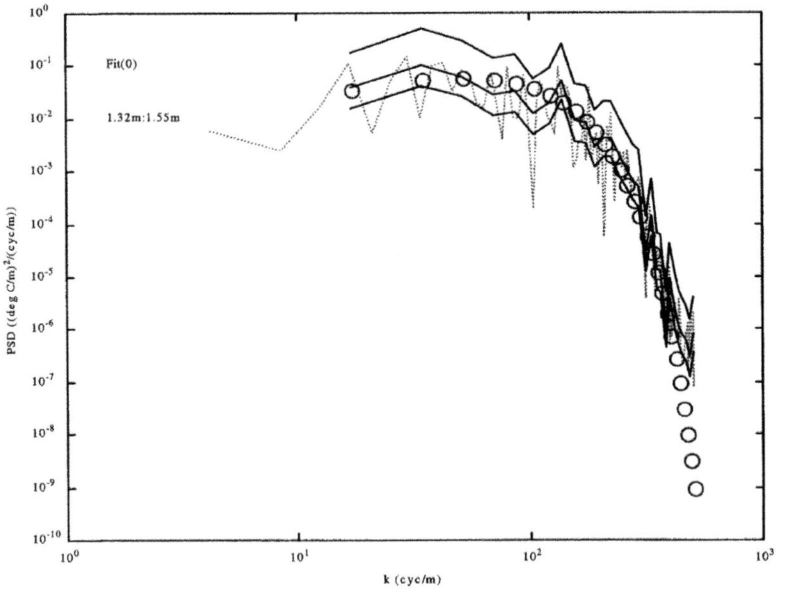

($\varepsilon = 7.5148*10^{-8}$ m^2/sec^3; $\chi_T = 5.5177*10^{-6}$ (degC)^2/sec)

Figure 9.5. A 'good fit' for a segment at depth between 1.32 and 1.55 m of profile 1210.

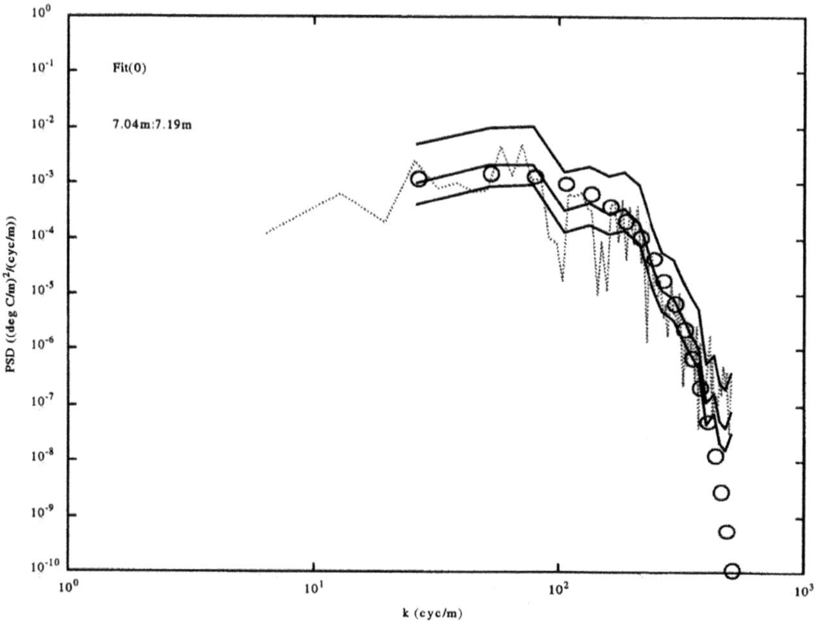

($\varepsilon = 8.9095*10^{-8}$ m^2/sec^3; $\chi_T = 1.5734*10^{-7}$ (degC)^2/sec)

Figure 9.6. A 'good fit' for a segment at depth between 7.04 and 7.19 m of profile 11349 - - -: Periodogram; —: Multitaper spectrum and 95% C.I.; o: Fitted Batchelor spectrum. Fit(I > 0): Unresolved spectrum; Fit(I < 0): 'Bad fit'; Fit(I = 0): 'Good fit'.

9.2.2. Estimated Values of ε and χ_T from Resolved Spectra. Estimated values of ε and χ_T in Figs. 9.7-9.9 and Figs. 9.10-9.12 are compared in Figs. 9.13-9.15 and Figs. 9.16-9.18, respectively. Only the estimated values from resolved spectra are considered in this analysis. Each estimated value is plotted with a segment between a pair of symbols which indicate the corresponding test used in the segmentation algorithm. In general, most of the estimated values at the same depth vary within two orders with different tests used in the algorithm. A few exceptions are shown in Figs. 9.13 and 9.16 at depth between 14 and 17 m (Profile 1111), in Fig. 9.14 at depth between 5 and 7 m (Profile 1210), and in Figs. 9.15 and 9.19 at depth between 18 and 19 m (Profile 1502).

9.2.3. Estimated Values of ε and χ_T from Profiles in the Same Lake. Estimated values of ε and χ_T from four profiles in Lake Calhoun are compared in Figs. 9.19 and 9.20. Estimated values of ε and χ_T from five profiles in Lake Waconia are compared in Figs. 9.21 and 9.22. The stationary segments are determined by the segmentation algorithm using test 5. Only the estimated values from resolved spectra are given in these plots. Each estimated value is plotted with a segment between a pair of symbols which indicate the corresponding profile. In general, most of the estimated values of ε in the same lake vary within two orders with different profiles (Figs. 9.19 and 9.21); while the estimated values of χ_T in the same lake vary much more with different profiles (Figs. 9.20 and 9.22).

Table IX.1. Ratios of unresolved, bad-fit and good-fit segments partitioned from profiles of temperature gradient microstructure. (U: Unresolved; B: Bad fit; G: Good fit.)

Lake	Profile	Test 1			Test 2			Test 4			Test 5		
		U	B	G	U	B	G	U	B	G	U	B	G
Calhoun	1111	36/61	13/61	12/61	3/18	7/18	8/18	19/44	10/44	15/44	30/55	14/55	11/55
	1127	41/65	13/65	11/65	5/20	5/20	10/20	24/49	8/49	17/49	23/40	8/40	9/40
	1228	48/68	8/68	12/68	12/32	9/32	11/32	41/62	8/62	13/62	38/53	5/53	10/53
	1455	20/34	4/34	10/34	3/12	7/12	2/12	20/32	6/32	6/32	18/31	4/31	9/31
Christmas	1210	9/28	6/28	13/28	0/12	2/12	10/12	5/26	4/26	17/26	9/24	5/24	10/24
	1239	5/23	5/23	13/23	1/8	2/8	5/8	10/26	5/26	11/26	9/27	5/27	13/27
	1256	5/28	8/28	15/28	2/17	6/17	9/17	5/23	6/23	12/23	7/27	6/27	14/27
	1326	4/22	5/22	13/22	1/11	4/11	6/11	4/23	4/23	15/23	3/23	7/23	13/23
	1349	6/27	4/27	17/27	2/13	2/13	9/13	5/27	7/27	15/27	5/24	6/24	13/24
Waconia	1502	26/57	19/57	12/57	4/12	5/12	3/12	21/43	12/43	10/43	30/57	17/57	10/57

9.2.4. Estimated Values of ε and χ_T from Different Lakes. Estimated values of ε and χ_T from one profile in each lake are compared in Figs. 9.23 and 9.24. The stationary segments are determined by the segmentation algorithm using test 5. Only the estimated values from resolved spectra are given in these plots. Each estimated value is plotted with a segment between a pair of symbols which indicate the corresponding profile and lake. Most of the estimated values of ε at the same depth vary within two orders with different lakes except at depth greater than 8 m (Fig. 9.23); while the estimated values of χ_T at the same depth vary much more with different lakes (Fig. 9.24). Generally, both estimated values of ε and χ_T decrease as the depth increases.

9.3. CONCLUSIONS ON SEGMENTATION OF TEMPERATURE GRADIENT PROFILES

High numbers of stationary segments are identified in the profiles of temperature gradient by the algorithm using tests 1, 4 and 5, which indicate strong intermittency in temperature gradient profiles. However, not all the stationary segments of temperature gradient profiles have spectra which fit the Batchelor spectra well.

Temperature gradient profiles taken from different lakes have different ratios of unresolved, 'bad fit' and 'good-fit' segments. In spite of using different tests in the segmentation algorithm, the profiles taken from the Calhoun lake and the Waconia lake

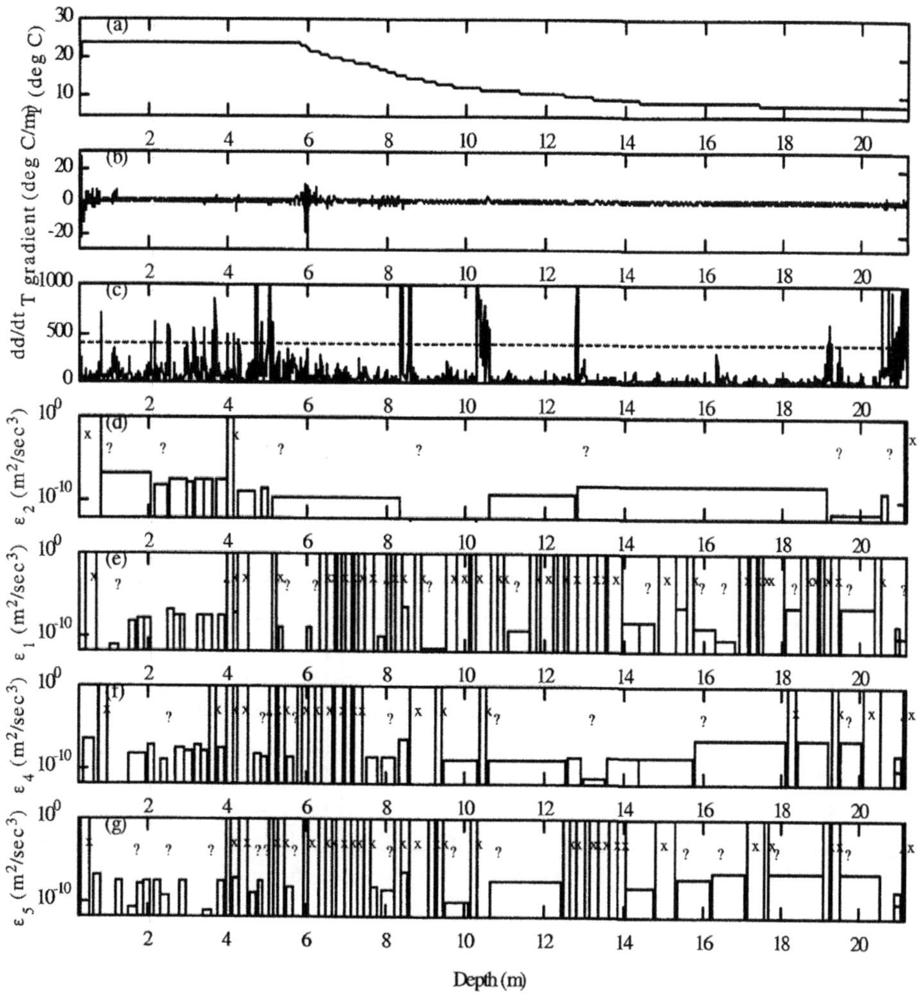

Figure 9.7. Estimated dissipation rate ε for profile 1111.
x: Unresolved spectrum; ?: 'Bad fit'.
Test used in the segmentation algorithm:
ε_1: test 1; ε_2: test 2; ε_4: test 4; ε_5: test 5.

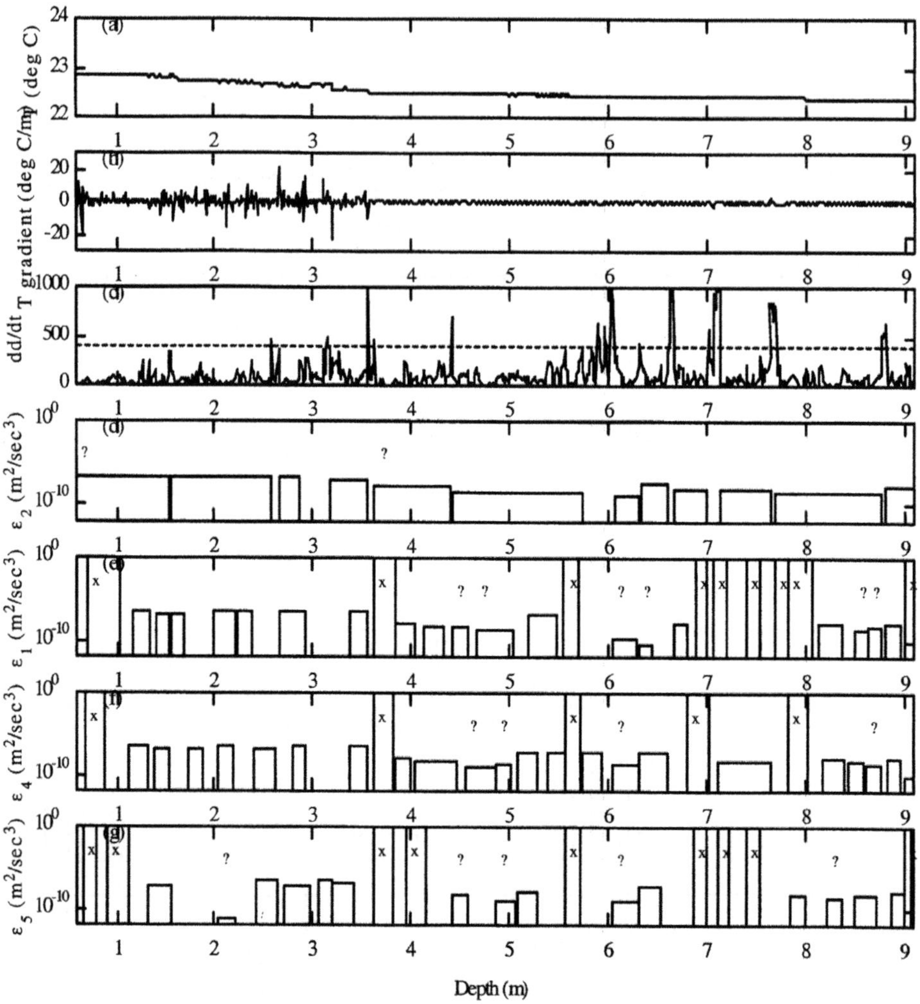

Figure 9.8. Estimated dissipation rate ε for profile 1210.
x: Non-Batchelor shaped or unresolved spectrum; ?: 'Bad fit'.
Test used in the segmentation algorithm:
ε_1: test 1; ε_2: test 2; ε_4: test 4; ε_5: test 5.

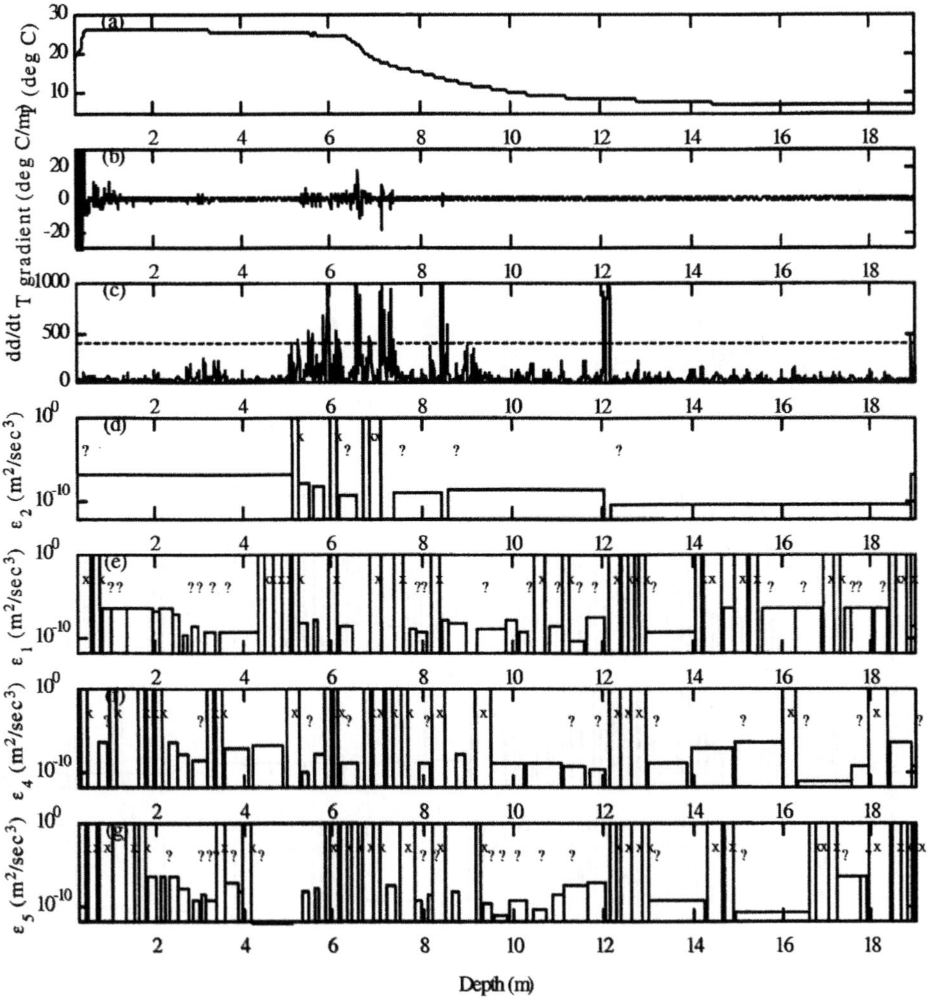

Figure 9.9. Estimated dissipation rate ε for profile 1502.
x: Non-Batchelor shaped or unresolved spectrum; ?: 'Bad fit'.
Test used in the segmentation algorithm:
ε_1: test 1; ε_2: test 2; ε_4: test 4; ε_5: test 5.

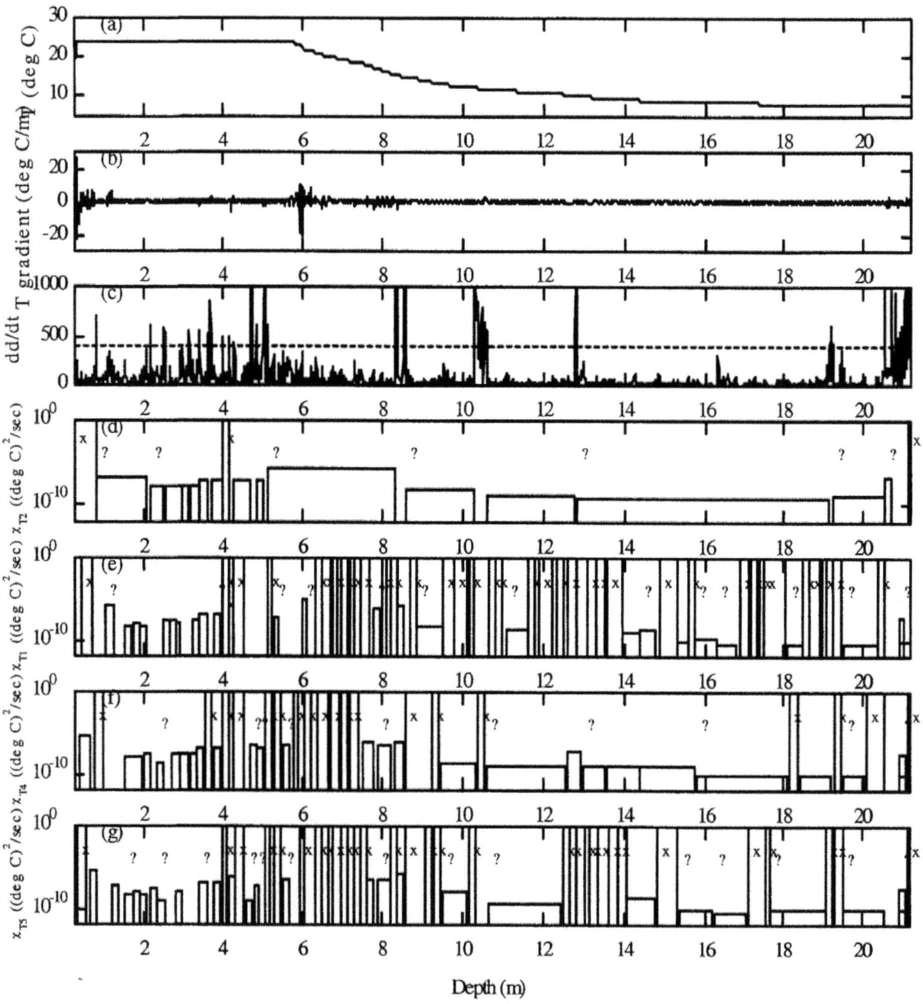

Figure 9.10. Estimated dissipation of temperature variance x_T for profile 1111.
x: Non-Batchelor shaped or unresolved spectrum; ?: 'Bad fit'.
Test used in the segmentation algorithm:
x_{T1}: test 1; x_{T2}: test 2; x_{T4}: test 4; x_{T5}: test 5.

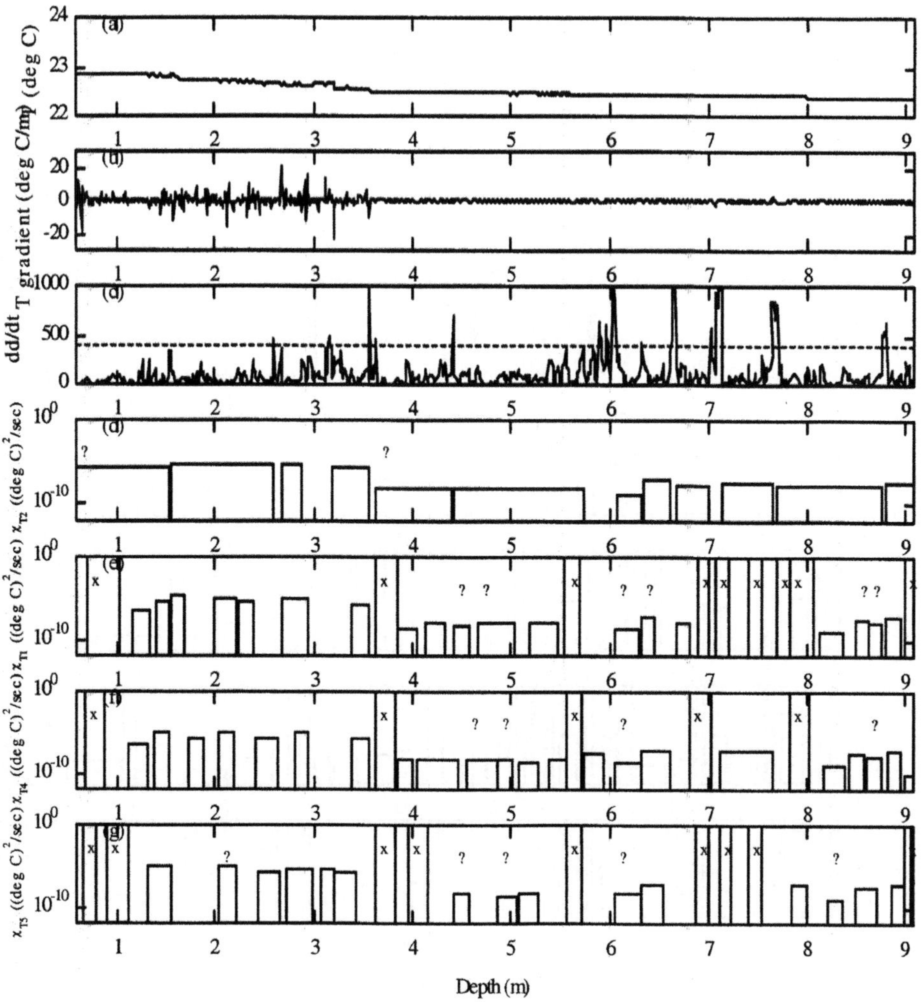

Figure 9.11. Estimated dissipation of temperature variance χ_T for profile 1210.
x: Non-Batchelor shaped or unresolved spectrum; ?: 'Bad fit'.
Test used in the segmentation algorithm:
χ_{T1}: test 1; χ_{T2}: test 2; χ_{T4}: test 4; χ_{T5}: test 5.

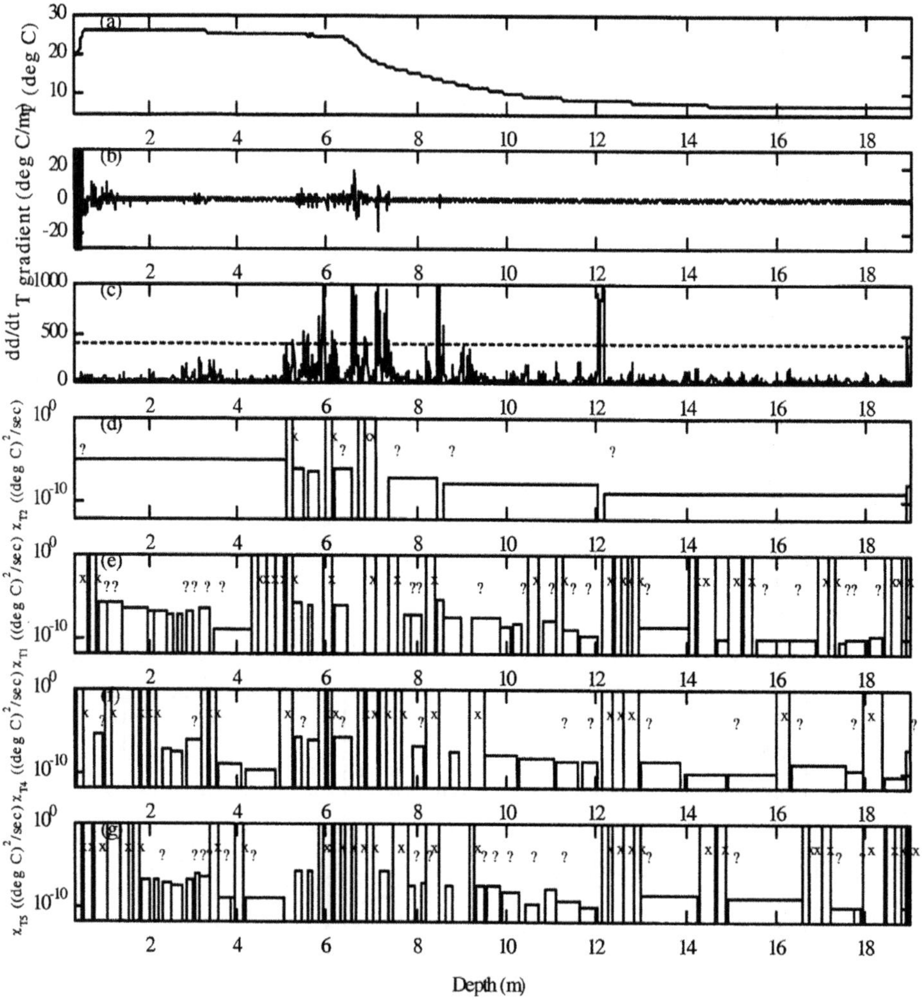

Figure 9.12. Estimated dissipation of temperature variance x_T for profile 1502.
x: Non-Batchelor shaped or unresolved spectrum; ?: 'Bad fit'.
Test used in the segmentation algorithm:
x_{T1}: test 1; x_{T2}: test 2; x_{T4}: test 4; x_{T5}: test 5.

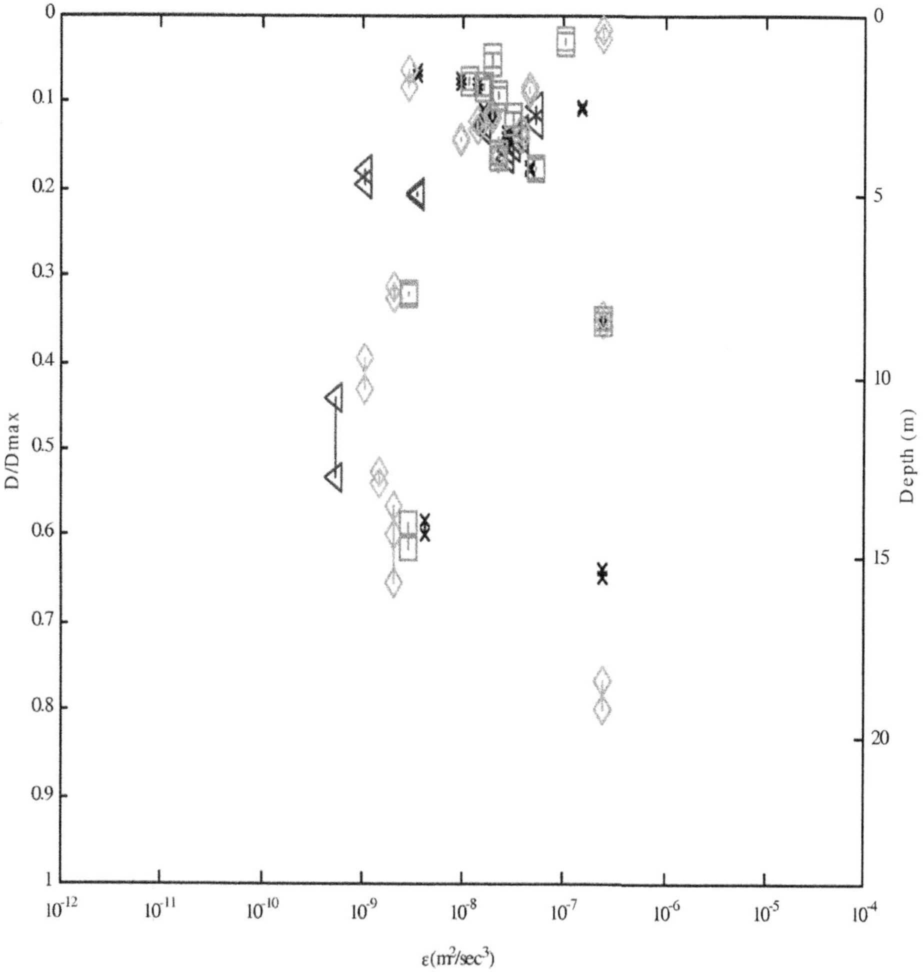

Figure 9.13. Estimated dissipation rate ε for profile 1111.
(×: test 1; ∇: test2; ◊: test 4; □: test5)

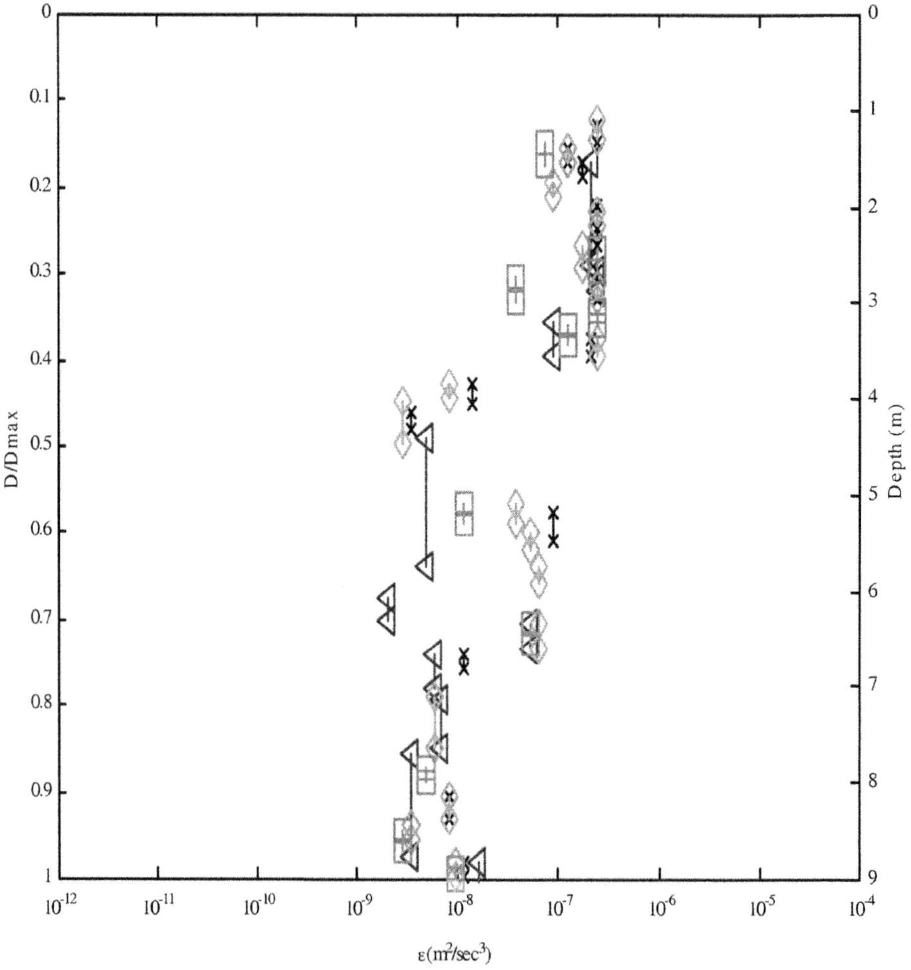

Figure 9.14. Estimated dissipation rate ε for profile 1210.
(×: test 1; ∇: test2; ◊: test 4; □: test5)

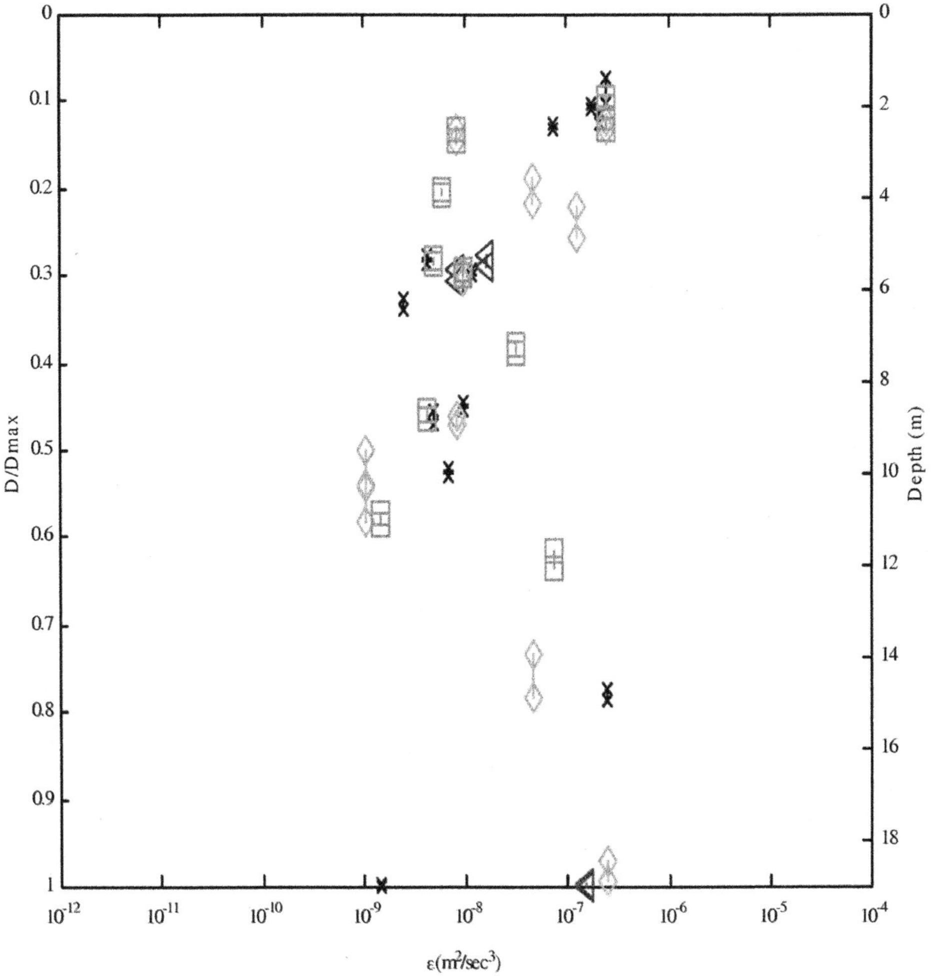

Figure 9.15. Estimated dissipation rate ε for profile 1502.
(×: test 1; ∇: test2; ◊: test 4; □: test5)

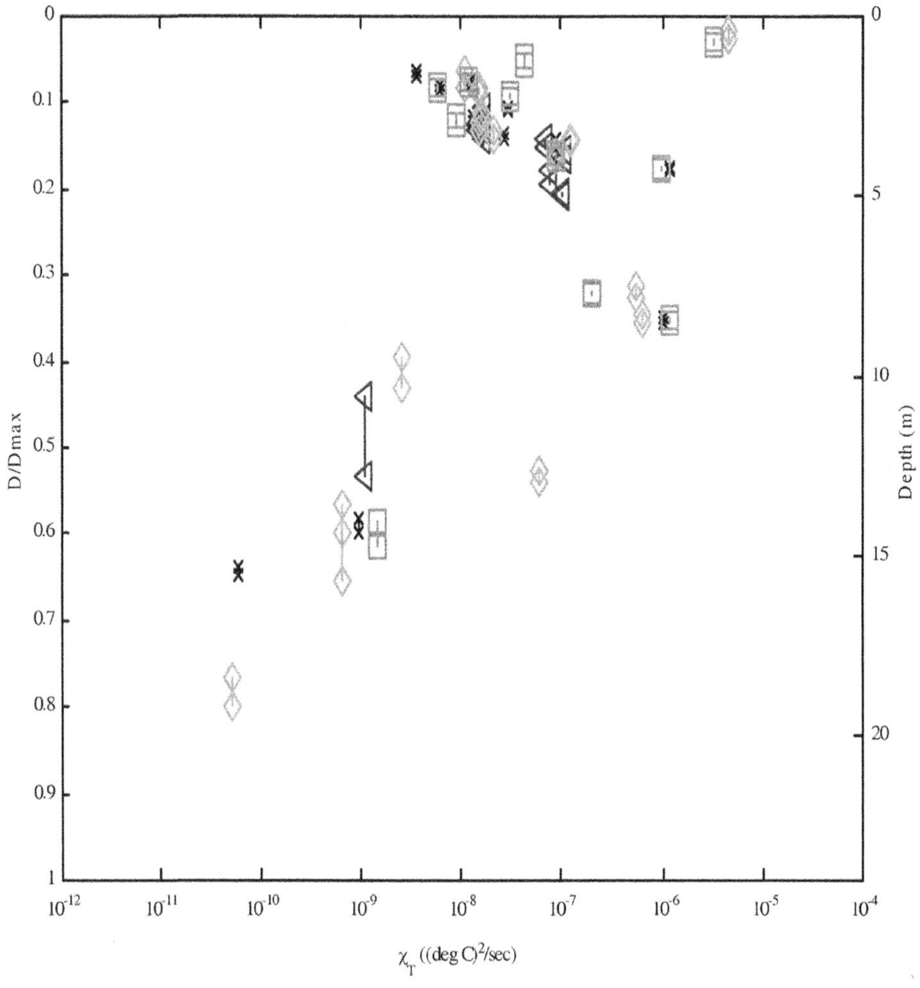

Figure 9.16. Estimated dissipation of temperature variance x_T for profile 1111. (×: test 1; ∇: test2; ◊: test 4; □: test5)

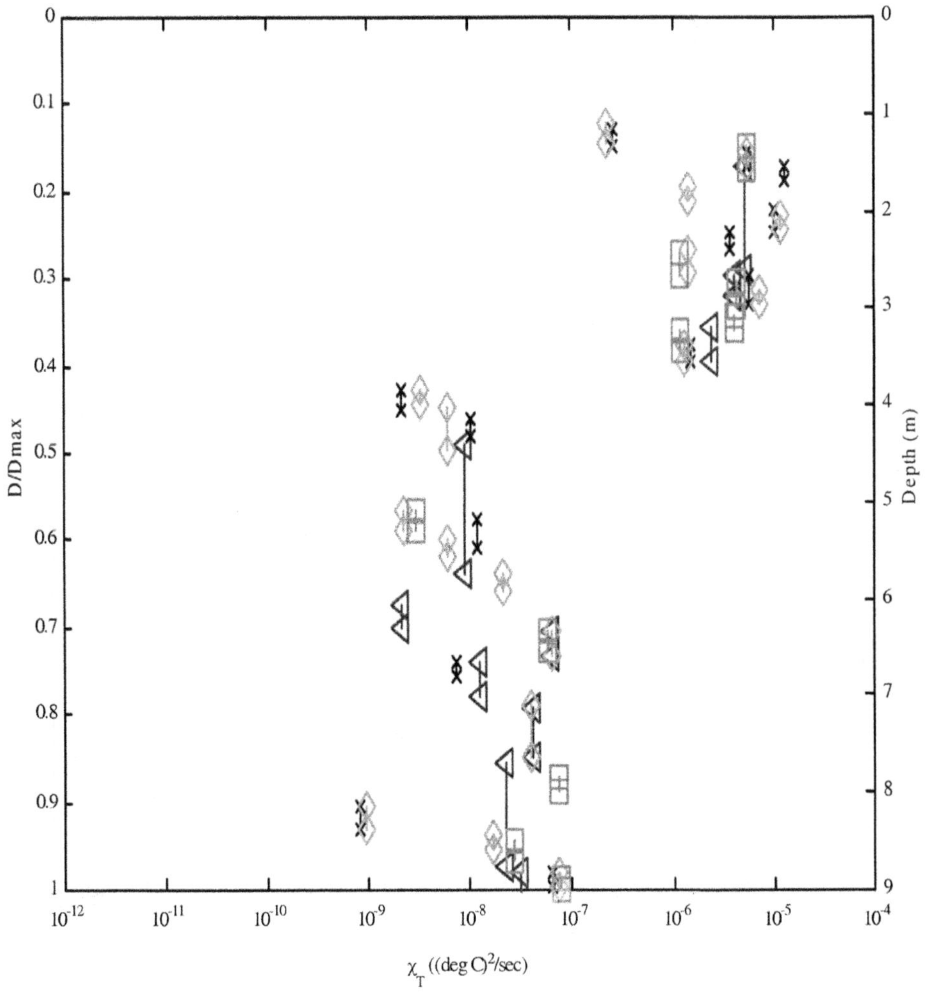

Figure 9.17. Estimated dissipation of temperature variance x_T for profile 1210. (×: test 1; ∇: test2; ◊: test 4; □: test5)

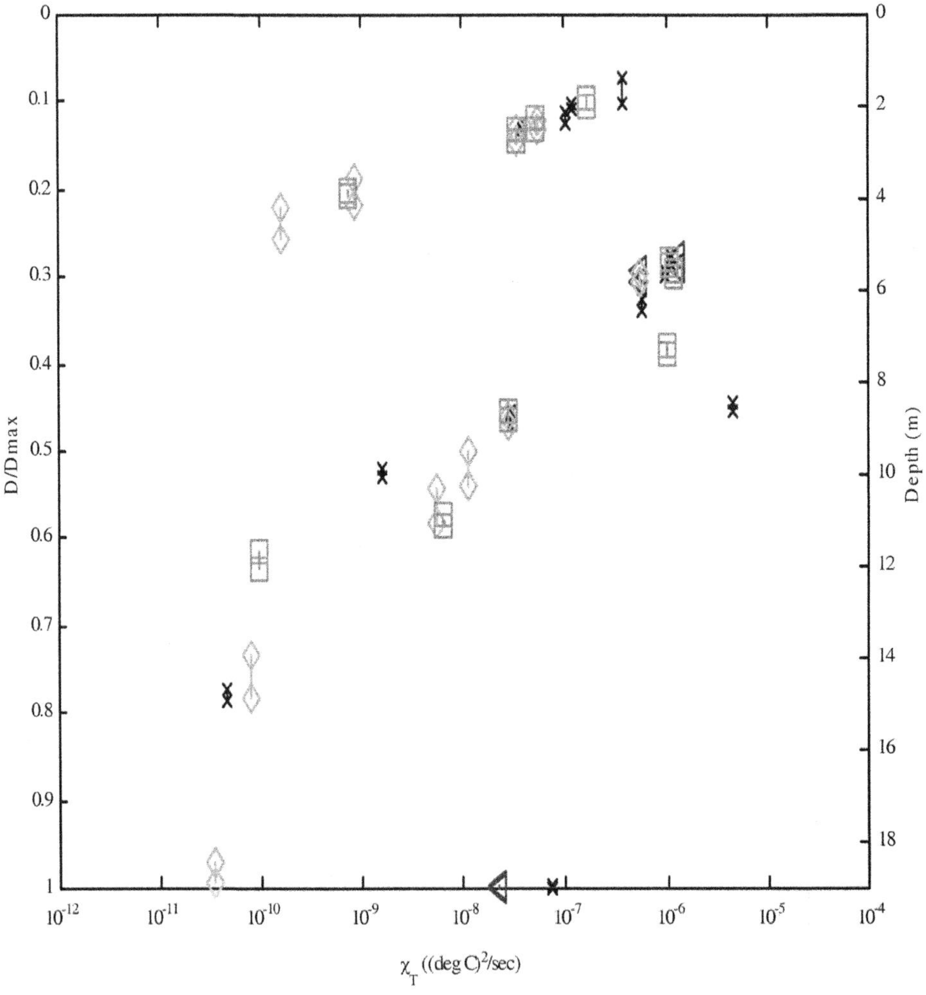

Figure 9.18. Estimated dissipation of temperature variance x_T for profile 1502. (×: test 1; ∇: test2; ◊: test 4; □: test5)

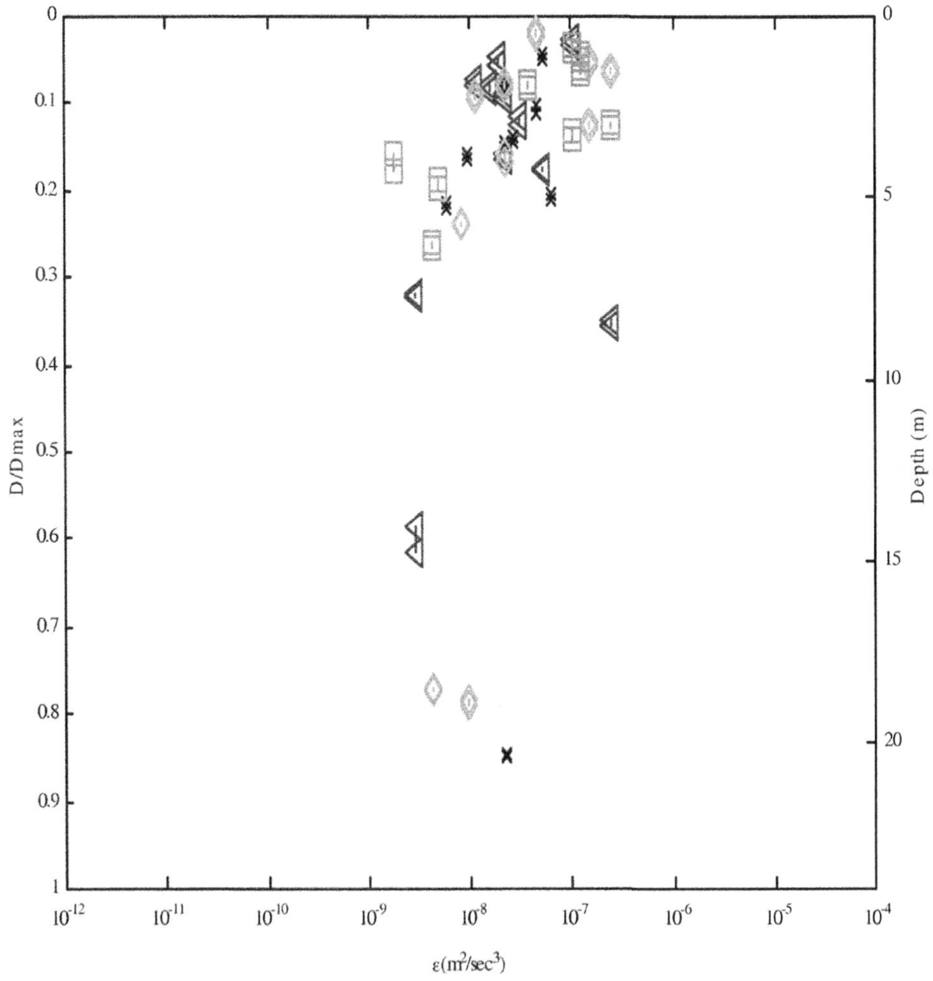

Figure 9.19. Estimated dissipation rate ε for profiles in Lake Calhoun.
(∇: Profile 1111; ×: Profile 1127; ◊: Profile 1228; □: Profile 1455)

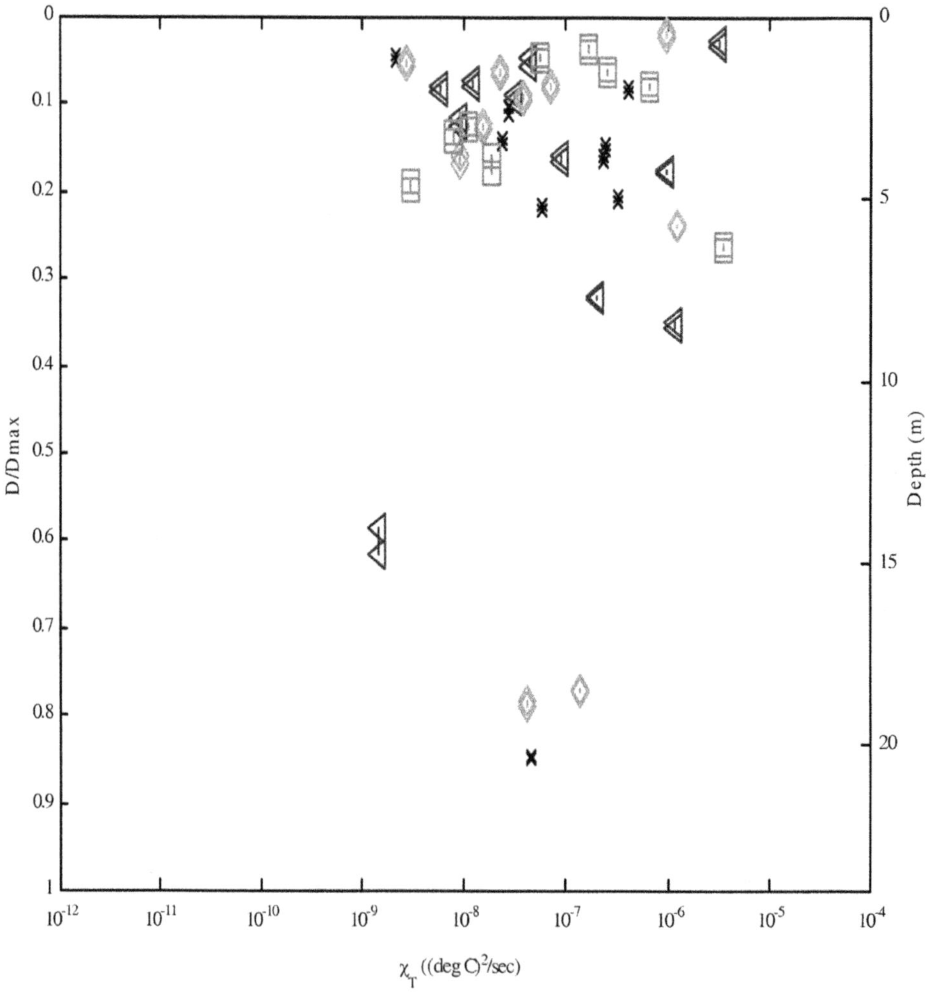

Figure 9.20. Estimated dissipation of temperature variance x_T for profiles in Lake Calhoun. (∇: Profile 1111; ×: Profile 1127; ◊: Profile 1228; □: Profile 1455)

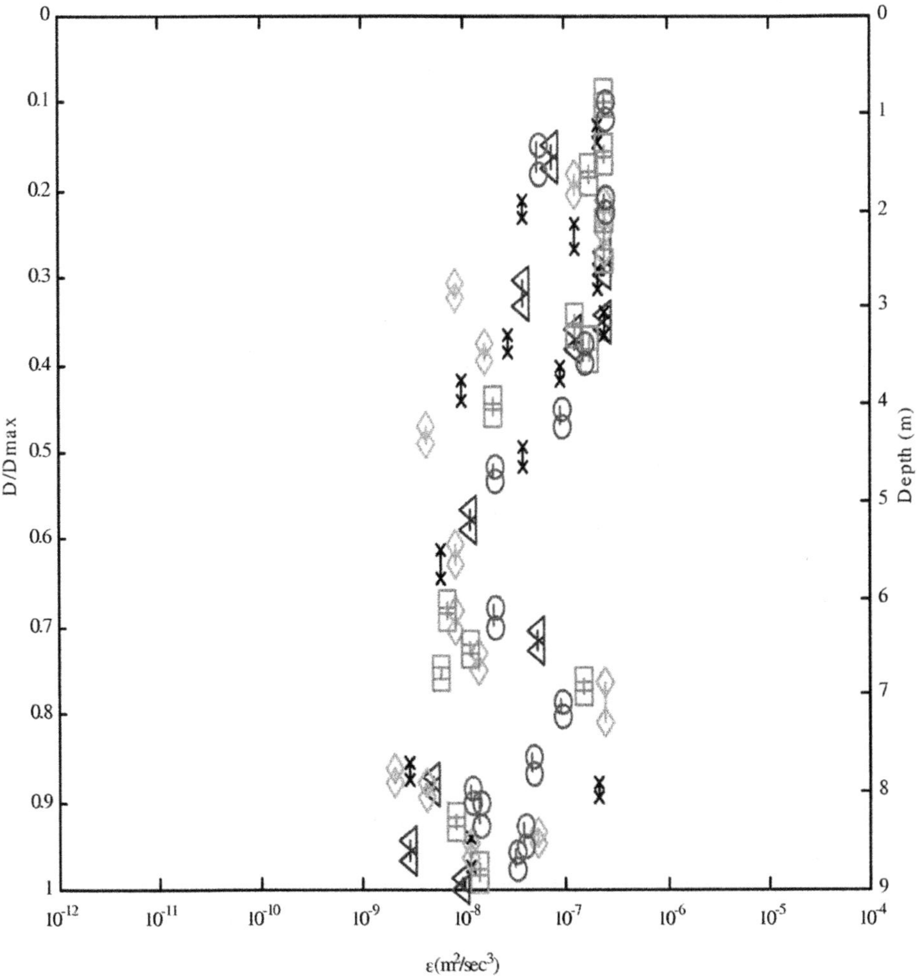

Figure 9.21. Estimated dissipation rate ε for profiles in Lake Waconia.
(∇: Profile 1210; ×: Profile 1239; ◊: Profile 1256; □: Profile 1326; o: Profile 1349)

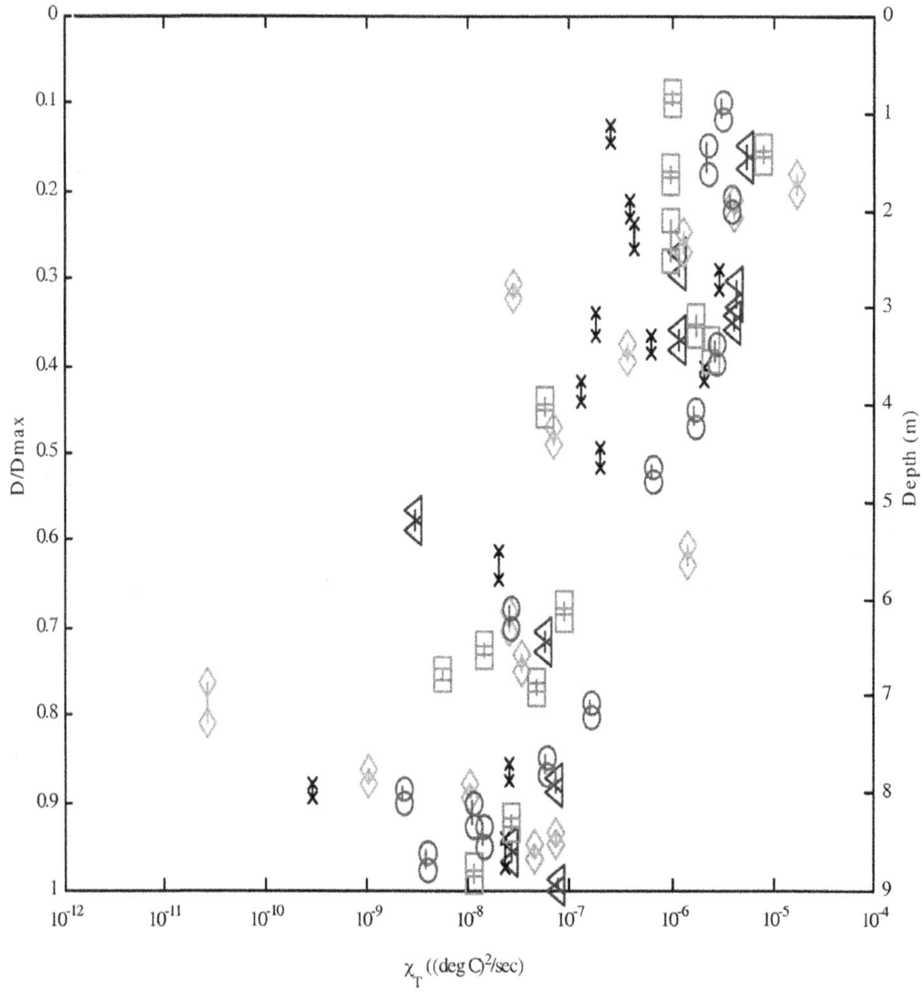

Figure 9.22. Estimated dissipation of temperature variance x_T for profiles in Lake Waconia.
(∇: Profile 1210; ×: Profile 1239; ◊: Profile 1256; □: Profile 1326; o: Profile 1349)

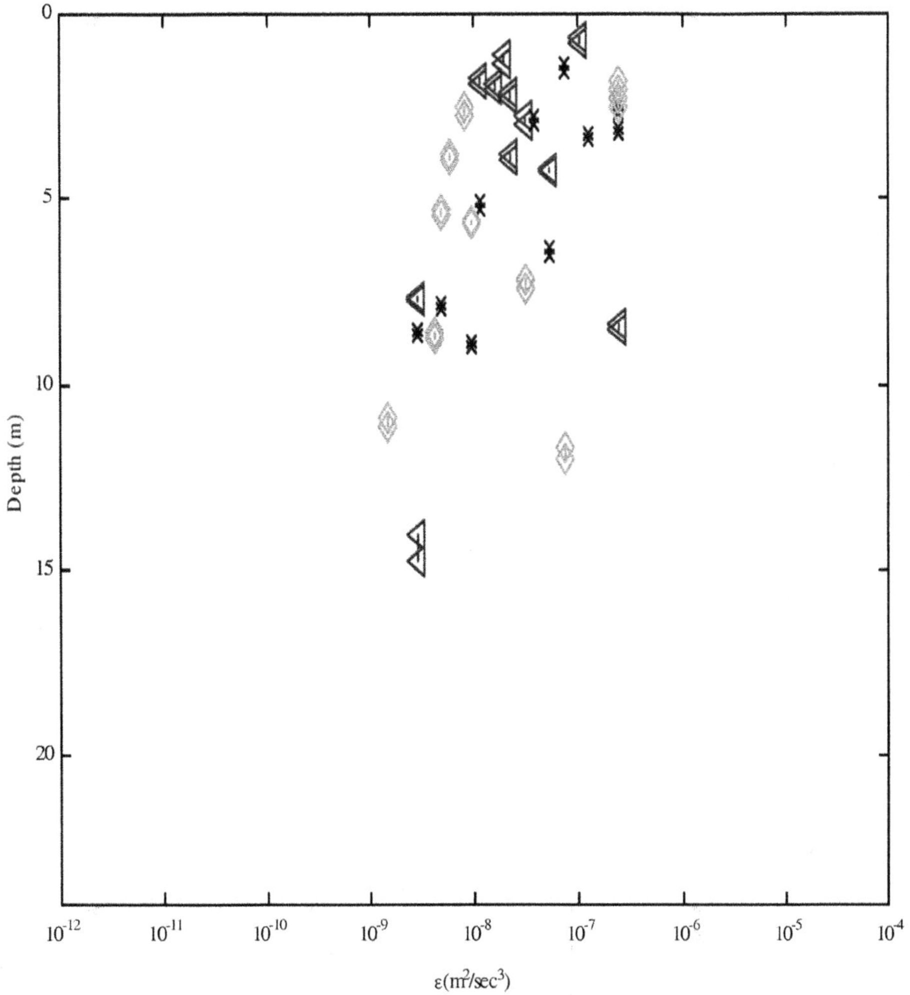

Figure 9.23. Estimated dissipation rate ε for profiles in different lakes. (∇: Profile 1111 in Lake Calhoun; $\times$: Profile 1210 in Lake Waconia; $\Diamond$: Profile 1502 in Lake Christmas)

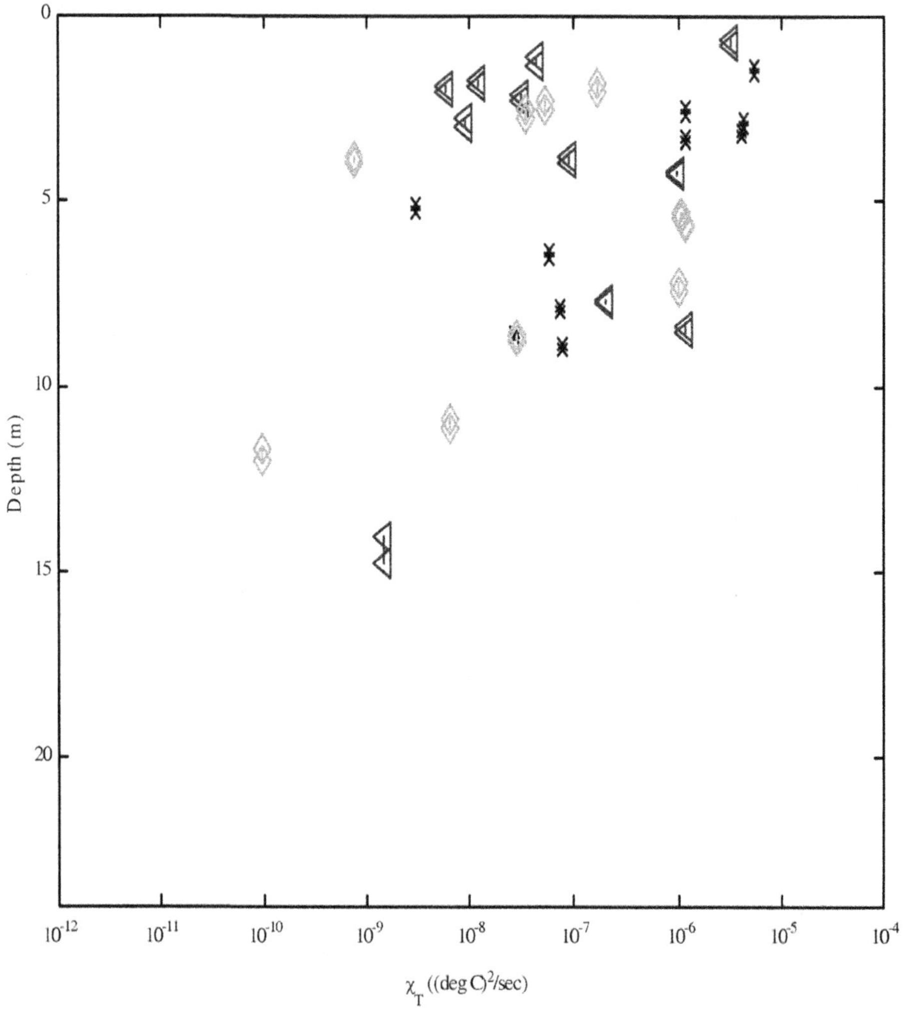

Figure 9.24. Estimated dissipation of temperature variance χ_T for profiles in different lakes.
(∇: Profile 1111 in Lake Calhoun; $\times$: Profile 1210 in Lake Waconia; $\Diamond$: Profile 1502 in Lake Christmas)

have more than 50% of stationary segments which do not have 'good-fit' spectra, while most of the profiles taken from the Christmas lake have less than 50% of those which do not have 'good fit' spectra.

Estimated results of ε and x_T from resolved spectra show that: (1) these estimated values of ε and x_T at the same depth of a temperature gradient profile vary by about two orders of magnitude with different tests used in the algorithm; (2) most of the estimated values of ε in the same lake vary within two orders of magnitude with different profiles; while the estimated values of x_T in the same lake vary more than two orders of magnitude with different profiles; (3) most of the estimated values of ε at the same depth vary within two orders of magnitude with different lakes; while the estimated values of x_T at the same depth vary much more with different lakes. Generally, both estimated values of ε and x_T decrease as the depth increases.

9.4. HYDROLOGIC SERIES

Four sets of hydrologic and climatic monthly series are analyzed in this study, including streamflow, temperature, precipitation and Palmer's drought severity index (PDSI) series (Chapter 2). The standardized series and the first difference (1-B) of the standardized series are partitioned, respectively, into stationary segments by the segmentation algorithm using tests 1, 4 and 5 (Chapter 7).

9.4.1. Stationary Segments from Hydrologic Series. The segmentation results corresponding to different types of hydrologic series are discussed as follows:

1. The numbers of stationary segments partitioned from differenced standardized and standardized monthly streamflow series are shown in Table IX.2. Segmentation results from monthly streamflow of Minnesota River at Clinton, IA are shown in Fig. 9.25. According to the results of the algorithm using test 1 (Table IX.2), the numbers of segments partitioned from both differenced standardized and standardized series are the same. According to the results of the algorithm using test 4 (Table IX.2), the numbers of segments partitioned from both differenced standardized and standardized series are different for two rivers. According to the results of the algorithm using test 5 (Table IX.2), the numbers of segments partitioned from both differenced standardized and standardized series are different for three rivers. The complete segmentation results show that these series (differenced standardized and standardized) are non-stationary, except the results for two rivers (Kalamazoo River and Wisconsin River) of which both differenced standardized and standardized series are identified as stationary by the algorithm using test 1.

2. The numbers of stationary segments partitioned from differenced standardized and standardized monthly temperature series are shown in Table IX.3. Segmentation results from monthly temperature of Urbana, IL are shown in Fig. 9.26. According to the results of the algorithm using test 1 (Table IX.3), the numbers of segments partitioned from both differenced standardized and standardized series are different for two stations. According to the results of the algorithm using test 4 (Table IX.3), the numbers of segments partitioned from both differenced standardized and

standardized series are different for one station. According to the results of the algorithm using test 5 (Table IX.3), the numbers of segments partitioned from both differenced standardized and standardized series are different for one station. The segmentation results of the algorithm using tests 1 and 5 show that both differenced standardized and standardized series of two stations (Aledo, IL and Ft. Wayne, IN) are stationary.

Table IX.2. Number of segments partitioned from monthly streamflow series. (DS: Differenced standardized series; S: Standardized series.)

Station	Test 1		Test 4		Test 5	
	DS	S	DS	S	DS	S
Minnesota River at Clinton, IA	3	3	3	4	4	3
White River near Alpine, IN	1	1	1	1	1	2
Kalamazoo River at Fennville, MI	1	1	2	2	2	2
Missouri River at Hermann, MO	2	2	2	3	2	2
Wisconsin River at Merrill, WI	1	1	3	3	2	3

Table IX.3. Number of segments partitioned from monthly temperature series. (DS: Differenced standardized series; S: Standardized series.)

Station	Test 1		Test 4		Test 5	
	DS	S	DS	S	DS	S
Urbana, IL	2	1	2	2	3	2
Aledo, IL	1	1	2	2	1	1
Ft. Wayne, IN	1	1	2	1	1	1
Evansville, IN	1	1	2	2	1	1
Minneapolis, MN	2	1	1	1	2	2

3. The numbers of stationary segments partitioned from differenced standardized and standardized monthly precipitation series are shown in Table IX.4. Segmentation results from monthly precipitation of Urbana, IL are shown in Fig. 9.27. According to the results of the algorithm using test 1 (Table IX.4), the numbers of segments partitioned from both differenced standardized and standardized series are the same. According to the results of the algorithm using test 4 (Table IX.4), the numbers of segments partitioned from both differenced standardized and standardized series are different for any one station. According to the results of the algorithm using test 5 (Table IX.4), the numbers of segments partitioned from both differenced standardized and standardized series are different for two stations. The segmentation results show that both

differenced standardized and standardized series of two stations (Urbana, IL and Aledo, IN) are identified as stationary by the algorithm using tests 1, 4 and 5. It is also shown that the algorithm using test 1 detects both differenced standardized and standardized series of one station (Indianapolis, IN) are non-stationary, and the algorithm using test 4 detects both differenced standardized and standardized series of another station (Waseca, MN) are non-stationary.

Table IX.4. Number of segments partitioned from monthly precipitation series. (DS: Differenced standardized series; S: Standardized series.)

Station	Test 1		Test 4		Test 5	
	DS	S	DS	S	DS	S
Urbana, IL	1	1	1	1	1	1
Aledo, IL	1	1	1	1	1	1
Ft. Wayne, IN	1	1	1	1	2	1
Indianapolis, IN	2	2	1	1	1	1
Minneapolis, MN	1	1	3	2	4	3

4. The numbers of stationary segments partitioned from differenced standardized and standardized monthly PDSI series are shown in Table IX.5. Segmentation results from monthly PDSI of Region 2, IL are shown in Fig. 9.28. According to the results of the algorithm using test 1 (Table IX.5), the numbers of segments partitioned from both differenced standardized and standardized series are the same for four regions. According to the results of the algorithm using test 4 (Table IX.5), the numbers of segments partitioned from both differenced standardized and standardized series are the same for one region. According to the results of the algorithm using test 5 (Table IX.5), the numbers of segments partitioned from both differenced standardized and standardized series are different for one region. Most of the results show non-stationarity in both differenced standardized and standardized series, except for one region (Region 7, IN) both differenced standardized and standardized series are identified as stationary by the algorithm using test 1.

Table IX.5. Number of segments partitioned from monthly PDSI series. (DS: Differenced standardized series; S: Standardized series.)

Station	Test 1		Test 4		Test 5	
	DS	S	DS	S	DS	S
Region 2, IL	1	1	3	1	2	3
Region 8, IL	1	2	4	3	4	4
Region 1, IN	1	1	4	4	3	4
Region 7, IN	1	1	3	5	1	4
Region 2, OH	1	2	4	3	2	4
Region 9, OH	1	1	2	4	1	3

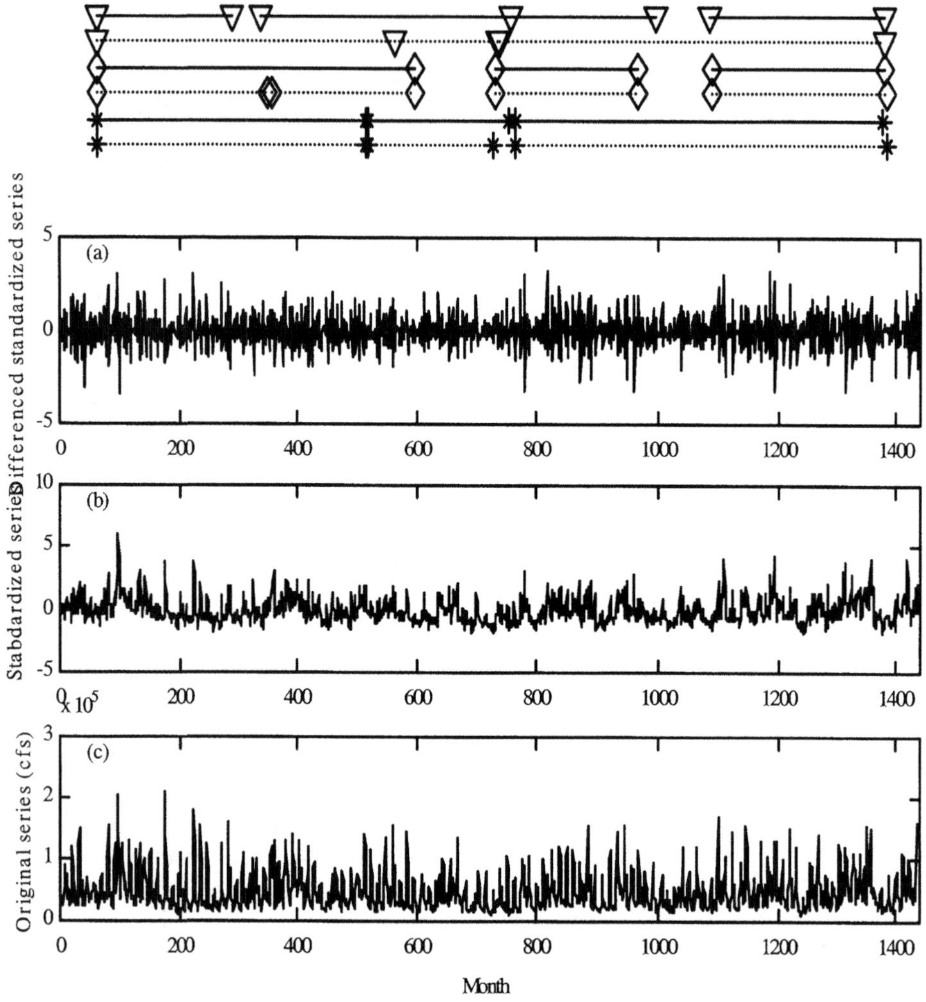

Figure 9.25. Segmentation results from monthly streamflow of Minnesota River at Clinton, IA.
Test used in the segmentation algorithm:
*: test 1; ◊: test 4; ∇: test5.
- - -: Segments from standardized series;
—: Segments from differenced standardized series

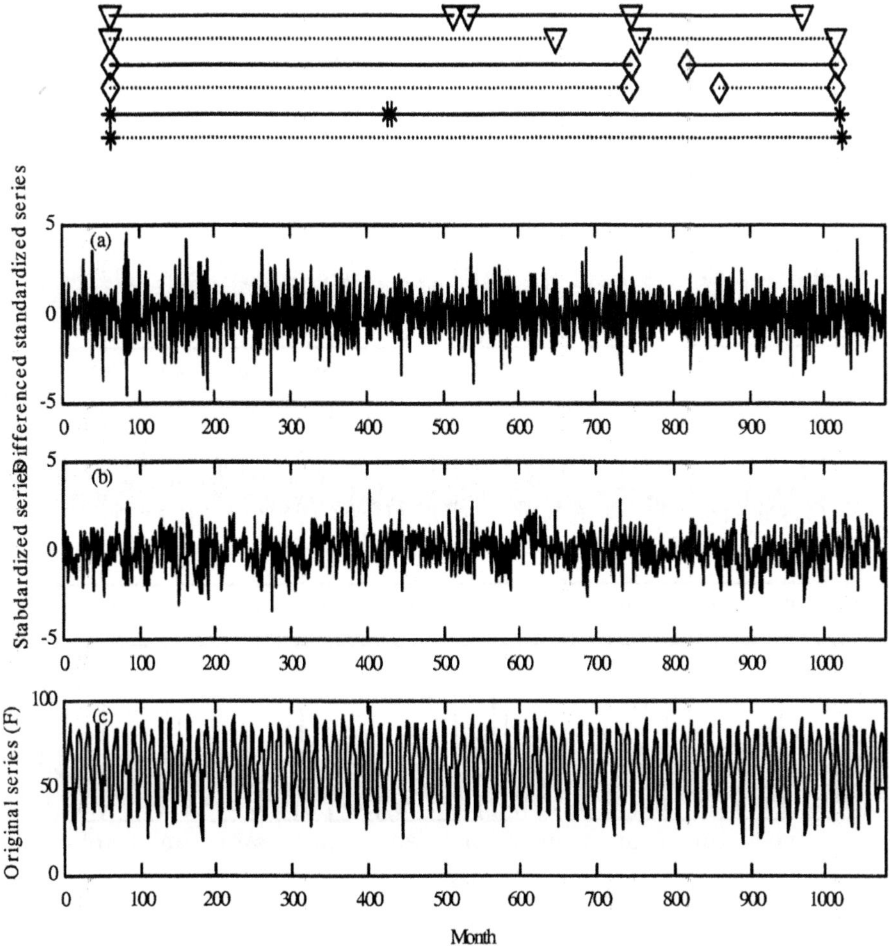

Figure 9.26. Segmentation results from monthly temperature of Urbana, IL.
Test used in the segmentation algorithm:
∗: test 1; ◊: test 4; ∇: test5.
- - -: Segments from standardized series;
—: Segments from differenced standardized series

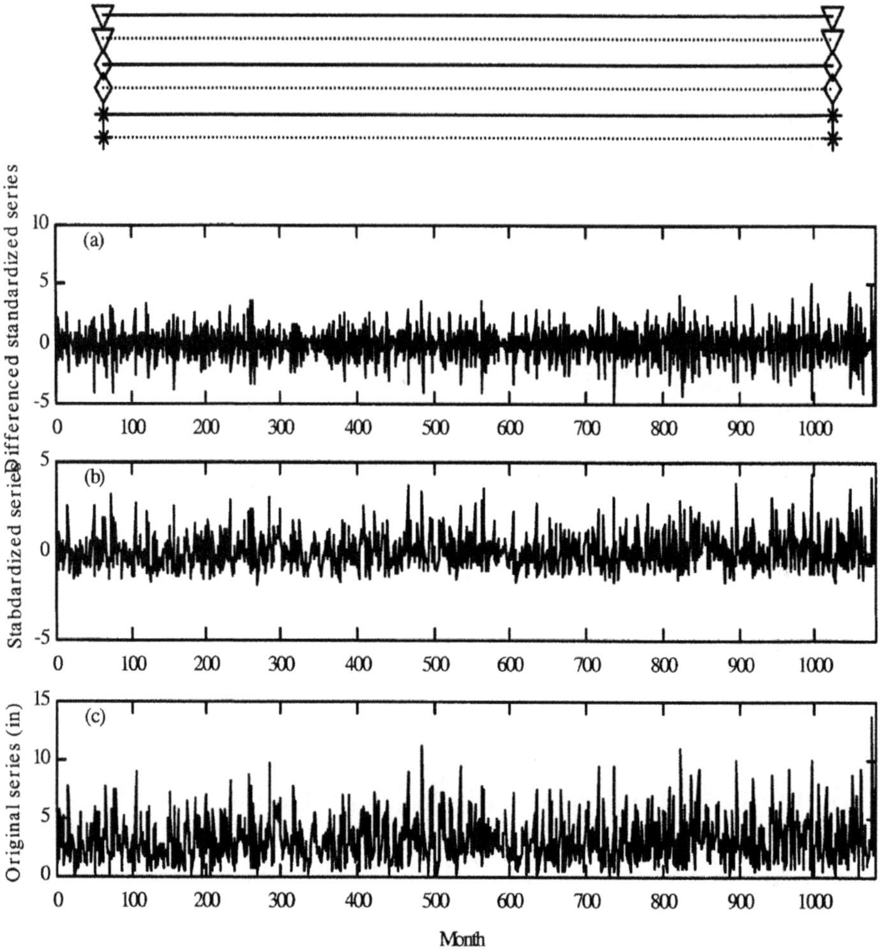

Figure 9.27. Segmentation results from monthly precipitation of Urbana, IL.
Test used in the segmentation algorithm:
*: test 1; ◊: test 4; ∇: test5.
- - -: Segments from standardized series;
—: Segments from differenced standardized series

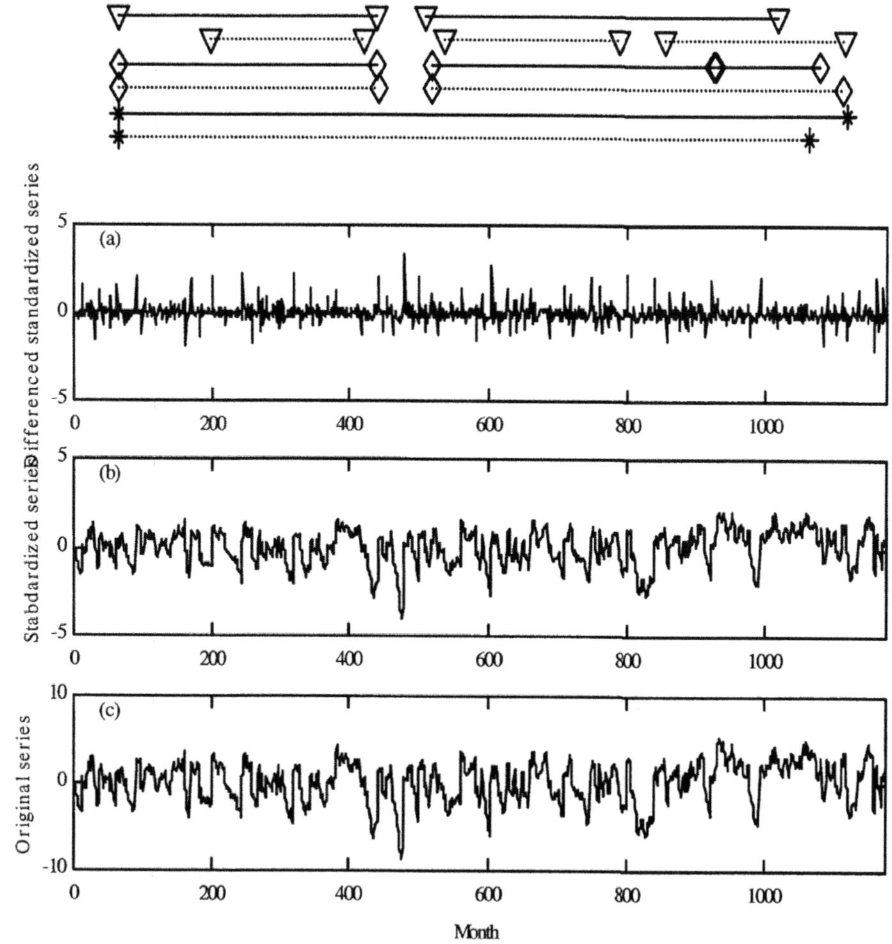

Figure 9.28. Segmentation results from monthly PDSI of region 2, IL.
Test used in the segmentation algorithm:
*: test 1; ◊: test 4; ∇: test5.
- - -: Segments from standardized series;
—: Segments from differenced standardized series

These results, in general, indicate that the segmentation algorithm using test 1 is less sensitive than that using tests 4 and 5. More series of monthly streamflow and PDSI are identified as non-stationary, while more series of monthly precipitation are stationary. These results also indicate that both differenced standardized and standardized series do not have the same segmentation results by using the same test in the segmentation algorithm. Consequently, the conventional assumption that the standarized hydrologic series are stationary are not always valid.

9.4.2. Change Points in Hydrologic Series. Stationary segments identified in differenced standardized and standardized monthly hydrologic series by the algorithm using test 5 are shown in Figs. 9.29 and 9.30, respectively. Each figure includes four subplots: (a) stationary segments of monthly streamflow series; (b) stationary segments of monthly temperature series; (c) stationary segments of monthly precipitation series; (d) stationary segments of monthly PDSI series. The starting boundaries (change points) of segments, excluding the first segment, are summarized in Tables IX.6 and IX.7.

The first commonly found starting boundary is between 1960 and 1970. It is shown in Fig. 9.29 and Table IX.6 that a starting boundary between 1960 and 1970 is found in 8 out of 21 differenced standardized monthly series, including three of streamflow series, two of temperature series, two of precipitation series and one of PDSI series. It is also shown in Fig. 9.30 and Table IX.7 that a starting boundary between 1960 and 1970 is found in 8 out of 21 standardized monthly series, including two of streamflow series, two of temperature series, one of precipitation series and three of PDSI series.

The second commonly found starting boundary is between 1930 and 1940. It is shown in Fig. 9.29 and Table IX.6 that a starting boundary between 1930 and 1940 is found in 6 out of 21 differenced standardized monthly series, including one streamflow series, one precipitation series and four PDSI series. It is also shown in Fig. 9.30 and Table IX.7 that a starting boundary between 1930 and 1940 is found in 8 out of 21 standardized monthly series, including one streamflow series, one precipitation series and six PDSI series.

Table IX.6. Starting boundaries (change points) detected in differenced standardized monthly series by the segmentation algorithm using test 5.

Series	1900-10	1911-20	1921-30	1931-40	1941-50	1951-60	1961-70	1971-80
Streamflow	1			1		1	3	
Temperature					1		2	
Precipitation		1		1			2	
PDSI				4		2	1	
Total	1	1		6	1	3	8	

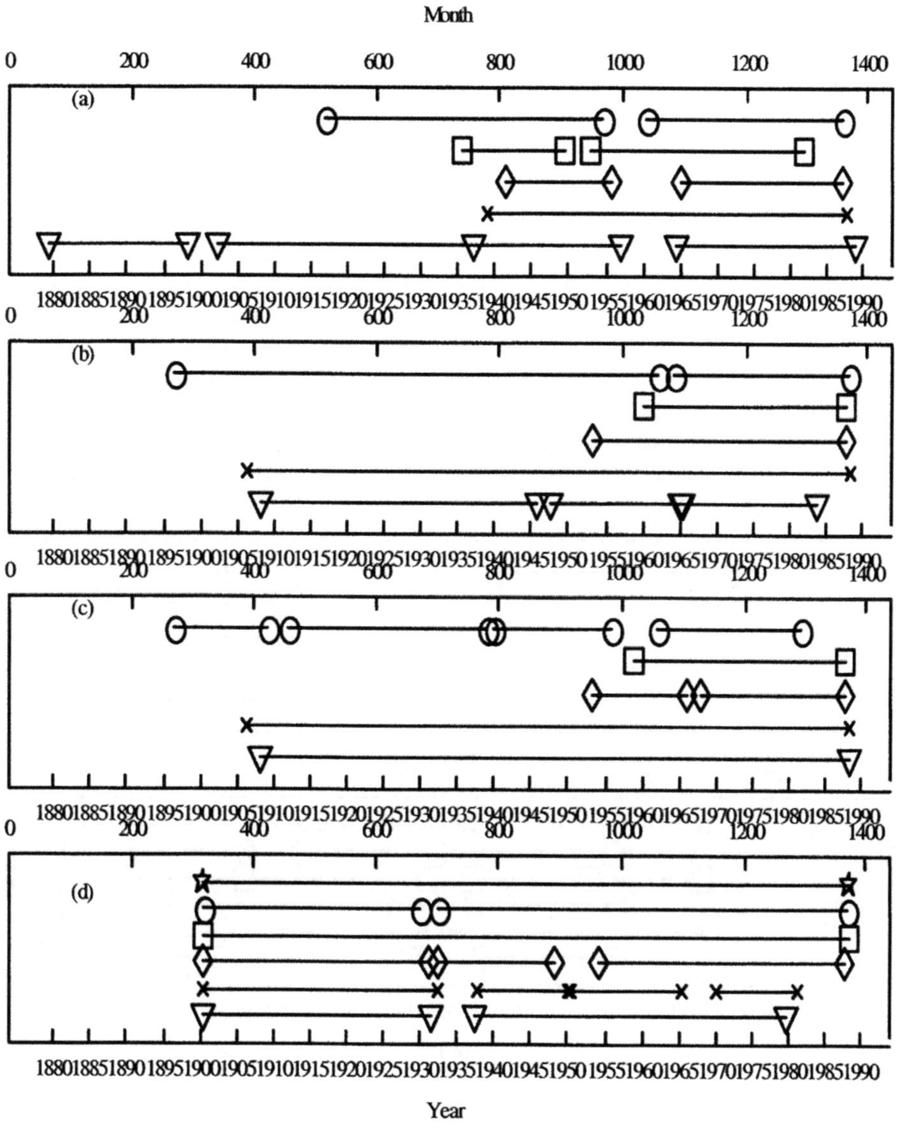

Figure 9.29. Stationary segments identified in differenced standardized monthly series by the segmentation algorithm using test 5.

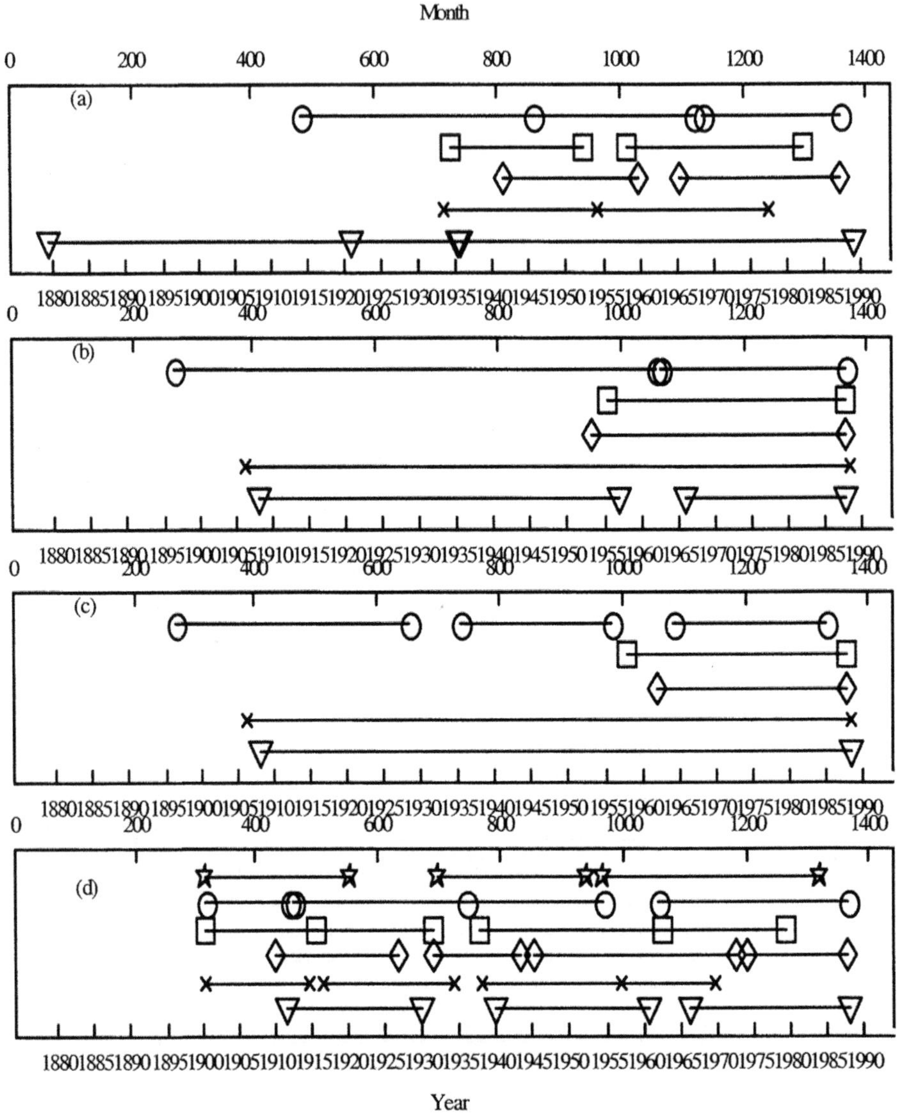

Figure 9.30. Stationary segments identified in standardized monthly series by the segmentation algorithm using test 5.

Table IX.7. Starting boundaries (change points) detected in standardized monthly series by the segmentation algorithm using test 5.

Series	1900-10	1911-20	1921-30	1931-40	1941-50	1951-60	1961-70	1971-80
Streamflow			1	1		2	2	
Temperature							2	
Precipitation				1			1	
PDSI		3		6	1	2	3	1
Total		3	1	8	1	4	8	1

These results show that change points in different types of hydrologic series, either differenced or not, are distributed in a similar pattern. It is possible that a common non-stationary mechanism has an influence on these hydrologic series. In general, standardized hydrologic monthly series, either differenced or not, are non-stationary.

9.5. CONCLUSIONS ON SEGMENTATION OF HYDROLOGIC SERIES

The results of segmentation of hydrologic series show that: (1) the segmentation algorithm using test 1 is less sensitive than that using tests 4 and 5; (2) more series of monthly streamflow and PDSI are identified as non-stationary, while more series of monthly precipitation are stationary; (3) both differenced standardized and standardized series do not have the same segmentation results by using the same test in the segmentation algorithm; (4) two starting boundaries (change points) are commonly observed in the segmentation results from differenced standardized and standardized series, one between 1960 and 1970 and the other between 1930 and 1940; and (5) change points in different types of hydrologic series, either differenced or not, are distributed in a similar pattern. In general, standardized hydrologic monthly series, either differenced or not, are non-stationary. Consequently, the conventional assumption that the standardized hydrologic series are stationary is not always valid.

10. Linearity and Gaussianity Analysis

10.1. INTRODUCTION

Linear models and Gaussian distributed variables are usually used in time series analysis mainly due to the convenience in studying relevant statistical properties. For models such as linear autoregressive (AR) and autoregressive-moving average (ARMA) models, procedures for model identification and parameter estimation have been well formalized based on linearity and Gaussianity (Box and Jenkins, 1976; Priestley, 1981; Hamilton, 1994). Both linearity and Gaussianity also play important roles in a recently developed signal processing technique — wavelet analysis. A considerable amount of linear wavelet families have been introduced in practice (Meyer, 1986; Mallat,1989; Daubechies, 1988; Chui,1992). Gaussian-distributed wavelet coefficients have been assumed in selective wavelet reconstruction (Donoho and Johnstone, 1994), detection of jumps and cusps (Hu, 1994; Wang, 1995; Ogden and Richwine, 1996), and segmentation of stationary time series (Chapter 7). However, non-linear mechanisms are often encountered in physical sciences. Non-Gaussian stationary time series may be generated as a result of some non-linear operation on a Gaussian input process. Therefore, non-linear modeling approaches have gained increasing attention from time series analysts (Subba Rao and Gabr, 1984; Priestley, 1988; Nikias and Petropulu, 1993).

Examples given in Subba Rao and Indukumar (1996) show that the second order cumulants (sample autocovariances) and spectra do not contain enough information to characterize non-linear or non-Gaussian time series. Standard whiteness tests cannot detect non-linear serial dependence in the residuals obtained from fitting a linear model, because these tests mostly rely on the second order cumulants. Recently, diagnostic tests based on the bispectrum (the third order spectrum) have been designed to detect non-linear or non-Gaussian time series. Subba Rao and Gabr (1980) detected non-linearity and non-Gaussianity based on a sample estimate of the bispectrum. Hinich (1982) modified Subba Rao and Gabr's (1980) approach by using the asymptotic variance of the sample bispectrum. It is shown in Ashley, Patterson and Hinich's (1982) simulations that Hinich's (1982) test has substantial power for many non-linear models. Both of their empirical and theoretical results also show that the test is equally powerful in detecting non-linearity in source or residual series.

In this chapter, the Hinich's linearity and Gaussianity tests are applied to the stationary segments of temperature gradient profiles and hydrologic series, respectively. These segments are partitioned by using the segmentation algorithm defined in Chapter 7.

10.2. TESTS FOR GAUSSIANITY AND LINEARITY (HINICH, 1982)

Suppose that $\{Y_t\}$ is linear, that is

$$Y_t = \sum_{j=0}^{\infty} b_j \varepsilon_{t-j} , \tag{10.1}$$

where $\{b_j\}$ are weights, and $\{\varepsilon_j\}$ are assumed to be i.i.d. The bispectrum of $\{Y_t\}$ is

$$S_{3Y}(f_1, f_2) = \mu_3 B(f_1) B(f_2) B^*(f_1 + f_2), \tag{10.2}$$

where $\mu_3 = E[\varepsilon_t^3]$,

$$B(f) = \sum_{j=0}^{\infty} b_j \exp(-2\pi i f j), \tag{10.3}$$

and $B^*(f)$ is its complex conjugate. A normalized bispectrum is defined in the form of the squared bicoherence,

$$\psi^2(f_1, f_2) = \frac{|S_{3Y}(f_1, f_2)|^2}{S_Y(f_1) S_Y(f_2) S_Y(f_1 + f_2)} = \frac{\mu_3^2}{\sigma_\varepsilon^6}, \tag{10.4}$$

where

$$S_Y(f) = \sigma_\varepsilon^2 |B(f)|^2 \tag{10.5}$$

is the spectrum of $\{Y_t\}$. The squared bicoherence is equivalent to the square of the skewness function of $\{Y_t\}$, which is constant if $\{Y_t\}$ is linear. If $\{Y_t\}$ is Gaussian distributed, the squared bicoherence is zero. An estimator of the bispectrum S_{3Y} is defined as follows. Let

$$F(j,k) = \frac{Y\left(\frac{j}{N}\right) Y\left(\frac{k}{N}\right) Y^*\left(\frac{j+k}{N}\right)}{N}, \tag{10.6}$$

where

$$Y\left(\frac{j}{N}\right) = \sum_{t=0}^{N-1} Y_t \exp\left(\frac{-2\pi i t j}{N}\right). \tag{10.7}$$

A consistent estimator is obtained by averaging $F(j,k)$ over M^2 adjacent frequency pairs as

$$\hat{S}_{3Y}(m,n) = \frac{\sum_{j=(m-1)M}^{mM-1} \sum_{k=(n-1)M}^{nM-1} F(j,k)}{M^2}, \tag{10.8}$$

where M is an integer greater than $(N)^{0.5}$. It follows from Eq. 10.4 that an estimator of the bicoherency is defined as

$$\hat{\psi}_{m,n} = \frac{\hat{S}_{3Y}(m,n)}{\left(\frac{N}{M^2}\right)^{\frac{1}{2}} \sqrt{\hat{S}_Y(g_m)\hat{S}_Y(g_n)\hat{S}_Y(g_{m+n})}}, \tag{10.9}$$

where

$$g_j = \frac{(2j-1)M}{2N}, \tag{10.10}$$

and $\hat{S}_Y$ is a smoothed estimator of the periodogram. $\hat{\psi}_{m,n}$ is complex Gaussian distributed with unit variance. Under the null hypothesis of Gaussianity, the bicoherence of $\{Y_t\}$ is zero:

$$\mathbf{H_{0G}}: \psi^2(f_1, f_2) = 0.$$

A test statistic defined as

$$T_\psi = 2\sum_m \sum_n \left|\hat{\psi}_{m,n}\right|^2 \tag{10.11}$$

is asymptotically chi-square distributed with $2P$ degrees of freedom, where P is the number of frequency pairs. Consequently, the decision rule is

$$\begin{cases} T_\psi < \chi_\alpha^2(2P), \text{accept } \mathbf{H_{0G}} \\ T_\psi \geq \chi_{1-\alpha}^2(2P), \text{do not accept } \mathbf{H_{0G}} \end{cases}.$$

Under the null hypothesis of linearity, the bicoherence of $\{Y_t\}$ is constant. If $\{Y_t\}$ is Gaussian (the bispectrum is zero), it cannot be concluded that whether or not the process is linear based on the bispectrum alone. Therefore, the hypothesis of linearity is tested only if $\{Y_t\}$ is non-Gaussian. Assume $\{Y_t\}$ is non-Gaussian, the null hypothesis of linearity is

$$\mathbf{H_{0L}}: \psi^2(f_1, f_2) \text{ is constant.}$$

Under $\mathbf{H_{0L}}$, $2\left|\hat{\psi}_{m,n}\right|^2$ are asymptotically non-central chi-square distributed ($\chi^2(2, \lambda_{m,n})$) with two degrees of freedom and noncentrality parameter

$$\lambda_{m,n} = \frac{2M^2}{N} \frac{\left|S_{3Y}(f_1, f_2)\right|^2}{S_Y(f_1)S_Y(f_2)S_Y(f_1+f_2)}. \tag{10.12}$$

A consistent estimator of $\lambda_{m,n}$ is

$$\hat{\lambda}_0 = \frac{T_\psi}{P} - 2. \tag{10.13}$$

A robust test statistic based on the interquartile range, $d\xi = \xi_3 - \xi_1$, is asymptotically Gaussian distributed as $N(d\xi, \sigma_0^2)$, where

$$\sigma_0^2 = \frac{\left(3f^{-2}(\xi_1) - 2f^{-1}(\xi_1)f^{-1}(\xi_3) + 3f^{-2}(\xi_3)\right)}{16P}, \tag{10.14}$$

and f is the density function of $\chi^2(2, \lambda_{m,n})$. Let $d\xi_s$ denote standardized sample estimate,

$$d\xi_s = \frac{d\hat{\xi} - d\xi}{\sigma_0}, \tag{10.15}$$

where $d\hat{\xi}$ is obtained from sample. The decision rule is

$$\begin{cases} N_{1-\frac{\alpha}{2}}(0,1) < d\xi_s < N_{\frac{\alpha}{2}}(0,1), \text{ accept } \mathbf{H_{0L}} \\ d\xi_s \le N_{1-\frac{\alpha}{2}}(0,1) \text{ or } d\xi_s \ge N_{\frac{\alpha}{2}}(0,1), \text{ do not accept } \mathbf{H_{0L}} \end{cases}.$$

The threshold is chosen corresponding to a given level of significance α.

10.3. TESTING FOR STATIONARY SEGMENTS

The algorithms for testing Gaussianity and linearity are composed of two stages of hypothesis testing. First, $\mathbf{H_{0G}}$ is tested. If $\mathbf{H_{0G}}$ holds, the assumption of Gaussianity is accepted. If the bispectrum is zero, it cannot be concluded that whether or not the process is linear based on the bispectrum alone. Consequently, $\mathbf{H_{0L}}$ is tested only if $\{Y_t\}$ is non-Gaussian. If $\mathbf{H_{0L}}$ holds, the assumption of linearity is accepted. The averaging parameter M (Eq. 10.8) is specified by rounding off $(N)^{0.5}$. Both hypotheses are tested at $\alpha = 0.05$ significance level.

10.3.1. Testing Temperature Gradient Profiles. Results of testing stationary segments partitioned from profiles of temperature gradient are shown in Table X.1. The denominator indicates the number of stationary segments of each profile. These results show that, in most of the profiles, only a small number of stationary segments are Gaussian. Generally, more segments are identified as non-linear than those identified as linear, regardless which test is used in the segmentation algorithm.
Tests 1-5 (Chapter 7) used in the segmentation algorithm are all based on the assumption of Gaussianity and linearity. If a series is not linear and Gaussian distributed, the test statistics derived based on these assumptions are no longer valid. Consequently, these results pose a question on using these tests in the segmentation algorithm applied to temperature gradient profiles.

10.3.2. Testing Hydrologic Series. Results of testing stationary segments partitioned from both the standardized and the differenced standardized hydrologic monthly series are discussed as follows:

1. The results of Gaussianity and linearity tests on stationary segments partitioned from differenced standardized and standardized monthly streamflow series are shown in Tables X.2 and X.3. According to the results of testing on differenced standardized series (Table X.2), most of the segments are either Gaussian or linear except one segment is non-linear. According to the results of testing on standardized series (Table X.3), two segments are Gaussian, and more segments are linear than those are non-linear.

Table X.1. Ratios of Gaussian, linear and non-linear segments partitioned from profiles of temperature gradient microstructure. (G: Gaussian; L: Linear; NL: Non-Linear.)

Lake	Profile	Test 1			Test 2			Test 4			Test 5		
		G	L	NL	G	L	NL	G	L	NL	G	L	NL
Calhoun	1111	12/61	26/61	23/61	1/18	7/18	10/18	2/44	13/44	29/44	7/55	23/55	25/55
	1127	12/65	18/65	35/65	0/20	7/20	13/20	6/49	24/49	19/49	3/40	20/40	17/40
	1228	2/68	18/68	48/68	1/32	10/32	21/32	6/62	24/62	32/62	6/53	16/53	31/53
	1455	3/34	12/34	19/34	0/12	6/12	6/12	1/32	20/32	11/32	3/31	10/31	18/31
Christmas	1210	0/28	13/28	15/28	1/12	4/12	7/12	0/26	10/26	16/26	0/24	13/24	11/24
	1239	0/23	13/23	10/23	0/8	2/8	6/8	0/26	9/26	17/26	1/27	9/27	17/27
	1256	1/28	13/28	14/28	0/17	7/17	10/17	0/23	8/23	15/23	1/27	12/27	14/27
	1326	0/22	8/22	14/22	0/11	6/11	5/11	0/23	8/23	15/23	0/23	16/23	7/23
	1349	0/27	9/27	18/27	0/13	5/13	8/13	0/27	12/27	15/27	0/24	11/24	13/24
Waconia	1502	9/57	25/57	23/57	0/12	3/12	9/12	5/43	16/43	22/43	10/57	21/57	26/57

2. The results of Gaussianity and linearity tests on stationary segments partitioned from differenced standardized and standardized monthly temperature series are shown in Tables X.4 and X.5. According to the results of testing on differenced standardized series (Table X.4), most of the segments are Gaussian except one segment is linear. According to the results of testing on standardized series (Table X.5), all segments are Gaussian. No segment is non-linear.

3. The results of Gaussianity and linearity tests on stationary segments partitioned from differenced standardized and standardized monthly precipitation series are shown in Tables X.6 and X.7. These results show that most of the segments are Gaussian except a few segments are linear. No segment is non-linear.

Table X.2. Ratios of Gaussian, linear and non-linear segments partitioned from differenced standardized monthly streamflow series.
(G: Gaussian; L: Linear; NL: Non-Linear.)

Station	Test 1			Test 4			Test 5		
	G	L	NL	G	L	NL	G	L	NL
Minnesota River at Clinton, IA	1/3	2/3		1/3	2/3		3/4	1/4	
White River near Alpine, IN		1/1			1/1				1/1
Kalamazoo River at Fennville, MI	1/1				2/2			2/2	
Missouri River at Hermann, MO		2/2			2/2			2/2	
Wisconsin River at Merrill, WI	1/1			1/3	2/3		1/2	1/2	

Table X.3. Ratios of Gaussian, linear and non-linear segments partitioned from monthly standardized streamflow series. (G: Gaussian; L: Linear; NL: Non-Linear.)

Station	Test 1			Test 4			Test 5		
	G	L	NL	G	L	NL	G	L	NL
Minnesota River at Clinton, IA		2/3	1/3		3/4	1/4		1/3	2/3
White River near Alpine, IN		1/1			1/1			2/2	
Kalamazoo River at Fennville, MI		1/1		1/2	1/2			2/2	
Missouri River at Hermann, MO		1/2	1/2		1/3	2/3		1/2	1/2
Wisconsin River at Merrill, WI	1/1				3/3			2/3	1/3

Table X.4. Ratios of Gaussian, linear and non-linear segments partitioned from differenced standardized monthly temperature series.
(G: Gaussian; L: Linear; NL: Non-Linear.)

Station	Test 1			Test 4			Test 5		
	G	L	NL	G	L	NL	G	L	NL
Urbana, IL	2/2			2/2			1/3	2/3	
Aledo, IL	1/1			2/2			1/1		
Ft. Wayne, IN	1/1			2/2			1/1		
Evansville, IN	1/1			2/2			1/1		
Minneapolis, MN	2/2			1/1			2/2		

Table X.5. Ratios of Gaussian, linear and non-linear segments partitioned from standardized monthly temperature series. (G: Gaussian; L: Linear; NL: Non-Linear.)

Station	Test 1			Test 4			Test 5		
	G	L	NL	G	L	NL	G	L	NL
Urbana, IL	1/1			2/2			2/2		
Aledo, IL	1/1			2/2			1/1		
Ft. Wayne, IN	1/1			1/1			1/1		
Evansville, IN	1/1			2/2			1/1		
Minneapolis, MN	1/1			1/1			2/2		

Table X.6. Ratios of Gaussian, linear and non-linear segments partitioned from differenced standardized monthly precipitation series.
(G: Gaussian; L: Linear; NL: Non-Linear.)

Station	Test 1			Test 4			Test 5		
	G	L	NL	G	L	NL	G	L	NL
Urbana, IL	1/1			1/1			1/1		
Aledo, IL	1/1			1/1			1/1		
Ft. Wayne, IN	1/1			1/1			2/2		
Indianapolis, IN	1/2	1/2		1/1			1/1		
Minneapolis, MN	1/1			2/3	1/3		2/4	2/4	

Table X.7. Ratios of Gaussian, linear and non-linear segments partitioned from standardized monthly precipitation series. (G: Gaussian; L: Linear; NL: Non-Linear.)

Station	Test 1			Test 4			Test 5		
	G	L	NL	G	L	NL	G	L	NL
Urbana, IL	1/1			1/1			1/1		
Aledo, IL	1/1			1/1			1/1		
Ft. Wayne, IN	1/1			1/1			1/1		
Indianapolis, IN		2/2		1/1			1/1		
Minneapolis, MN	1/1			2/2			1/3	2/3	

4. The results of Gaussianity and linearity tests on stationary segments partitioned from differenced standardized and standardized monthly PDSI series are shown in Tables X.8 and X.9. According to the results of testing on differenced standardized series (Table X.8), all the segments are either Gaussian or linear. According to the results of testing on standardized series (Table X.9), most of the segments are either Gaussian or linear, except four segments are detected as non-linear.

10.4. CONCLUSIONS ABOUT TESTING THE HYDROLOGIC SERIES

These results of testing Gaussianity and linearity on monthly hydrologic series show that: (1) only one segment of differenced standardized series is detected as non-linear, while 15 segments of standardized series (without differencing) are detected as non-linear; (2) most of the segments of temperature and precipitation series, either differenced standardized or standardized (without differencing), are Gaussian distributed. No non-linear segment is detected in temperature and precipitation series; and (3) compared to temperature and precipitation series, a number of segments from standardized (without differencing) monthly streamflow and PDSI series are non-Gaussian or non-linear. Therefore, the conventional assumption of linear models of Gaussian-distributed variables may not be valid for all standardized monthly streamflow and PDSI series.

Table X.8. Ratios of Gaussian, linear and non-linear segments partitioned from differenced standardized monthly PDSI series.
(G: Gaussian; L: Linear; NL: Non-Linear.)

Station	Test 1			Test 4			Test 5		
	G	L	NL	G	L	NL	G	L	NL
Region 2, IL		1/1		2/3	1/3		1/2	1/2	
Region 8, IL	1/1			2/4	2/4		3/4	1/4	
Region 1, IN	1/1			3/4	1/4		2/3	1/3	
Region 7, IN	1/1			3/3			1/1		
Region 2, OH	1/1			3/4	1/4		2/2		
Region 9, OH	1/1			2/2			1/1		

Table X.9. Ratios of Gaussian, linear and non-linear segments partitioned from standardized monthly PDSI series. (G: Gaussian; L: Linear; NL: Non-Linear.)

Station	Test 1			Test 4			Test 5		
	G	L	NL	G	L	NL	G	L	NL
Region 2, IL		1/1		1/2	1/2			2/3	1/3
Region 8, IL	1/2	1/2		1/3	1/3	1/3	1/4	2/4	1/4
Region 1, IN	1/1			1/4	3/4		1/4	3/4	
Region 7, IN	1/1			2/5	3/5		1/4	3/4	
Region 2, OH	2/2			1/3	2/3		1/4	2/4	1/4
Region 9, OH	1/1			2/4	2/4		2/3	1/3	

11. Bayesian Detection of Shifts in Hydrologic Time Series

11.1. INTRODUCTION

Detection of shifts in a hydrologic time series is an important problem because shifts can indicate climatic changes. Climatic changes can manifest as changes in the characteristics of hydrologic time series. These characteristics include the mean, variance and other statistical properties of time series.

Hydrologists have dealt with the problems of shifts in hydrologic time series for a long time. Shifts in hydrologic time series, for example, have been shown to account for the "Hurst effect". At the operational level, a common reason for observed significant changes in rainfall series is moving raingages from one location to another. The data with shifts are adjusted by using mass curves and double mass curves. In the mass curve adjustment, data from only the affected station are used. In the double mass curve technique, data from affected as well as the neighboring unaffected stations are used. However, these methods are very simple arithmetic adjustments made to the data and have only an empirical basis. Significance of changes in the data characteristics cannot be inferred by these procedures.

There have been several methods extant which are used to investigate changes in the characteristics of hydrologic time series. Prominent among these are intervention analysis and methods based on spectral analysis. In intervention analysis, times at which the characteristics of the time series have changed must be specified. Models are fitted to segments of data before and after the observed change. Inferences about changes in time series are based on these models. Because of the necessity to prespecify times of changes, Intervention analysis may be of limited utility, especially when observed changes are small.

The inferences from Intervention analysis are model dependent. The robustness of decisions depends on the length of data and the appropriateness of the model. If there are multiple changes in the data, with only a few observations between changes, or if the changes are small, inferences from Intervention analysis would not be robust.

Methods based on evolutionary spectra of data also have drawbacks. The basic idea behind the method based on evolutionary spectra is to keep track of changes in spectral characteristics of the data. These methods may also be model based. For example, the procedure developed by Rao and Shapiro (1970) and used by Rao and Yu (1984) is based on an exponential moving average model. The parameters of the model are updated whenever a significant change is encountered in the spectrum of data.

There are well known problems in the estimation of spectra of nonstationary series. The models used in these spectrum-based procedures may not be unique. Sufficient information is generally not available about the dependence of inferences on the model used. Consequently, quite frequently, results from the methods based on evolutionary spectra may not be robust.

Examples of the differences in the inferences obtained by using Intervention analysis and evolutionary spectra are shown in Table XI.1. The first eight series are of long-term annual rainfall in the East Coast in the US., the ninth is a series of particulate data in Chicago and the last series is the Colorado river flows at Compact Point. There is only one case (NEWH) where the same change point has been detected by both the methods. In other cases, there is no correspondence between the results from the two methods. Results such as those in Table XI.1 provide the motivation to search for better and robust methods to investigate the significance of changes in the characteristics of hydrologic time series.

Table XI.1. Change point locations according to previous results.

Data Name	Intervention Analysis (Potter, 1976, 1979)	Evolutionary Spectrum (Rao & Yu, 1984)
Eastport (EAST)	4,69	77,87
New Bedford (NEWB)	93	85,91,159
Woodstock (WOVT)	56	82
Fitchburg (FITC)	61	51,97
Taunton (TAUN)	49	55,57,98
New Haven (NEWH)	74	73
Moorestown (MOOR)	61	30,31,112
West Chester (WEST)	28,49	100
Chicago (CHIC)		74,75,109,137
Colorado (COLO)		209,233,281,297,316,357, 377,389,425,429,497,533,569 581,605,629,653

Box and Tiao (1962) were the first to develop a procedure to detect changes in a time series by using a Bayesian inferential procedure which provides a theory of statistical influence. Box and Tiao (1962,1964a,b) used the Bayesian approach to compare two means and two variances from two samples drawn from specific populations. A parent distribution characterized by a location parameter θ, a scale parameter σ and a non-normality parameter β was chosen by them for analysis. As the Bayesian method is successful in other applications, it may be used to investigate shifts in hydrologic and climatologic time series. More specifically, the aspects discussed in this chapter are to detect whether observed changes in the mean values of a series are statistically significant or not, to locate the change point by using the Bayesian inferential procedure and to examine the behavior of the distributions of mean, standard deviation and shape parameters.

11.2. DATA USED IN THIS CHAPTER

The data used in this chapter consist of eight series of annual precipitation from the northeastern U.S., a series of monthly average suspended particulates and a series of monthly river flow data. The annual precipitation data are measured at Eastport, New Bedford, Woodstock, Fitchburg, Taunton, New Haven, Moorestown and West Chester. These data were originally collected by Potter (1979). The period, length of records and statistics of the data are shown in Table XI.2. Figure 11.1 and 11.2 show the annual precipitation data of Eastport and New Bedford.

The monthly average suspended particulate data were collected in Chicago (CHIC) between 1964 and 1977 for a period of 14 years. The monthly suspended particulate data series is shown in Fig. 11.3. In January 1970 (corresponding to the 73th observation), a law banning the use of high sulfur coal went into effect in the city of Chicago. The particulate concentration changed because of this law. The statistics of the series before this intervention date are as follows: the mean is 155.1, variance is 1042, and the skewness is 0.36. After the intervention date, the mean is 100.7, the

Table XI.2. Rainfall characteristics.

Name	Period	Record Length	Mean (in)	Variance (in^2)	Skewness Coeff.
Eastport (EAST)	1874-1976	103	39.8	67.6	0.52
New Bedford (NEWB)	1814-1976	163	44.5	59.8	0.50
Woodstock (WOVT)	1870-1976	107	39.0	27.2	0.29
Fitchburg (FITC)	1865-1976	112	42.7	54.0	0.22
Taunton (TAUN)	1875-1976	102	45.6	47.3	0.16
New Haven (NEWH)	1873-1976	104	45.0	47.6	0.20
Moorestown (MOOR)	1864-1976	113	44.2	39.0	0.39
West Chester (WEST)	1860-1976	117	47.5	58.7	0.11

variance is 414, and the skewness is 0.26. The level of suspended particulates dropped significantly after January 1970 as industries complied with the new law.

The monthly flows from the Colorado river (COLO) between 1914 and 1969 measured at Compact Point for a period of 56 years are used in this study. The flows are measured in second-feet. The mean of the series is 16000, the variance is 3.5×10^8, and the skewness is 2.5. The Colorado river flows have a periodicity of 12 months and there are obvious changes in the series at several points. Figure 11.4 through 11.6 show the monthly Colorado river flows of the first 100 and last 172 observations.

All these data show evidence of changes in level, nonstationarity, periodicity and other features. These series have been investigated in the past for the changes in level or shifts and the results are shown in Table XI.1.

11.3. A BAYESIAN METHOD TO DETECT SHIFTS IN DATA

11.3.1. Theory. Hsu (1982) developed a robust Bayesian inference procedure to detect parameter shifts in time series. Hsu's method is briefly discussed below to investigate changes in parameter estimates and to detect changes in time series. Hsu's procedure is based upon the development by Box and Tiao (1973). The Bayesian procedure discussed herein applies to cases where the normal distribution assumption may not be valid. This method employs a family of distributions called the exponential power distributions to investigate changes in series which may not be normally distributed.

A general multiple linear regression model which represents the relationship between the observations y_j and the regressors $X_{j1}, X_{j2}, \ldots, X_{jp-1}$ is written as in Eq. (11.1), and is the basis of this method.

$$y_j = \theta_0 + \theta_1 X_{j1} + \theta_2 X_{j2} + \ldots + \theta_{p-1} X_{jp-1} + \varepsilon_j \tag{11.1}$$

Equation (11.1) can be written in matrix form as in Eq. (11.1a),

$$y_j = \mathbf{X}_j^T \boldsymbol{\theta} + \varepsilon_j \qquad j = 1,2,\ldots,N \tag{11.1a}$$

where j defines the time period, N defines the sample size and p is the total number of regression parameters. $\mathbf{X}_j$ and $\boldsymbol{\theta}$ are vectors defined by

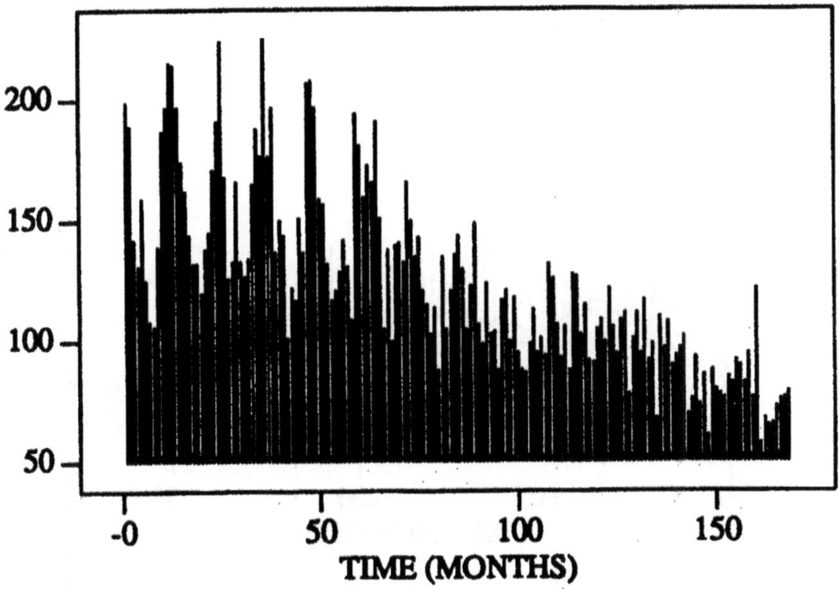

Figure 11.1. Eastport annual precipitation (1874-1976)

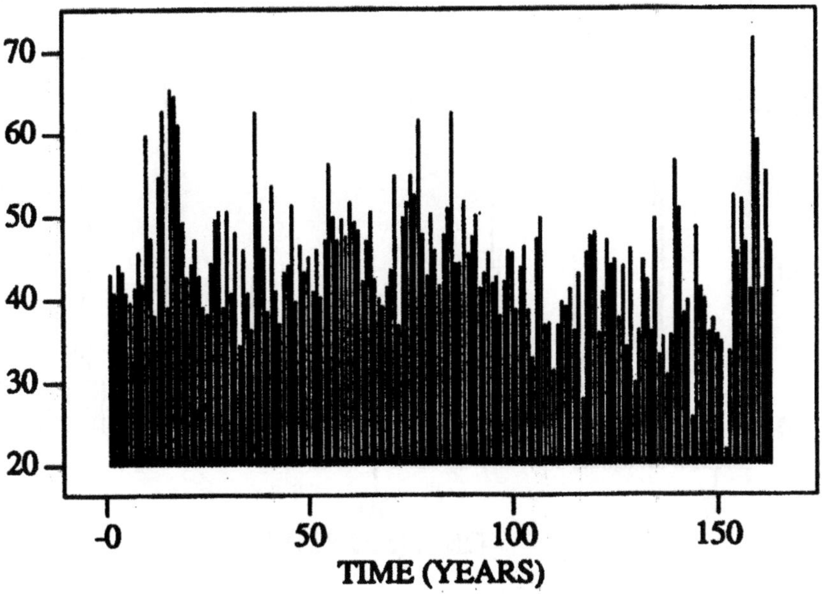

Figure 11.2. New Bedford annual precipitation (1814-1976)

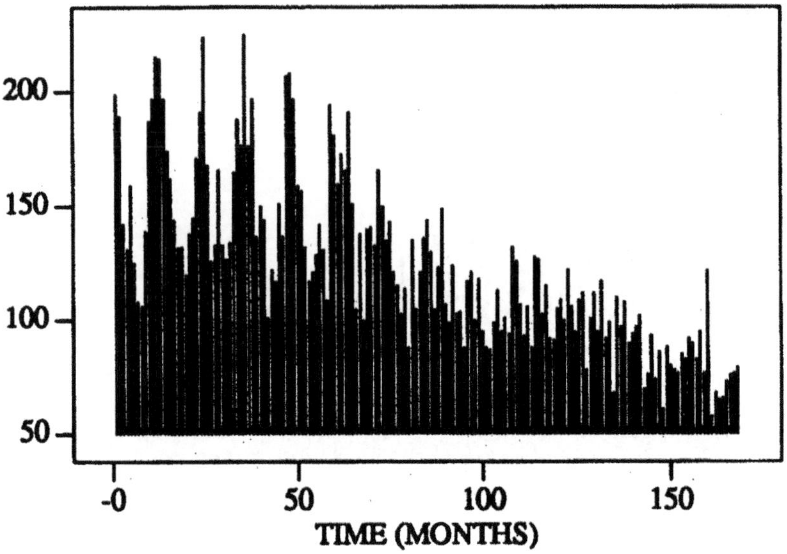

Figure 11.3. Chicago monthly suspended particulates (1964 – 1977)

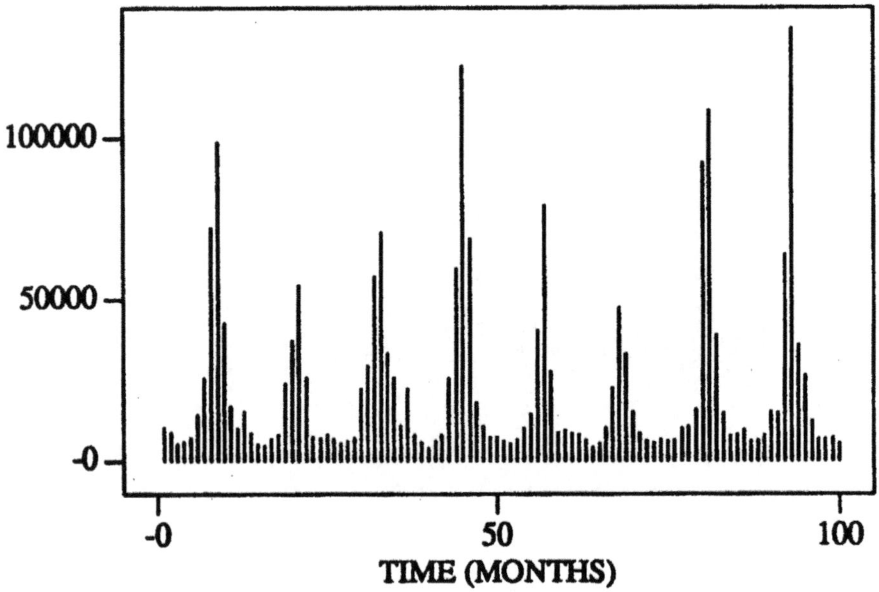

Figure 11.4. Colorado River monthly flows at compact point (1914 – 1969)

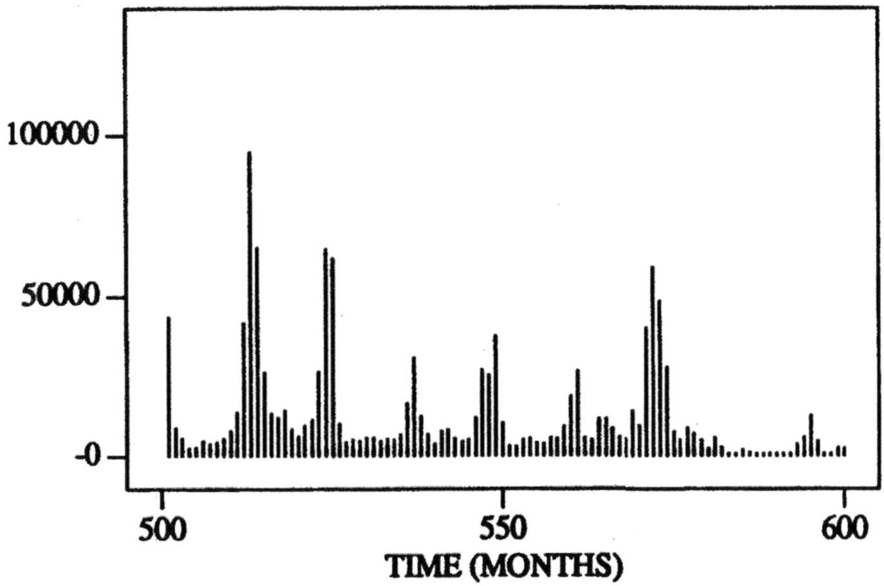

Figure 11.5. Colorado River monthly flows at compact point (1914 – 1969)

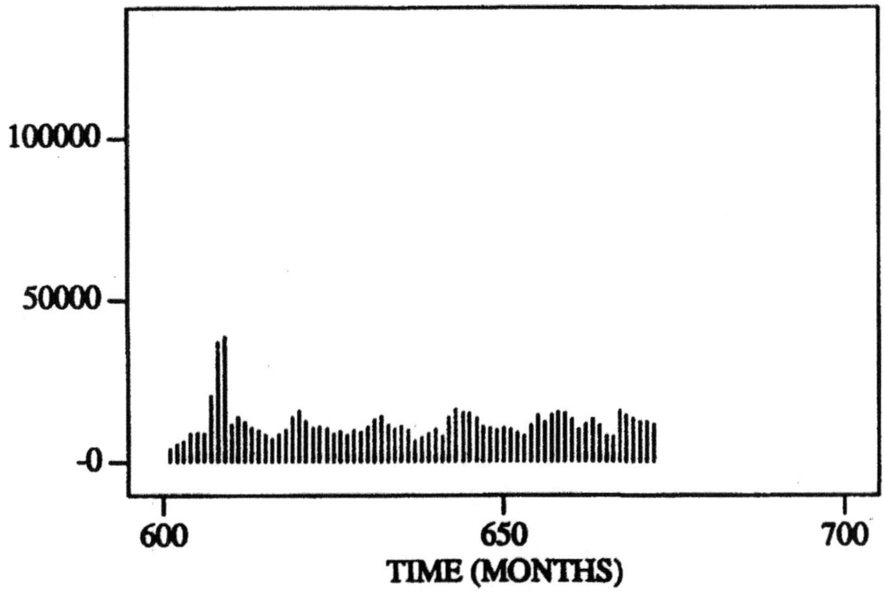

Figure 11.6. Colorado River monthly flows at compact point (1914 – 1969)

$$\boldsymbol{X}_j^{T} = \left[1, \boldsymbol{X}_{j1}, \boldsymbol{X}_{j2}, \ldots, \boldsymbol{X}_{jp-1}\right] \tag{11.1b}$$

and

$$\boldsymbol{\theta} = \left[\theta_0, \theta_1, \theta_2, \ldots, \theta_{p-1}\right]^{T} \tag{11.1c}$$

In Eq. (11.1), ε_j are independent identically distributed random variables whose density function is given in Eq. (11.2),

$$p(y \mid \boldsymbol{\theta}, \sigma, \beta) = \omega(\beta)\, \sigma^{-1} \exp\left[-c(\beta) \left| \frac{y - \boldsymbol{\theta}}{\sigma} \right|^{2/(1+\beta)} \right] \qquad -\infty < y < \infty \tag{11.2}$$

$$\sigma > 0, \qquad -\infty < \boldsymbol{\theta} < \infty, \qquad -1 < \beta \le 1$$

where $c(\beta)$ and $\omega(\beta)$ are defined below.

$$c(\beta) = \left\{ \frac{\Gamma[(3/2)(1+\beta)]}{\Gamma[(1/2)(1+\beta)]} \right\}^{1/(1+\beta)} \tag{11.2a}$$

$$\omega(\beta) = \frac{\{\Gamma[(3/2)(1+\beta)]\}^{1/2}}{(1+\beta)\{\Gamma(1/2)(1+\beta)]\}^{3/2}} \tag{11.2b}$$

In Eq. (11.2), θ and σ are respectively the mean and the standard deviation of the distribution. β is an index of peakedness of the distribution (Box and Tiao, 1973). The probability density function in Eq. (11.2) is quite versatile and describes a variety of situations. When $-1 < \beta < 0$, the distribution is short tailed and level, for $0 < \beta \le 1$, it is long tailed and more peaked. Special cases are when $\beta = 1$ (double exponential distribution), $\beta = 0$ (normal distribution) and $\beta \to -1$ (uniform distribution). Some examples of the distribution in Eq. (11.2) for various values of β are shown in Fig. 11.7.

Let a sequence of observations be of length N with the first n_1 values distributed with parameters $\boldsymbol{\theta}_1, \sigma_1,$ and β_1 and the remaining n_2 values where $n_2 = N - n_1$ with parameters $\boldsymbol{\theta}_2, \sigma_2,$ and β_2. The n_1-th point in the observation sequence is the unknown change point. The parameters of the distribution and the change point are shown in Fig. 11.8. The parameters can be further examined by analyzing the difference in the shape parameter, $\delta = \beta_2 - \beta_1$, the ratio of the variance, $\upsilon = \sigma_2^2/\sigma_1^2$, and the difference between the regression equation parameters, $\gamma = \boldsymbol{\theta}_2 - \boldsymbol{\theta}_1$.

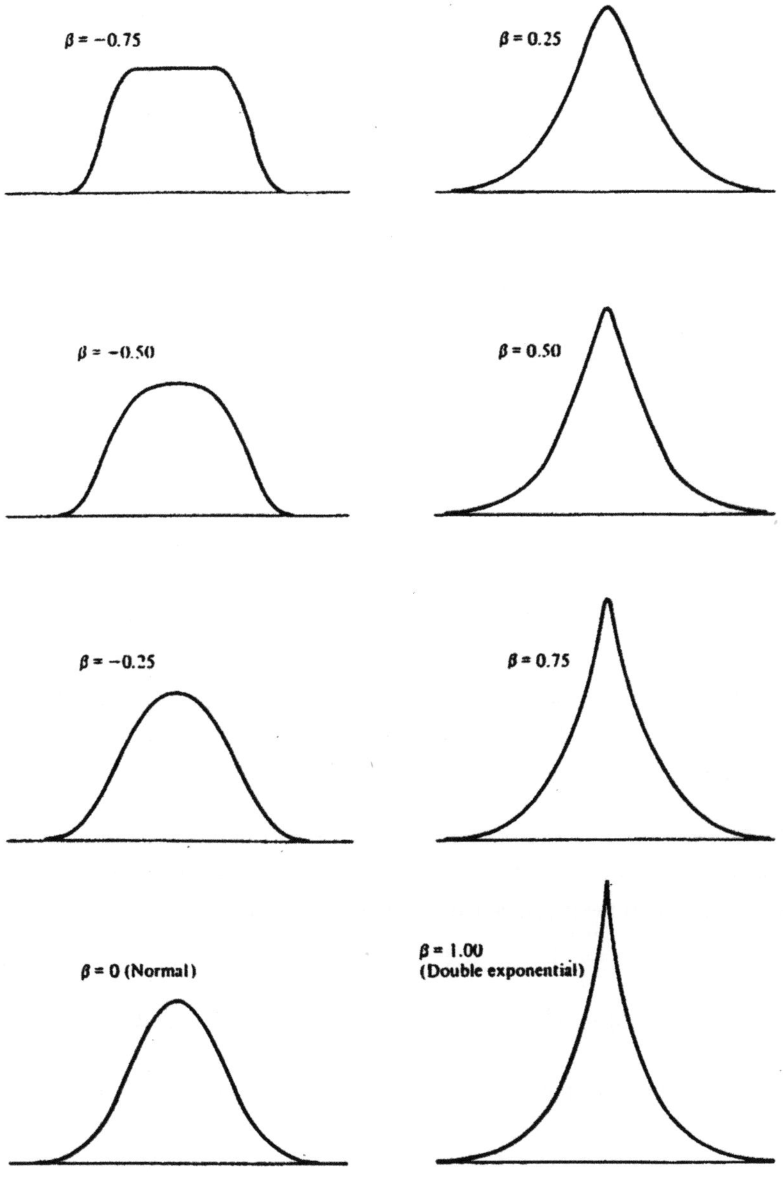

Figure 11.7. Exponential power distribution for various values of β from Box and Tiao, 1973)

The posterior distribution of the change point n_1 is determined below to find the location of the change points. The posterior distribution of parameters δ, υ and γ unconditional on the change point is examined to see if there is a difference between the distribution based on the first n_1 and the remaining n_2 data points.

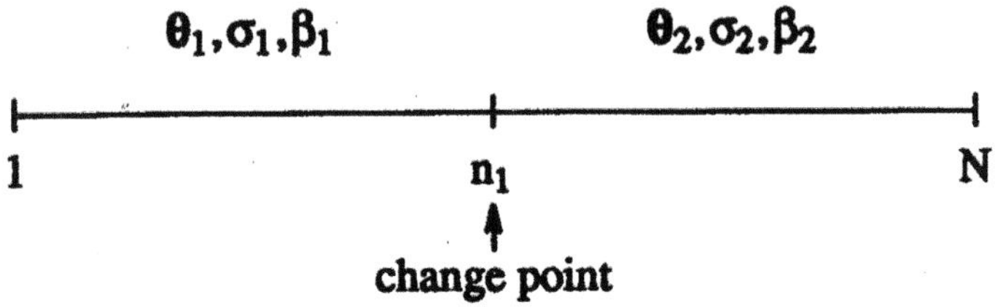

Figure 11.8. Parameters of the distribution and the change point location

If the differences in the posterior distributions are found, the posterior distributions of β_i, σ_i^2 and θ_i conditional on the change point are examined. These aspects are separately discussed below.

11.3.1.1. The Posterior Distribution of the Change Point n_1 The prior distribution of the change point $p(n_1)$ is assumed to be uniform over the range $1 \le n_1 \le N-1$ and the prior distribution of the parameter β_i is given by Eq. (11.3),

$$p(\beta_1) = \Gamma(2a)[\Gamma(a)]^{-2} 2^{-(2a-1)} (1-\beta_i^2)^{a-1} \qquad a \ge 1 \qquad (11.3)$$

where a is an adjustable parameter of the change point n_1 given by Eq. (11.4).

$$p(n_1 \mid y) \propto \prod_{i=1}^{2} [n_i 2^{n_i})^{-1} \int_{-1}^{+1} A_i p(\beta_i) d\beta_i] p(n_1) \qquad 1 \le n_1 \le N-1 \qquad (11.4)$$

where,

$$A_i = \frac{\Gamma(v_i)/2\Gamma\left[1+\frac{n_i}{2}(1+\beta_i)\right]}{|\mathbf{D}_i|^{1/2}\Gamma\left[\frac{1}{2}(v_i+p)\right]\left\{\Gamma\left[1+\frac{1}{2}(1+\beta_i)\right]\right\}^{n_i}\left[n_i s_i\left(\beta_i,\hat{\boldsymbol{\theta}}_i\right)\right]^{v_i/2}} \tag{11.5}$$

In. Eq. (11.5),

$$v_i = n_i(1+\beta_i) - p \ . \tag{11.6}$$

Also

$$s_i(\beta_i,\hat{\boldsymbol{\theta}}_i) = \min_{\boldsymbol{\theta}_i} \{s_i(\beta_i,\boldsymbol{\theta}_i)\} \tag{11.7}$$

and

$$s_i(\beta_i,\boldsymbol{\theta}_i) = \frac{1}{n_i}\sum_{j=1}^{n_i} \left|y_{ij} - \mathbf{X}_{ij}^T\boldsymbol{\theta}_i\right|^{2/(1+\beta_i)} \ . \tag{11.8}$$

In Eq. (11.5), $\mathbf{D}_i$ is a scaling matrix and is determined at a given value of β_i such that the condition in eq. (11.9) is met.

$$\mathbf{M}(\boldsymbol{\theta}) \approx \mathbf{M}(\boldsymbol{\theta}_i) + \left(\boldsymbol{\theta}_i - \hat{\boldsymbol{\theta}}_i\right)^T \mathbf{D}_i \left(\boldsymbol{\theta}_i - \hat{\boldsymbol{\theta}}_i\right) \tag{11.9}$$

where

$$\mathbf{M}(\boldsymbol{\theta}_i) = \sum_{j=1}^{n_i} \left|y_{ij} - \mathbf{X}_{ij}^T\boldsymbol{\theta}_i\right|^{2/(1+\beta_i)} \tag{11.10}$$

and

$$\mathbf{M}\left(\hat{\boldsymbol{\theta}}_i\right) = \min_{\boldsymbol{\theta}_i} \{\mathbf{M}(\boldsymbol{\theta}_i)\} \ . \tag{11.11}$$

The posterior probability density function of the change point $p(n_1 \mid y)$ given by Eq. (11.4) is used to locate the positions of the change points by plotting it against n_1.

11.3.1.2. The Unconditional Posterior Distributions of δ, υ, and γ. The distribution of $\delta = \beta_2 - \beta_1$ conditional on the change point n_1 is given by Eq. (11.12).

$$p(\delta \mid n_1, \mathbf{y}) = \int_{-1}^{+1} p(\beta_1, \beta_2 \mid n_1, \mathbf{y})\, d\beta_1$$

$$= \int_{-1}^{+1} p(\beta_1) p(\beta_2) p_u(\beta_1, \beta_2 \mid n_1, \mathbf{y}) d\beta_1 \qquad -2 < \delta < 2 \tag{11.12}$$

In. Eq. (11.12), $P_u(\cdot)$ is the probability based on uniform prior distribution.

$$p_u(\beta_1, \beta_2 \mid n_1, \mathbf{y}) \propto \prod_{i=1}^{2} p_u(\beta_i \mid n_i, \mathbf{y}_i) \qquad -1 < \beta_i \le 1 \tag{11.13}$$

$$p_u(\beta_i \mid n_i, \mathbf{y}_i) \propto \Gamma\left[1 + \frac{1}{2} n_i (1+\beta_i)\right] \left\{\Gamma\left[1 + \frac{1}{2}(1+\beta_i)\right]\right\}^{-n_i} J(\beta_i) \propto A_i \tag{11.14}$$

$$J(\beta_i) = \int_{-\infty}^{\infty} \cdots \int_{-\infty}^{\infty} [\mathbf{M}(\boldsymbol{\theta}_i)]^{-n_i(1+\beta_i)/2} d\theta_{i0} d\theta_{i1} \ldots d\theta_{ip-1} \tag{11.15}$$

The unconditional distribution of δ is obtained by multiplying the conditional distribution of δ given by Eq. (11.12) with the posterior distribution of the change point n_1 given by Eq. (11.4) and summing over the observations.

$$p(\delta \mid \mathbf{y}) = \sum_{n_1=1}^{N-1} p(\delta \mid n_1, \mathbf{y}) p(n_1 \mid \mathbf{y}) \qquad -2 < \delta < 2 \tag{11.16}$$

The distribution of the variance ratio $\upsilon = \sigma_2^2 / \sigma_1^2$ conditional on the change point n_1 is given by Eq. (11.17),

$$p(\upsilon \mid \beta_1, \beta_2, n_1, \mathbf{y}) = \int_0^{\infty} \sigma_1^2 f_2(\upsilon \sigma_1^2 \mid \beta_2, n_2, \mathbf{y}_2) f_1(\sigma_1^2 \mid \beta_1, n_1, \mathbf{y})\, d\sigma_1^2 \tag{11.17}$$

where

$$f_i(\sigma_i^2 \mid \beta_i, n_i, \mathbf{y}) = \frac{\left[n_i s_i(\beta_i, \hat{\boldsymbol{\theta}}_i)c(\beta_i)\right]^{v_i/2}}{(1+\beta_i)\Gamma(v_i/2)} (\sigma_i^2)^{-\{[1/(1+\beta_i)](v_i/2)+1\}}$$

$$\times \exp\left[-\frac{n_i s_i(\beta_i, \hat{\boldsymbol{\theta}}_i)c(\beta_i)}{(\sigma_i^2)^{1/(1+\beta_i)}}\right] . \qquad (11.18)$$

The joint probability distribution of β_1 and β_2 is given in Eq. (11.19),

$$p(\beta_1, \beta_2 \mid n_1, \mathbf{y}) = p(\beta_1 \mid n_1, \mathbf{y}_1)p(\beta_2 \mid n_2, \mathbf{y}_2) \qquad (11.19)$$

where

$$p(\beta_i \mid n_i, y_i) \propto p_u(\beta_i \mid n_i, y_i)p(\beta_i) \qquad -1 < \beta_i \le 1 . \qquad (11.20)$$

The posterior distribution of υ is obtained by multiplying the conditional distribution of υ given by Eq. (11.17) by the joint distribution of β_1 and β_2 given by Eq. (11.19) and integrating over β_1 and β_2.

$$p(\upsilon \mid n_1, y) = \int_{-1}^{+1}\int_{-1}^{+1} p(\upsilon \mid \beta_1, \beta_2, n_1, y)p(\beta_1, \beta_2 \mid n_1, y)d\beta_1 d\beta_2 \qquad \upsilon > 0 \qquad (11.21)$$

The unconditional posterior distribution of υ is obtained by multiplying the conditional distribution of υ given by Eq. (11.21) with the posterior distribution of the change point n_1 given by Eq. (11.4) and summing over the observations.

$$p(\upsilon \mid \mathbf{y}) = \sum_{n_1=1}^{N-1} p(\upsilon \mid n_1, \mathbf{y})p(n_1 \mid \mathbf{y}) \qquad (11.22)$$

The posterior distribution of $\boldsymbol{\gamma} = \boldsymbol{\theta}_2 - \boldsymbol{\theta}_1$ conditional on the change point n_1 is given by

$$p(\gamma \mid n_1, \mathbf{y}) = \int_{-1}^{+1}\int_{-1}^{+1} p(\gamma \mid \beta_1, \beta_2, n_1, \mathbf{y})p(\beta_1 \mid n_1, \mathbf{y}_1)p(\beta_2 \mid n_2, \mathbf{y}_2)\, d\beta_1\, d\beta_2 \qquad (11.23)$$

$$-\infty < \gamma_k < \infty \qquad k = 0, 1, \ldots, p-1$$

where γ approximately follows a Student's t distribution $t(\mu(s^2, \upsilon)$ with mean μ, scale parameter s, and υ degrees of freedom,

$$p(\gamma_k \mid \beta_1, \beta_2, n_1, y) \sim t(\hat{\theta}_{2k} - \hat{\theta}_{1k}, a_k^2(\tau_{1k}^2 + \tau_{2k}^2), b_k) \tag{11.24}$$

where τ_{ik}^2 is the k-th diagonal element of $[\mathbf{M}(\hat{\theta}_i)/v_i]\mathbf{D}_i^{-1}, i = 1,2$. The quantities a_k and b_k are given below.

$$a_k^2 = \left[\frac{b_k - 2}{b_k}\right] f_1 \qquad b_k = 4 + \frac{f_1^2}{f_2} \tag{11.25}$$

where

$$f_1 = \left[\frac{v_2}{v_2 - 2}\right]\cos^2\varphi + \left[\frac{v_1}{v_1 - 2}\right]\sin^2\varphi \tag{11.26}$$

and

$$f_2 = \frac{v_2^2}{(v_2 - 2)^2(v_2 - 4)}\cos^4\varphi + \frac{v_1^2}{(v_1 - 2)^2(v_1 - 4)}\sin^4\varphi \tag{11.27}$$

$$\cos^2\varphi = \frac{\tau_{2k}^2}{\tau_{1k}^2 + \tau_{2k}^2} \tag{11.28}$$

The unconditional posterior distribution of γ is obtained by multiplying the conditional distribution of γ given by Eq. (11.23) with the posterior distribution of the change point n_1 given by Eq. (11.4) and summing over the observations.

$$p(\gamma \mid y) = \sum_{n_1=1}^{N-1} p(\gamma \mid n_1, y) p(n_1 \mid y) \tag{11.29}$$

11.3.1.3. The Conditional Posterior Distributions of β_i, σ_i^2 and $\boldsymbol{\theta}_i$. Taking the possible change point n_1 as the correct change point and analyzing the first and second portions of data, the posterior distribution of β conditional on the change point n_1 is obtained by using Eq. (11.20).

When β_1 and β_2 are found to be unequal, further investigation is conducted on the posterior distributions of σ_1^2 and σ_2^2 for several selected values of β_1 and β_2 based on the probability density function of σ_1^2 given by Eq. (11.18). The unconditional distributions of σ_1^2 and σ_2^2 are obtained by combining the posterior distributions of σ_i^2 for various values of β_i. To obtain the distribution of σ_i^2 conditional on the change point n_1, a weight is assigned to each of the posterior distributions of σ_i^2 according to the posterior distribution of β_i given by Eq. (11.20). A well separated distribution of σ_1^2 and σ_2^2 indicates that σ_1^2 and σ_2^2 are unequal.

The posterior distribution of the mean $\boldsymbol{\theta}_i$ for different values of β_i conditional on the change point is given by

$$p_u(\boldsymbol{\theta}_i \mid n_i, \mathbf{y}_i) = \int_{-1}^{+1} p_u(\boldsymbol{\theta}_i \mid \beta_i, n_i, \mathbf{y}_i) p_u(\beta_i \mid n_i, \mathbf{y}_1) p(\beta_i) d\beta_i \tag{11.30}$$

$$-\infty < \theta_{ik} < \infty \qquad k = 0, 1, \ldots, p-1$$

where

$$p_u(\boldsymbol{\theta}_i \mid \beta_i, n_i, y_i) = [J(\beta_i)]^{-1} [\mathbf{M}(\boldsymbol{\theta}_i)]^{-n(1+\beta_i)/2} \ . \tag{11.31}$$

To obtain the distribution of $\boldsymbol{\theta}_i$ conditional on the change point n_i, a weight is assigned to each of the posterior distributions of $\boldsymbol{\theta}_i$ according to the posterior distribution of β_i given by Eq. (11.20).

11.3.2. Computational Sequences. The observed data are scaled down by a factor such that the processed data are between 0 and 1. Once a change point is located within the specified range, additional information regarding the posterior distributions of the mean, standard deviation and shape parameters are investigated. The computational procedure is outlined below.

(a) The Posterior Distributions of the Change Point n_1

1. The series of data to be analyzed is read in. β_i is assumed to range from –9 to +9. Define a range within the data series for the change point search. A change point is assumed to be within the search range.
2. The series is divided into two parts, the part before and after the change point. Each part is analyzed separately.
3. A first order autoregressive model $y_t = a + by_{t-1}$ is assumed for the data y_t, and parameters a and b of the model are estimated. These parameters are used to find $M(\hat{\boldsymbol{\theta}}_i)$ given by Eq. (11.11) which is used in Eq. (11.9) to

find the scaling matrix $\mathbf{D}_i$. The quantity A_i is calculated by using Eq. (11.5). The posterior distribution of the change point $p(n_1|y)$, is calculated by using Eq. (11.4) with the prior distribution of the parameter β_i given by Eq. (11.3). As the prior distribution of β_i is not known. A uniform prior distribution is assumed herein.

4. Step 3 is repeated by assuming the next point in the search range. After all points within the range have been examined, the probability of the change point is normalized.

(b) The Unconditional Posterior Distributions of $\delta, \upsilon,$ *and* γ

1. For each point within the search range, the probability of the shape parameter β_1 conditional on the change point is calculated by using Eq. (11.14) and is normalized. The joint probability of β_1 and β_2 conditional on the change point is calculated by using Eq. (11.13). The conditional distribution of the difference in the shape parameter $\delta = \beta_2 - \beta_1$ is calculated by using Eq. (11.12) by varying δ from –2 to +2. As the prior probability distribution of β_1 is not known, a uniform prior is assumed. The unconditional distribution of δ is calculated by using Eq. (11.16) by multiplying the conditional distribution of δ with the posterior distribution of the change point $p(n_1|y)$ and summing over the observations.
2. The distribution of the variance σ_i^2 conditional on the change point is calculated by using Eq. (11.18) and the conditional distribution of the variance ratio $\upsilon = \sigma_2^2/\sigma_i^2$ is calculated by using Eq. (11.17). The joint probability distribution of β_1 and β_2 is calculated by using Eq. (11.19). To eliminate the β_1 parameters, this distribution is multiplied by the joint distribution and integrated over the β_1 values as given in Eq. (11.21). The conditional posterior distribution of υ is calculated by using Eq. (11.22) by multiplying the conditional distribution of υ with the posterior distribution of the change point $p(n_1|y)$ and summing over the observations.
3. The posterior distribution of the difference in the mean $\gamma = \theta_2 - \theta_1$ conditional on the change point is calculated by using Eq. (11.23). The unconditional distribution of γ is calculated by using Eq. (11.29) by multiplying the conditional distribution of γ with the posterior distribution of the change pint $p(n_1|y)$ and summing over the observations.

(c) The Conditional Posterior Distribution of β_i, σ_i^2 and $\boldsymbol{\theta}_i$

1. The posterior distribution of the shape parameter β_i conditional on the change point is calculated in step 1 of the previous section (unconditional posterior distribution of δ).
2. The posterior distribution of the variance σ_i^2 conditional on the change point is calculated in step 2 of the previous section.
3. The posterior distribution of the mean $\boldsymbol{\theta}_i$ conditional on the change point is calculated by using Eq. (11.30).

11.4. DISCUSSION OF RESULTS

11.4.1. The Posterior Distribution of the Change Point n_1. The data are classified to be nonhomogeneous or probably nonhomogeneous by using the results of Intervention analysis of Potter (1979). Previous studies of the same data are conducted by Potter (1976, 1979) and Rao and Yu (1984). Based on these studies it was found that for some of the data, more than one change point is found. Results of these studies are presented in Table XI.1.

The posterior probability of the change point $p(n_1|y)$ is calculated by using Eq. (11.4). The change points are located by analyzing a range of data near the change point. The number of observations is also changed to study the variation in the results. When the mean of the series changes significantly, then a change point will be characterized by a large probability within the range. The results are given in Table XI.3.

Some of the results in Table XI.3 show correspondence with previous results given in Table XI.1. Similar results indicate that a change point might be in the vicinity. Contradictory results indicate that there might not be a change point or that the change in the mean of the series is too small to be detected. For example, the 68th point in EAST corresponds to the 69th point obtained by Intervention analysis but not with the result of the spectrum method. The 99th point in NEWB corresponds to both Intervention analysis (93th) and spectrum method (91th), suggesting a possible change point. On the other hand, WOVT and FITC do not have any change point that corresponds to previous results. The 54th point in TAUN corresponds to the spectrum method (55th). NEWH, MOOR and WEST do not have any corresponding change points. The 74th point in CHIC corresponds to the results obtained from the spectrum method, as do the 281th, 428th and 609th points in COLO. This shows that EAST 68th, NEWB 99th, TAUN 54th, CHIC 74th, COLO 281th, 428th and 609th points are possible locations of change points. The changes in mean in the WOVT, FITC, NEWH, MOOR and WEST do not appear to be significant.

11.4.2. The Unconditional Posterior Distributions of δ, υ and γ. Taking a change point in Table XI.1 as the correct change point, the distribution of ($\delta = \beta_2 - \beta_1$) conditional on the change point is calculated by using Eq. (11.12). The probabilities of δ being less than zero are given in Table XI.4. A small probability implies the equality of β_1 and β_2, while a large probability indicates otherwise. In general, the conditional probability of δ being less than zero is small (less than 0.5). Consequently the conditional probability of β_1 being equal to β_2 is quite large. Therefore the part of data before and after the change point have distributions with similar shape parameters β_i. An exception is COLO in the range of 601-640, which suggests that shape parameters β_1 and β_2 are definitely different.

The posterior density function of ($\delta = \beta_1 - \beta_2$), unconditional on the change point is found by multiplying the conditional distribution of δ with the posterior distribution of the change point $p(n_1|y)$ and the summing over the observations as described in Eq. (11.16). The probabilities of δ being less than zero are given in Table XI.5. The results are very similar to that of the conditional probabilities in Table XI.4. These results further suggest that the part of data before and after the change point have distributions with similar shape parameters β_i. The posterior density function of $\upsilon = \sigma_2^2/\sigma_i^2$ conditional on the change point is calculated by using Eq. (11.17). The probabilities of υ being less than one are given in Table XI.6. A small probability indicates that the part of data before and after the change point have different variances σ_i^2. Almost all the conditional probabilities of υ are small (less than 0.5). WOVT, CHIC and COLO have probabilities that are virtually zero, which indicates that variances σ_i^2 are unequal. To eliminate the influence of the change point, the posterior distribution of υ with the posterior distribution of the change point $p(n_1|y)$ and summing over all observations as given in Eq. (11.22). The probabilities of υ being less than one are also given in Table XI.6. Similar results are obtained for the unconditional distribution of υ. This result suggests that the part of data before and after the change point have different variances.

The posterior distribution of $\gamma = \theta_2 - \theta_l$ given the change point is calculated by using Eq. (11.23). The results for the Colorado river data are given in Table XI.7. For both the precipitation and the particulate data, almost all of the conditional probabilities of γ are close to 0.5 because the variance of γ is found to be quite large. However, the COLO the variance of γ is not as large and the probabilities of γ being greater than zero are given in Table XI.7. A small probability value in Table XI.7 indicates that the parameter θ_i for the data before and after the change point are not significantly different. The unconditional distribution of γ is obtained by multiplying the conditional distribution of γ with the posterior probability function of the change point $p(n_1|y)$ and summing over all observations as described in Eq. (11.29). The probabilities of γ being less than one are also given in Table XI.7. Similar results are obtained for the unconditional distribution of υ. This further suggest that data before and after the change point have different θ_i parameters.

11.4.3. The Conditional Posterior Distributions of β_i, σ_i^2 and θ_i. Taking the change points in Table XI.1 as correct and isolating the first part of the data from the second, the posterior density functions of β_1 and β_2 are calculated by using Eq. (11.20). In general, both density functions are located in the negative interval.

Taking the change point in Table XI.1 as being correct, the posterior distributions of σ_1^2 and σ_2^2 conditional on β_1 and β_2 are calculated by using Eq. (11.18). By weighing the posterior distribution of σ_1^2 according to the posterior distribution of β_1 and β_2, unconditional distributions are obtained. In general, the results show that the posterior distribution of σ_1^2 and σ_2^2 overlap each other, suggesting that the values of σ_1^2 and σ_2^2 are approximately the same.

Table XI.3. Change points.

Name	Range	Change Points	$P(n_1 \mid y)$
EAST	66-75	68	0.535
	56-81	68	0.429
NEWB	86-100	99	0.112
	71-110	102	0.080
WOVT	46-65	65	0.219
	60-85	65	0.113
FITC	51-70	70	0.097
	31-70	70	0.081
TAUN	41-60	54	0.117
	36-75	65	0.081
NEWH	66-80	80	0.117
	51-82	82	0.138
MOOR	41-60	43	0.107
	26-65	43	0.088
WEST	51-70	53	0.147
	21-60	47	0.119
CHIC	41-80	74	0.201
	101-140	101	0.289
COLO	41-80	44	0.179
	81-120	118	0.113
	121-160	130	0.054
	161-200	200	0.088
	201-240	224	0.100
	241-280	269	0.053
	281-320	281	0.204
	321-360	329	0.157
	361-400	368	0.223
	401-440	428	0.180
	441-480	453	0.170
	481-520	515	0.139
	521-560	525	0.214
	561-600	574	0.403
	601-640	609	0.608

Table XI.4. Conditional distribution of $\delta = (\beta_2 - \beta_1)$.

Name	Range	$p(\beta_2 - \beta_1 < 0 \mid n_1, y)$
EAST	66-75	0.298
	56-81	0.298
NEWB	86-100	0.396
	71-110	0.423
WOVT	46-65	0.552
	60-85	0.552
FITC	51-70	0.420
	31-70	0.420
TAUN	41-60	0.542
	36-75	0.610
NEWH	66-80	0.389
	51-82	0.416
MOOR	41-60	0.348
	26-65	0.348
WEST	51-70	0.612
	21-60	0.608
CHIC	41-80	0.281
	101-140	0.050
COLO	41-80	0.156
	81-120	0.639
	121-160	0.696
	161-200	0.284
	201-240	0.356
	241-280	0.387
	281-320	0.403
	321-360	0.502
	361-400	0.260
	401-440	0.388
	441-480	0.177
	481-520	0.466
	521-560	0.583
	561-600	0.477
	601-640	0.999

Table XI.5. Unconditional distribution of $\delta = (\beta_2 - \beta_1)$.

Name	Range	$p(\beta_2 - \beta_1 < 0 \mid y)$
EAST	66-75	0.289
	56-81	0.280
NEWB	86-100	0.407
	71-110	0.438
WOVT	46-65	0.518
	60-85	0.532
FITC	51-70	0.411
	31-70	0.407
TAUN	41-60	0.564
	36-75	0.633
NEWH	66-80	0.429
	51-82	0.440
MOOR	41-60	0.373
	26-65	0.375
WEST	51-70	0.640
	21-60	0.625
CHIC	41-80	0.367
	101-140	0.295
COLO	41-80	0.258
	81-120	0.707
	121-160	0.812
	161-200	0.336
	201-240	0.410
	241-280	0.439
	281-320	0.536
	321-360	0.668
	361-400	0.359
	401-440	0.508
	441-480	0.251
	481-520	0.564
	521-560	0.690
	561-600	0.530
	601-640	0.994

Table XI.6. Distributions of $\upsilon = \sigma_2^2/\sigma_1^2$.

Name	Range	$p(\upsilon < 1 \mid n_1, y)$	$p(\upsilon < 1 \mid y)$
EAST	66-75	0.348	0.352
	56-81	0.348	0.349
NEWB	86-100	0.023	0.072
	71-110	0.013	0.051
WOVT	46-65	0.001	0.009
	60-85	0.001	0.007
FITC	51-70	0.048	0.119
	31-70	0.048	0.124
TAUN	41-60	0.011	0.022
	36-75	0.001	0.006
NEWH	66-80	0.327	0.361
	51-82	0.255	0.285
MOOR	41-60	0.560	0.454
	26-65	0.560	0.420
WEST	51-70	0.409	0.422
	21-60	0.481	0.466
CHIC	41-80	0.000	0.000
	101-140	0.000	0.000
COLO	41-80	0.000	0.000
	81-120	0.000	0.000
	121-160	0.000	0.000
	161-200	0.000	0.000
	201-240	0.000	0.000
	241-280	0.000	0.000
	281-320	0.000	0.000
	321-360	0.000	0.000
	361-400	0.000	0.000
	401-440	0.000	0.000
	441-480	0.000	0.000
	481-520	0.000	0.000
	521-560	0.000	0.000
	561-600	0.000	0.000
	601-640	0.000	0.000

Table XI.7. Distributions of $\gamma = \theta_2 - \theta_1$.

Name	Range	$p(\gamma > 0 \mid n_1, y)$	$p(\gamma > 0 \mid y)$
COLO	41-80	0.211	0.525
	81-120	0.958	0.933
	121-160	0.988	0.968
	161-200	0.092	0.150
	201-240	0.035	0.151
	241-280	0.966	0.785
	281-320	0.788	0.611
	321-360	0.595	0.553
	361-400	0.090	0.289
	401-440	0.026	0.141
	441-480	0.914	0.779
	481-520	0.998	0.963
	521-560	0.971	0.057
	561-600	0.998	0.997
	601-640	0.469	0.470

Finally, the θ_i parameters which give the central locations of the distribution for the first portion and the remainder of the data are analyzed. The posterior distribution of θ_1 and θ_2 conditional on the change point are calculated by using Eq. (11.30). By weighing each distribution of various values of β for the Eastport (EAST) precipitation data are given in Figures 11.9 through 11.15. The data range analyzed here is from 56th to the 81st point. Figure 11.9 shows that the dominant change point is located at n_1 = 68. Taking the 68th time point as the correct change point, the distribution of θ_1 and θ_2 are obtained. In general, the distributions of θ_i do not overlap each other as shown in Figure 11.15. An example of the distributions of the change point, the mean parameter θ, the standard deviation parameter σ, and the shape parameter β for the Eastport (EAST) precipitation data are given in Figures 11.9 through 11.15. The data range analyzed here is from 56th to the 81st point. Figure 11.9 shows that the dominant change point is located at $n_1 = 68$. Taking the 68th time point as the correct change point, the distribution of $\delta = \beta_2 - \beta_1$ conditional on the change point is shown in Figure 11.10. The posterior density function is located more to the right of the zero point, suggesting that β_1 and β_2 are equal. The unconditional distribution of $\delta = \beta_2 - \beta_1$ is shown in Figure 11.11. Similar to Figure 11.10, the posterior density function is located more to the right of the zero point, suggesting that β_1 is equal to β_2. The distribution of $\upsilon = \sigma_1^2 / \sigma_1^2$ are approximately equal. The distributions of β_1 and β_2 conditional on the change point are shown in Figure 11.13. Both distributions are located to the left of the zero point and are short tailed. The distributions of σ_1^2 and σ_2^2 given the change point are shown in Figure 11.14. Since both density functions overlap each other, there is a possibility that σ_1^2 might be equal

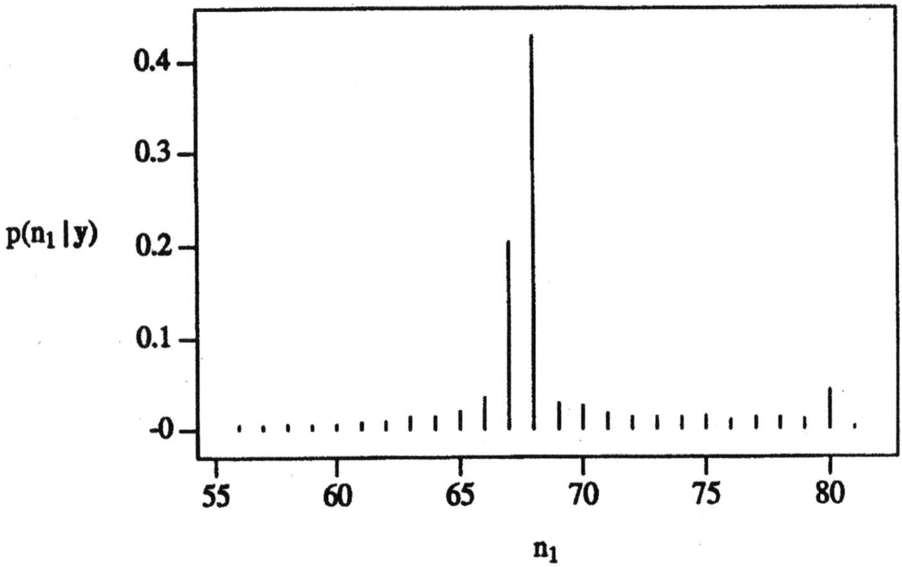

Figure 11.9. Posterior distribution of the change point $p(n_1,y)$ for EAST

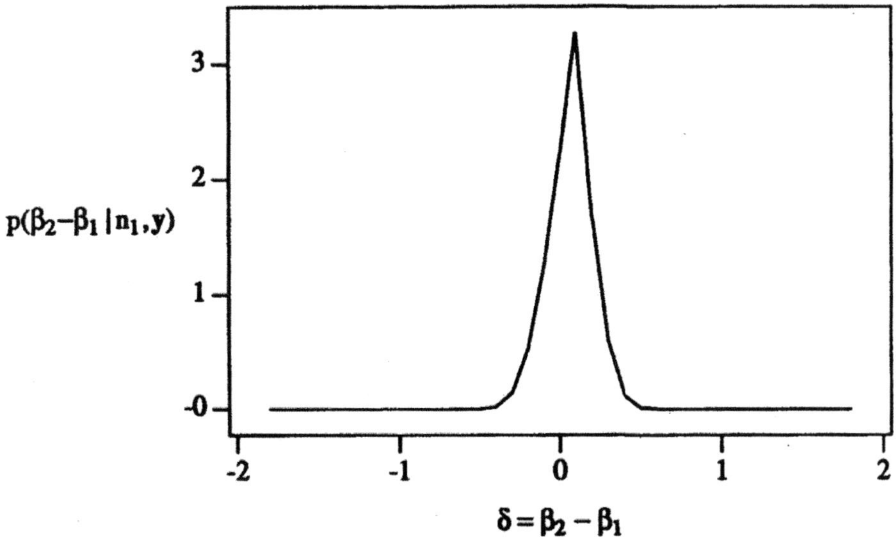

Figure 11.10. Distribution of ($(\beta_2 - \beta_1 \mid n_1 y)$ for EAST

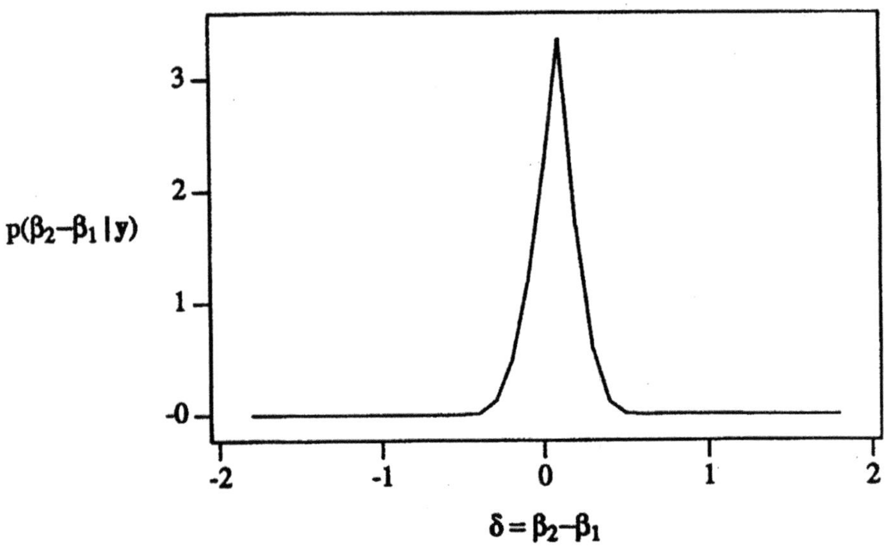

Figure 11.11. Unconditional distribution of $(\beta_2 - \beta_1 | \mathbf{y})$ for EAST

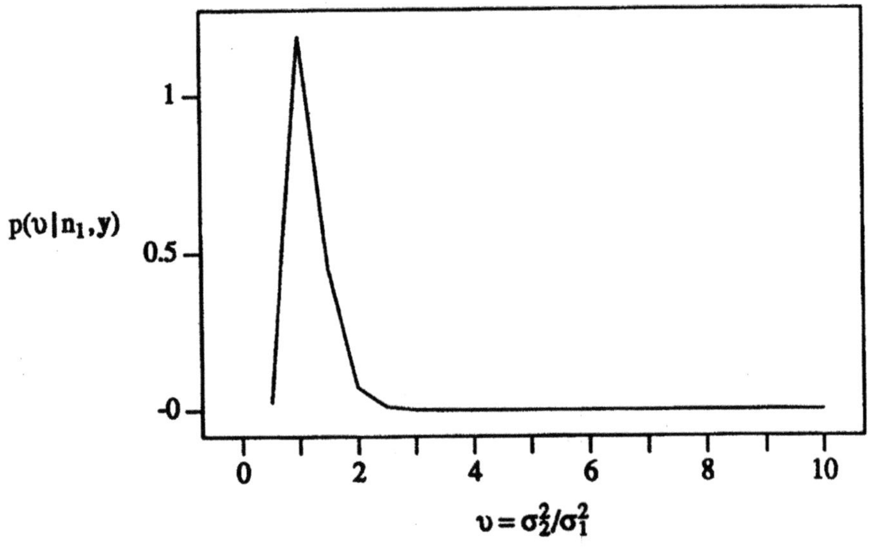

Figure 11.12. Distribution of $p(\upsilon | n_1, y)$ for EAST

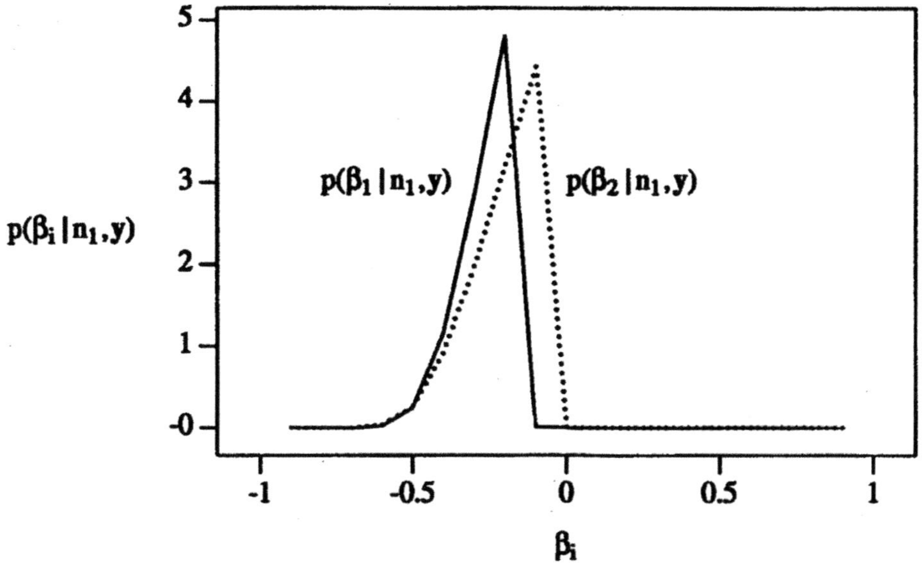

Figure 11.13. Unconditional distribution of $(\beta_i \mid n_1, y)$ for EAST

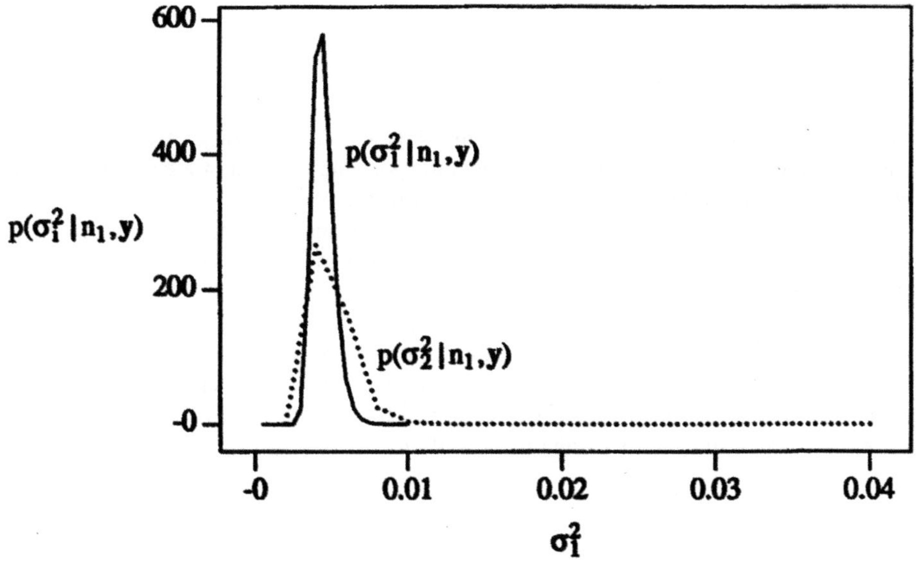

Figure 11.14. Distribution of $(\sigma_i^2 \mid n \;\; , y)$ for EAST

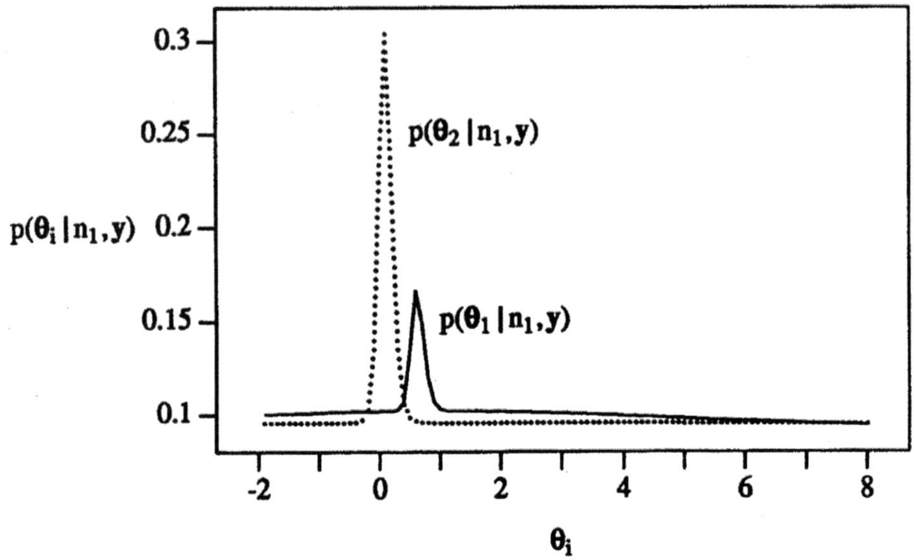

Figure 11.15. Distribution of $(p(\theta_i \mid n_1, y)$ for EAST

to σ_2^2. The distributions of θ_1 and θ_2 conditional on the change point is shown in Figure 11.15. Since the distributions are not overlapping each other, it is likely that θ_1 and θ_2 are not equal to each other.

11.5. CONCLUSIONS

The Bayesian method is very useful to investigate the possibility of changes in the characteristics of a series. In order to detect the possible location of a change point, the characteristics of the series have to change significantly.

The Bayesian method is better than Intervention analysis and methods based on spectral analysis because the times at which the change occurs need not be specified. Also the results are not model dependent.

The conclusions from this particular study are as follows: (1) The statistical significance of the observed changes in the mean values of a series can be investigated by using this method. (2) The Bayesian inferential procedure is useful in locating the change point in time series. (3) The behavior of the statistical characteristics of a time series can be investigated by using this procedure.

12. References

Appel, U. and A. V. Brandt. (1983) Adaptive sequential segmentation of piecewise stationary time series. *Information Sciences*, 29, 27-56.

Ashley, R. A., D.M. Patterson and M.J. Hinich. (1982) A diagnostic test for nonlinear serial dependence in time series fitting errors. *J. Time Series Anal.*, 7(3), 165-178.

Batchelor, G. K. (1959) Small-scale variation of convected quantities like temperature in turbulent field. *J. Fluid Mech.*, 5, 113-133.

Beran, M.A., and N.W. Arnell. (1996) *Climate change and hydrological disasters. In: Hydrology of Disasters,* V.P. Singh, editor, Water Science and Technology Library V24, pp. 41-62, Kluwer Academic Publishers, Dordrecht, The Netherlands.

Berger, A., J.L. Mélice and I. Van Der Mersch. (1990) Evolutive spectral analysis of sunspot data over the past 300 years. *Phil. Trans. Royal Soc.* Lond., A330, pp. 529-541.

Bevington, P. R. and D.K. Robinson. (1969) *Data Reduction and Error Analysis for the Physical Sciences*, McGraw-Hill, New York.

Bhattacharya, D. (1997) *Memory, Periodicity and Heteroskedasticity in Hydrologic Time Series*, Ph.D. Thesis, Purdue University, West Lafayette.

Boden, T.A., D.P. Kaiser, R.J. Sepanski and F.W. Stoss. (1994) *Trends 93: A Compendium of Data on Global Change.* ORNL/CDIAC-65. Carbon Dioxide Information Analysis Center, Oak Ridge National Laboratory, Oak Ridge, Tenn. USA.

Box, G.E.P. and G.C. Tiao. (1962) A Further Look at Robustness via Bayes's Theorem , *Biometrika* 49 (3 and 4), 419-432.

Box, G.E.P. and G.C. Tiao. (1964a) A Bayesian Approach to the Importance of Assumptions Applied to the Comparison of Variances , *Biometrika* 51 (1 and 2), 153-167.

Box, G.E.P. and G.C. Tiao. (1964b) A Note on Criterion Robustness and Inference Robustness , *Biometrika* 51 (1 and 2), 169-173.

Box, G.E.P. and G.C. Tiao. (1973) *Bayesian Inference in Statistical Analysis , Chapter 3 and 4*, Addison-Wesley, Reading, Massachusetts.

Box, G. E. P. and G.M. Jenkins. (1976) *Time Series Analysis: Forecasting and Control, 2nd ed.*, Holden-Day, San Francisco.

Brockwell, P.J. and R.A. Davis. (1991) *Time Series: Theory and Methods. Second Ed.*, Springer-Verlag, New York, NY.

Buishand, T.A. (1982) Some methods for testing the homogeneity of rainfall records. *J. Hydrol.*, 58, pp 11-27.

Burg, J.P. (1968) *A new analysis technique for time series data.* NATO Advanced Study Institute on Signal Processing with Emphasis on Underwater Acoustics. Reprinted in Childers (1978).

Burg, J.P. (1975) Maximum Entropy Spectral Analysis. Ph.D. Dissertation, Dept. Of Geophysics, Stanford University.

Burn, D.H. (1994) Hydrologic effects of climatic change in west central Canada. *J. Hydrology*, 160, pp. 53-70.

Burroughs, W.J. (1992) *Weather Cycles: Real or Imaginary.* Cambridge University Press.

Caldwell, D. R., T.M. Dillon, J.M. Brubaker, P.A. Newberger and C.A. Paulson. (1980). The scaling of vertical temperature gradient spectra. *J. Geophys. Res.*, 85(C4), 1917-1924.

Carlson, R. F., McCormick, A.J.A., and Watts, D.G. (1970). "Application of linear models to four annual streamflow series." *J. Water Resour. Res.*, 6, 4, pp. 1070-1078.

Carter, G. D. and J. Imberger. (1986) Vertically rising microstructure profiler. *J. Atmos. Oceanic Technol.*, 3, 462-471.

Chen, H.-L. (1999) *Stochastic Characteristics of Synthetic and Observed Time Series, Technical* Report CE-EHE-99-3, Purdue University, West Lafayette, IN.

Chen, H.-L. and A.R. Rao. (1998) Periodicity in midwest US climatic data: The Hale cycle or the luni-solar signal? *Stochastic Hydrol. Hydraul.*, 12(4), 205-222.

Chiew, F.H.S., T.A. McMahon, and T. Piechota. (1994) El-Niño/Southern Oscillation and streamflow patterns in South-East Australia *Australian Civil Eng. Trans* .Vol. CE36 ,No. 4 , pp. 285-291.

Childers, D.G. (1978) *Modern Spectrum Analysis.* IEEE Press.

Chui, C.K. 1992 *An Introduction to Wavelets.* Academic Press.

Combes, J.M., A. Grossmann, Tchamitchian, Ph. eds. (1990) *Wavelets, Time-Frequency Methods, and Phase Space.* Springer.

Cox, D.R., and A. Stewart. (1955) Some quick sign tests for trend in location and dispersion. *Biometrika,* 42, pp. 80-95.

Cramér, H. (1940) On the theory of stationary random processes. *Ann. Math.* 41, pp. 215-230.

Currie, R. (1974) Solar cycle in surface air temperature. *Journal of Geophysical Research*, Vol. 79, No. 36, pp 5657-5660.

Currie, R. (1981) Evidence for 18.6-year M_N Signal in temperature and drought conditions in North America since AD 1800. *Journal of Geophysical Research*, Vol. 86, No. C11, pp. 11,055-11,064.

Currie, R. (1984) On bistable phasing of 18.6-year nodal induced flood in India. *Geophysical Research Letters,* Vol. 11, No. 1, pp. 50-53.

Currie, R. (1987) On bistable phasing of 18.6-year induced drought and flood in the Nile record since AD 650. *J. Climatology*, Vol. 7, pp. 373-389.

Currie, R. and D.P. O'Brien (1988) Periodic 18.6-year and cyclic 10 to 11-year signals in Northeastern United States precipitation data. *J. Climatology*, Vol. 8, pp. 255-281.

Currie, R.G. and D.P. O'Brien. (1990) Deterministic signals in precipitation records from American Corn Belt. *International Journal of Climatology*, Vol. 10, pp. 179-189.

Daubechies, I. (1988) Orthonormal bases of compactly supported wavelets. *Commun. Pure Appl. Math.*, 41, 909-996.

Daubechies, I. (1992) *Ten Lectures on Wavelets.* SIAM, Philadelphia, PA

Daubechies, I. (1993) *Wavelet transforms and orthogonal wavelet bases.* Proc. Of Symposia in Applied Math., Volume 47, 1:33.

Davis, R. A., D. Huang and Y-C. Yao. (1995) Testing for a change in the parameter values and order of an aotoregressive model. *Ann. Statist.,* 21, 671-680.

De Souza, P. V. and P.J. Thomson. (1982) LPC distance measures and statistical tests with particular reference to the likelihood ratio. *IEEE Trans. on ASSP.*, 30(2), 304-315.

Denman, K. L. and A.E. Gargett (1995) Biological-physical interactions in the upper ocean: The role of vertical and small scale transport processes. *Ann. Rev. Fluid Mech.*, 27, 225-255.

Dillon, T. M., and D.R. Caldwell. (1980) The Batchelor spectrum and dissipation in the upper ocean. *J. Geophys. Res.*, 85(C4), 1910-1916.

Donoho, D. L. and I.M. Johnstone (1994) Ideal spatial adaptation via wavelet shrinkage. *Biometrika*, 81, 425-455.

EarthInfo. (1993) *Earthinfo CD^2 USGS Daily Values*, Technical Report, EarthInfo Inc., Boulder.

Farge, M. (1992) Wavelet transforms and their applications to turbulence. *Ann. Rev. Fluid Mech.*, 24, 395-457.

Farge, M., N. Kevlahan, V. Perrier and U. Goirand (1996) *Wavelets and turbulence.* Proc. IEEE, 84(4), 639-669.

Feller, W. (1966) *An Introduction to Probability Theory and its Applications. 2nd Edition*, Vol. 1, John Wiley, New York, NY.

Fozdar, F. M., G.J. Parker and J. Imberger (1985) Matching temperature and conductivity sensor response characteristics. *J. Phys. Oceanogr.*, 15, 1557-1569.

Gabor, D. (1946) Theory of communication. *Journ. IEE*, 93, pp. 429-457.

Garbrecht, J. and G.P. Fernandez (1994) Visualization of trends and fluctuations in climatic records. *Water Resources Bulletin*, 30(2), pp 297-306.

Georgiadi, A.G. (1991) *The change of the hydrological cycle under the influence of global warming. In: Hydrology for the Water Management of Large River Basins*, Proceedings of the Vienna Symposium, August, 1991, IAHS Publ. No. 201-1991, pp. 119-128.

Ghil, M. And S. Childress. (1987) *Topics in Fluid Dynamics: Atmosphere Dynamics, Dynamo Theory, and Climatic Dynamics*. Springer-Verlag, NewYork, NY.

Grigoriu, M. (1995). *Applied Non-Gaussian Processes: Examples, Theory, Simulation, Linear random Vibration, and MATLAB Solutions*. Prentice Hall, Englewood Cliffs, NJ.

Groisman, P.Ya. and D.R. Easterling. (1994) *Area average precipitation over the contiguous U.S., Alaska, and Canada.* In: Boden, T.A., Kaiser, D.P., Sepanski, R.J., and Stoss, F.W. (Eds.), Trends 93: A Compendium of Data on Global Change. ORNL/CDIAC-65, Carbon Dioxide Information Analysis Center, Oak Ridge National Lab., Oak Ridge, TN. pp. 786-798.

Guckenheimer, J. and P. Holmes. (1983) *Non-Linear Oscillations, Dynamical Systems, and Bifurcations of Vector Fluids.* Springer-Verlag, New York, NY.

Haar, A. (1910) Zur Theorie der orthogonalen Funktionensysteme. *Mathematische Annalen*, 69, 331-371.

Hamed, K. H. (1997) *Investigation of Stationarity of Hydrologic and Climatic Time Series,* Ph.D. Thesis, Purdue University, West Lafayette.

Hamed, K.H. and A.R. Rao. (1998) A modified Mann-Kendall trend test for autocorrelated data. *Journal of Hydrology*, 204: 182-196.

Hamilton, J. D. (1994) *Time Series Analysis*, Princeton University Press, Princeton.

Hazen, A. (1914) Discussion on 'Flood Flows' by W.E. Fuller. *Trans. ASCE, 77*, pp 626-632.

Hinich, M. J. (1980) Testing for Gaussianity and linearity of a stationary time series. *J. Time Series Anal.,* 3(3), 169-176.

Hipel, K.W., A.I. McLeod, and R.R. Weiler. (1988) Data analysis of water quality time series in Lake Erie. *Water Resources Bulletin*, 24(3), pp. 533-544.

Hirsch, R.M., R.B. Alexander and R.A. Smith (1991) Selection of methods for the detection and estimation of trends in water quality. *Water Resources Research*, 27(5), pp. 803-813.

Hirsch, R.M., and J.R. Slack. (1984) Nonparametric Trend Test for Seasonal Data with Serial Dependence. *Water Resources Research*, Vol. 20, No. 6, pp. 727-732.

Hondzo, M. and D. Lyn. (1999) Quantified small-scale turbulence inhibits the growth of a green alga. *Freshwater Biology*, 41(1), 51-61.

Howell, J. F. and L. Mahrt. (1994) *An adaptive decomposition: Application to turbulence. Wavelets in Geophysics,* edited by Foufoula-Georgiou, E., and Kumar, P., Academics Press, San Diego, 107-128.

Hu, Y-S. (1994) *Wavelet Approach to Change-Point Detection with Application to Density Estimation,* Ph.D. Thesis, Texas A&M University, College Station.

Hudgins, L. H., M.E. Mayer, M. E. and C.A. Friehe. (1993) *Fourier and wavelet analysis of atmospheric turbulence. Progress in Wavelet Analysis and Applications*, edited by Meyer, Y., and Roques, S., Editions Frontiers, Paris, 491-498.

Hurst, H.E. (1951) Long term storage capacity of reservoirs." Trans. Amer. Soc. Civil Engrs., 116, pp. 770-799.

Imberger, J. and G.N. Ivey. (1991) On the nature of turbulence in a stratified fluid. Part II: Applications to lakes. *J. Phys. Oceanogr.*, 21(5), 659-679.

Ivanov, P.Ch., M.G. Rosenblum, C.K. Peng,, J. Mietus, S. Havlin,, H.E. Stanley and A.L. Goldberger (1996) Scaling behavior of heartbeat intervals obtained by wavelet-based time-series analysis. Letters to Nature, *Nature Magazine*, Vol. 383 (26 Sept. 1996), pp. 323-327.

Jackson, B.B. (1975) "Markov mixture models for drought lengths." *Water Resour. Res.*, 11, 1, pp. 64-74.

Jenkins, G.M. and D.G. Watts (1968) *Spectral Analysis and Its Applications*. Holden-Day, San Francisco, CA.

Jones, D.D., T,M.L. Wigley, and K.R. Briffa. (1994) *Global and hemispheric temperature anomalies - Land and marine instrumental records.* In: Boden, T.A., Kaiser, D.P., Sepanski, R.J., and Stoss, F.W.(Eds.), Trends 93: A Compendium of Data on Global Change. ORNL/CDIAC-65, Carbon Dioxide Information Analysis Center, Oak Ridge National Lab., Oak Ridge, TN. pp. 603-608.

Kaiser, G. (1994) *A Friendly Guide to Wavelets*. Birkhauser, Boston.

Karl, T.R., D.R. Easterling, and P.Ya Groisman. (1994b) *United States Historical Climatology Network - National and regional estimates of monthly and annual precipitation.* In: Boden, T.A., Kaiser, D.P., Sepanski, R.J., and Stoss, F.W. (Eds.), Trends 93: A Compendium of Data on Global Change. ORNL/CDIAC-65, Carbon Dioxide Information Analysis Center, Oak Ridge National Lab., Oak Ridge, TN. pp. 830-905.

Kay, S.M. (1988) *Modern Spectral Estimation, Theory and Application*. Prentice-Hall Signal processing Series, Alan V. Oppenheim, Series Editor. Prentice-Hall, N.J.

Kendall, M.G. (1970) *Rank Correlation Methods*. Griffin, London.

Klemeš, V. (1978) Physically based stochastic hydrologic analysis. *In: Advances in Hydroscience,* V.T. Chow (Ed.), Academic Press, New York, NY.

Klemeš, V. and I. Klemeš. (1988) Cycles in finite samples and cumulative processes of higher orders. *Water resources Research*, 24(1), pp. 93-104.

Kottegoda, N.T. (1985) Assessment of non-stationarity in annual series through evolutionary spectra. *Journal of Hydrology*, 76, pp. 391-402.

Kumar, P. and E. Foufoula-Georgiou (1994) *Wavelet analysis in geophysics: An introduction. Wavelets in Geophysics*, edited by Foufoula-Georgiou, E., and Kumar, P., Academics Press, San Diego, 1-43.Labitzke, K., and VanLoon, H., 1990. Association between the 11-year Solar cycle, the Quasi-Biennial Oscillation, and the atmosphere: A summary of recent work. Phil. Trans. Royal Society Lond., A330, pp. 577-589.

Labitzke, K., and H. Van Loon (1990) Association between the 11-year Solar cycle, the Quasi-Biennial Oscillation, and the atmosphere: A summary of recent work." Phil. Trans. Royal Society Lond., A330, pp. 577-589.

Lall, U., and M. Mann (1995) The Great Salt Lake: A barometer of low-frequency climatic variability. *Water Resources Research*, 31(10), pp. 2503-2515.

Legates, D.R. (1991) An evaluation of procedures to estimate monthly precipitation probabilities. *J. Hydrology*, 122, pp. 129-140.

Lettenmaier, D.P. (1976) Detection of trends in water quality data from records with dependent observations. *Water Resources Research*, 12(5), pp. 1037-1046.

Lettenmaier, D.P. and S. J. Burges (1977) Operational assessment of hydrologic models of long-term persistence." *Water. Resour. Res.*, 13, 1, pp. 113-124.

Lovell, B. and B. Boashash. (1987) *Evaluation of criteria for detection of changes in nonstationary signals.* IASTED Int. Sym. on Signal Proc. and its App., Brisbane, 1, 291-296.

Luketina, D. A. (1987) *Frontogenesis of Freshwater Overflows*, Ph.D. Thesis, University of Western Australia, Nedlands..

Luketina, D. A. and J. Imberger. (1998) Determining kinetic energy dissipation from Batchelor curve fitting. *J. Atmos. Oceanic Technol.*, submitted.

Mallat, S. (1996) *Wavelet Signal Processing*. Academic Press, New York, NY.

Mallat, S. G. (1989) A theory for multiresolution signal decomposition: The wavelet representation. IEEE Trans. Pattern Anal. *Machine Intell.*, 11, 674-693.

Mandelbrot, B.B., and J.R. Wallis (1968) Noah, Joseph and operational hydrology. *Water Resour. Res.*, 7,6, pp. 1460-1468.

Matalas, N.C. (1967) Mathematical assessment of synthetic hydrology." *Water Resour. Res.*, 3, 4, pp. 937-945.

Matalas, N.C. and J.R. Wallis (1971) Statistical properties of multivariate fractional noise processes. *Water Resour. Res.*, 7,6, pp. 1460-1468.

McGilchrist, C.A. and K.D. Woodyer. (1975) Note on distribution-free CUMSUM technique. *Technometrics*, 17(3), pp. 321-325.

Meneveau, C. (1991) Analysis of turbulence in the orthonormal wavelet representation. *J. Fluid Mech.*, 332, 469-520.

Meyer, Y. (1986) *Ondelettes et functions splines.* Sem. Equations aux Derivees Partielles, Ecole Polytechnique, Paris.

Meyer, Y. (1993) *Wavelets: Algorithms and Applications.* SIAM.

Misiti, M., Y. Misiti, G. Oppenheim and J.M. Poggi. (1996) *Wavelet Toolbox for use with MATLAB.* The Mathworks, Inc., Natick, Mass.

Mitra K., and S.N. Dutta. (1992) 18.6-year Luni-Solar nodal an 10-11-year Solar signals in rainfall in India. *Int. J. of Climatology,* 12, pp. 839-851.

Mitra, K., S. Mukherji and S.N. Dutta. (1991) Some indication of 18.6 year Luni-Solar and 10-11 year Solar cycles in rainfall in North-West India, the plains of Uttar Pradesh and North-Central India. *Int. J. of Climatology,* 11, 645-652.

National Climatic Data Center (NCDC). (1991) *Data selections from the disk resident database (DRD964x).* Data Files, National Climatic Data Center, Asheville, NC.

National Oceanographic and Atmospheric Administration (NOAA). (1997) *International Tree-ring Data Bank (ITRDB).* Data Files, National Geophysical Data Center (NGDC), Boulder, CO.

Nelson, T.R. and T.A. Boden. (1993) *CDIAC's numeric data package collection: Selected data sets relevant to studies of greenhouse gases and climate.* Data Files, Version 1.02, Vols. 1 and 2, Carbon Dioxide Information Analysis Center, Oak Ridge National Lab., Oak Ridge, TN.

Newberger, P. A. and D.R. Caldwell (1981) An initial subrange in microstructure spectra. *J. Geophys. Res.,* 86(C5), 4265-4268.

Newbold, P. (1980) A note on relations between seasonally adjusted variables. *J. Time Series Analysis,* 1(1), pp. 31-55.

Nikias, C. L. and A.P. Petropulu. (1993) *Higher-Order Spectra Analysis,* Prentice-Hall, Englewood Cliffs.

Oakey, N. S. (1982) Determination of the rate of dissipation of turbulent energy from simultaneous temperature and velocity shear microstructure measurements. *J. Phys. Oceanogr.,* 12, 256-271.

O'Connell, P.E. (1971) A simple stochastic modeling of Hurst's law. In: Mathematical models in hydrology, Warsaw Symposium, (IAHS Publ. 100, 1974), 1, pp.169-187.

Ogden, R. T. and J. Richwine (1996) *Wavelets in Bayesian change-point analysis,* Technical Report, University of South Carolina, Columbia.

O'Melia, C. R. and C.L. Tiller (1993) *Physico chemical aggregation and deposition in aquatic environments.* Environmental Particles 1, edited by Buffle, J. H. P., and Leeuwin, V., Lewis Publishers, Boca Raton, 303-386.

Percival, D.B. and A.T. Walden. (1993) *Spectral Analysis for Physical Applications: Multi-taper and Conventional Univariate Techniques.* Cambridge Univ. Press, Great Britain.

Percival, D. B. and P. Guttorp. (1994) *Long-memory processes, the Allan variance and wavelets. Wavelets in Geophysics,* edited by Foufoula-Georgiou, E., and Kumar, P., Academics Press, San Diego, 325-344.

Philander, S.G. (1990) *El Niő, La Niă ,and the Southern Oscillations.* Academic Press , New York, NY., 293 pp.

Potter, K.W. (1976) Evidence for Nonstationarity as a Physical Explanation of the Hurst Phenomenon , *Water Resources Research* 12(5), 1047-1052.

Potter, K.W. (1979) Annual Precipitation in the Northeast United States: Long Memory, Short Memory, or No Memory?, *Water Resources Research,* 15(2), 340-346.

Priestley, M.B. (1965) Evolutionary Spectra and non-stationary processes. *J. Roy. Statist. Soc.,* Ser. B, 27, pp. 204-237.

Priestley, M.B. (1984) *Spectral Analysis and Time Series. Third Printing,* Academic Press, London.

Priestley, M.B. (1988) *Non-linear and Non-stationary Time Series Analysis.* Academic Press, London.

Priestley, M.B. and T. Subba Rao. (1969) A test for nos-stationarity of time series. *J. Royal Statist. Soc.,* 31(1), pp. 140-149.

Proakis, J. G. and D.G. Manolakis. (1996) *Digital Signal Processing: Principles, Algorithms, and Applications, 3rd ed.,* Prentice-Hall, Upper Saddle River.

Rao, A.G. and A Shapiro (1970). Adaptive Smoothing Using Evolutionary Spectra , *Management Science,* Vol. 17, No. 3, 208-218.

Rao., A.R. and G.H. Yu (1989) Hale cycle and Indian drought and flood area indexes. *ASCE J. Irr. And Drainage Eng.,* 115(3), pp. 315-333.

Rao, A.R. and G.H. Yu. (1984) *Modeling Nonstationary Hydrologic Time Series , Hydraulic and Systems Engineering,* Purdue University, Tech. Rep. No. CE-HSE-84-1.

Ruddick, B., A. Anis and K. Thompson (1999) Maximum likelihood spectral fitting: The Batchelor spectrum. *J. Atmos. Oceanic Technol.*, submitted.

Salas, J.D., J.W. Delleur, V. Yevjevich and W.L. Lane (1988) Applied Modeling of Hydrologic Time series. *Water Resources Publications*, Littleton, CO.

Sherman, J. T. and R.E. Davis (1985) Observations of temperature microstructure in NATRE. *J. Phys. Oceanogr.*, 25, 1913-1929.

Slepian, D. (1978). Prolate Spheroidal Wave Functions, Fourier analysis and uncertainty -V: The discrete case. *Bell Syst. Tech. J.*, 57, pp. 1371-1429.

Sneyers, R. (1975) *Sur l'analyse statistique des series d'observations.* WMO technical note 143, World Meteorological Organization, Geneva.

Strang, G. and T. Nguyen (1996) *Wavelets and Filter banks.* Wellesley -Cambridge Press, Wellesley, MA.

Subba Rao, T. and M.M. Gabr. (1980) A test for linearity of stationary time series. *J. Time Series Anal.*, 1(1), 145-158.

Subba Rao, T. and K.C. Indukumar. (1996) Spectral and wavelet methods for the analysis of nonlinear and nonstationary time series. *J. Franklin Inst.*, 333(B3), 425-452.

Sudler, C.E. (1927) Storage required for the regulation of streamflow. *Trans. Amer. Soc. Civil Engrs.* 91, pp. 622-660.

Tewfik, A. H. and M. Kim. (1992) Correlation structure of the discrete eavelet coefficients of Fractional Brownian Motion. *IEEE Trans. Inform. Theory*, 38, 904-909.

Thomas, H.A. and Fiering, M.B. (1962) Mathematical synthesis of streamflow sequences for the analysis of river basins by simulation. In: design of water resource systems (A. Mass et al., eds.) pp. 459-493, Cambridge, Massachusettes, Harvard University Press.

Thomson, D.J. (1982) *Spectrum estimation and harmonic analysis.* Proc. IEEE, 70, pp. 1055-1096.

Thomson, D.J. (1990a) Time series analysis of Holocene climate data. *Phil. Trans. Roy. Soc. London, Series A, 330, pp. 601-616.*

Thomson, D.J. (1990b) Quadratic-inverse spectrum estimates; Applications to paleoclimatology. *Phil. Trans. Roy. Soc. London, Series A*, 332, pp. 593-597.

Trevino, G. and E.L. Andreas (1996) On wavelet analysis of nonstationary turbulence. *Boundary-Layer Meteorol.*, 65, 357-379.

Tsay, R. S. (1988) Outliers, level shifts, and variance changes in time series. *J. Forecasting*, 7, 1-20.

Ulrych, T.J. and M. Ooe (1983) *Autoregressive and mixed Autoregressive-Moving Average models and spectra. In: Nonlinear Methods of Spectral Analysis (Second Edition),* edited by S. Haykin, Springer-Verlag, Berlin. Pp. 73-125.

United States Geological Survey (USGS) (1992) *USGS daily values. Data Files, Computer Laser Optical Disks*, EarthInfo, Boulder, CO.

Wilson, H., and J.H. Hansen (1994) *Global and hemispheric temperature anomalies from instrumental surface air temperature records.* In: Boden, T.A., Kaiser, D.P., Sepanski, R.J., and Stoss, F.W. (Eds.), Trends 93: A Compendium of Data on Global Change. ORNL/CDIAC-65, Carbon Dioxide Information Analysis Center, Oak Ridge National Lab., Oak Ridge, TN. pp. 609-614.

WMO (1995) WMO statement on the status of the global climate in 1994. WMO Series, No. 826, Geneva, Switzerland.

World Meteorological Organization (WMO). (1988) *Analyzing long time series of hydrological data with regard to climatic variability.* WCAP-3, WMO/TD-No. 224, World Meteorologiacl Organization, Geneva. 12pp.

Worsley, K.J. (1979) On the likelihood ratio test for a shift in location of normal populations. *J. Amer. Statist. Assoc.*, 74, 365-367.

Yevjevich, V. (1965). The application of surplus, deficit and range in hydrology. Hydrology paper 10, Colorado State University, Fort Collins, CO.

Yule, G.U. (1926) Why do we sometimes get nonsense-correlations between time series? A study in sampling and the nature of time series. *J. Roy. Statist. Soc.*, 89, pp. 1-63.

Wang, Y. (1995) Jump and sharp cusp detection by wavelets. *Biometrika,* 82, 385-397.

World Meteorological Organization (WMO). (1995) *Statement on the status of the global climate in 1994.* WMO, Geneva, Switzerland.

Wornell, G. W. (1996) *Signal Processing with Fractals: A Wavelet-Based Approach.* Prentice-Hall, Upper Saddle River.

Zetterqvist, L. (1991) Statistical estimation and interpretation of trends in water quality time series. *Water Resources Research*, 27(7), pp. 1637-1648.

13. Index

Water Science and Technology Library

1. A.S. Eikum and R.W. Seabloom (eds.): *Alternative Wastewater Treatment*. Low-Cost Small Systems, Research and Development. Proceedings of the Conference held in Oslo, Norway (7–10 September 1981). 1982 ISBN 90-277-1430-4
2. W. Brutsaert and G.H. Jirka (eds.): *Gas Transfer at Water Surfaces*. 1984 ISBN 90-277-1697-8
3. D.A. Kraijenhoff and J.R. Moll (eds.): *River Flow Modelling and Forecasting*. 1986 ISBN 90-277-2082-7
4. World Meteorological Organization (ed.): *Microprocessors in Operational Hydrology*. Proceedings of a Conference held in Geneva (4–5 September 1984). 1986 ISBN 90-277-2156-4
5. J. Němec: *Hydrological Forecasting*. Design and Operation of Hydrological Forecasting Systems. 1986 ISBN 90-277-2259-5
6. V.K. Gupta, I. Rodríguez-Iturbe and E.F. Wood (eds.): *Scale Problems in Hydrology*. Runoff Generation and Basin Response. 1986 ISBN 90-277-2258-7
7. D.C. Major and H.E. Schwarz: *Large-Scale Regional Water Resources Planning*. The North Atlantic Regional Study. 1990 ISBN 0-7923-0711-9
8. W.H. Hager: *Energy Dissipators and Hydraulic Jump*. 1992 ISBN 0-7923-1508-1
9. V.P. Singh and M. Fiorentino (eds.): *Entropy and Energy Dissipation in Water Resources*. 1992 ISBN 0-7923-1696-7
10. K.W. Hipel (ed.): *Stochastic and Statistical Methods in Hydrology and Environmental Engineering*. A Four Volume Work Resulting from the International Conference in Honour of Professor T. E. Unny (21–23 June 1993). 1994
    10/1: Extreme values: floods and droughts ISBN 0-7923-2756-X
    10/2: Stochastic and statistical modelling with groundwater and surface water applications ISBN 0-7923-2757-8
    10/3: Time series analysis in hydrology and environmental engineering ISBN 0-7923-2758-6
    10/4: Effective environmental management for sustainable development ISBN 0-7923-2759-4
    Set 10/1–10/4: ISBN 0-7923-2760-8
11. S.N. Rodionov: *Global and Regional Climate Interaction: The Caspian Sea Experience*. 1994 ISBN 0-7923-2784-5
12. A. Peters, G. Wittum, B. Herrling, U. Meissner, C.A. Brebbia, W.G. Gray and G.F. Pinder (eds.): *Computational Methods in Water Resources X*. 1994 Set 12/1–12/2: ISBN 0-7923-2937-6
13. C.B. Vreugdenhil: *Numerical Methods for Shallow-Water Flow*. 1994 ISBN 0-7923-3164-8
14. E. Cabrera and A.F. Vela (eds.): *Improving Efficiency and Reliability in Water Distribution Systems*. 1995 ISBN 0-7923-3536-8
15. V.P. Singh (ed.): *Environmental Hydrology*. 1995 ISBN 0-7923-3549-X
16. V.P. Singh and B. Kumar (eds.): *Proceedings of the International Conference on Hydrology and Water Resources* (New Delhi, 1993). 1996
    16/1: Surface-water hydrology ISBN 0-7923-3650-X
    16/2: Subsurface-water hydrology ISBN 0-7923-3651-8

16/3: Water-quality hydrology ISBN 0-7923-3652-6
16/4: Water resources planning and management ISBN 0-7923-3653-4
Set 16/1–16/4 ISBN 0-7923-3654-2

17. V.P. Singh: *Dam Breach Modeling Technology*. 1996 ISBN 0-7923-3925-8
18. Z. Kaczmarek, K.M. Strzepek, L. Somlyódy and V. Priazhinskaya (eds.): *Water Resources Management in the Face of Climatic/Hydrologic Uncertainties*. 1996 ISBN 0-7923-3927-4
19. V.P. Singh and W.H. Hager (eds.): *Environmental Hydraulics*. 1996 ISBN 0-7923-3983-5
20. G.B. Engelen and F.H. Kloosterman: *Hydrological Systems Analysis*. Methods and Applications. 1996 ISBN 0-7923-3986-X
21. A.S. Issar and S.D. Resnick (eds.): *Runoff, Infiltration and Subsurface Flow of Water in Arid and Semi-Arid Regions*. 1996 ISBN 0-7923-4034-5
22. M.B. Abbott and J.C. Refsgaard (eds.): *Distributed Hydrological Modelling*. 1996 ISBN 0-7923-4042-6
23. J. Gottlieb and P. DuChateau (eds.): *Parameter Identification and Inverse Problems in Hydrology, Geology and Ecology*. 1996 ISBN 0-7923-4089-2
24. V.P. Singh (ed.): *Hydrology of Disasters*. 1996 ISBN 0-7923-4092-2
25. A. Gianguzza, E. Pelizzetti and S. Sammartano (eds.): *Marine Chemistry*. An Environmental Analytical Chemistry Approach. 1997 ISBN 0-7923-4622-X
26. V.P. Singh and M. Fiorentino (eds.): *Geographical Information Systems in Hydrology*. 1996 ISBN 0-7923-4226-7
27. N.B. Harmancioglu, V.P. Singh and M.N. Alpaslan (eds.): *Environmental Data Management*. 1998 ISBN 0-7923-4857-5
28. G. Gambolati (ed.): *CENAS. Coastline Evolution of the Upper Adriatic Sea Due to Sea Level Rise and Natural and Anthropogenic Land Subsidence*. 1998 ISBN 0-7923-5119-3
29. D. Stephenson: *Water Supply Management*. 1998 ISBN 0-7923-5136-3
30. V.P. Singh: *Entropy-Based Parameter Estimation in Hydrology*. 1998 ISBN 0-7923-5224-6
31. A.S. Issar and N. Brown (eds.): *Water, Environment and Society in Times of Climatic Change*. 1998 ISBN 0-7923-5282-3
32. E. Cabrera and J. García-Serra (eds.): *Drought Management Planning in Water Supply Systems*. 1999 ISBN 0-7923-5294-7
33. N.B. Harmancioglu, O. Fistikoglu, S.D. Ozkul, V.P. Singh and M.N. Alpaslan: *Water Quality Monitoring Network Design*. 1999 ISBN 0-7923-5506-7
34. I. Stober and K. Bucher (eds): *Hydrogeology of Crystalline Rocks*. 2000 ISBN 0-7923-6082-6
35. J.S. Whitmore: *Drought Management on Farmland*. 2000 ISBN 0-7923-5998-4
36. R.S. Govindaraju and A. Ramachandra Rao (eds.): *Artificial Neural Networks in Hydrology*. 2000 ISBN 0-7923-6226-8
37. P. Singh and V.P. Singh: *Snow and Glacier Hydrology*. 2001 ISBN 0-7923-6767-7
38. B.E. Vieux: *Distributed Hydrologic Modeling Using GIS*. 2001 ISBN 0-7923-7002-3

Water Science and Technology Library

39. I.V. Nagy, K. Asante-Duah and I. Zsuffa: *Hydrological Dimensioning and Operation of Reservoirs*. Practical Design Concepts and Principles. 2002 ISBN 1-4020-0438-9
40. I. Stober and K. Bucher (eds.): *Water-Rock Interaction*. 2002 ISBN 1-4020-0497-4
41. M. Shahin: *Hydrology and Water Resources of Africa*. 2002 ISBN 1-4020-0866-X
42. S.K. Mishra and V.P. Singh: *Soil Conservation Service Curve Number (SCS-CN) Methodology*. 2003 ISBN 1-4020-1132-6
43. C. Ray, G. Melin and R.B. Linsky (eds.): *Riverbank Filtration*. Improving Source-Water Quality. 2003 ISBN 1-4020-1133-4
44. G. Rossi, A. Cancelliere, L.S. Pereira, T. Oweis, M. Shatanawi and A. Zairi (eds.): *Tools for Drought Mitigation in Mediterranean Regions*. 2003 ISBN 1-4020-1140-7

Kluwer Academic Publishers – Dordrecht / Boston / London

GPSR Compliance

The European Union's (EU) General Product Safety Regulation (GPSR) is a set of rules that requires consumer products to be safe and our obligations to ensure this.

If you have any concerns about our products, you can contact us on ProductSafety@springernature.com

In case Publisher is established outside the EU, the EU authorized representative is:

Springer Nature Customer Service Center GmbH
Europaplatz 3
69115 Heidelberg, Germany

Batch number: 10755284

Printed by Printforce, the Netherlands